AF902730

# Diesel Engines and Biodiesel Engines Technologies

*Edited by Freddie L. Inambao*

Published in London, United Kingdom

Diesel Engines and Biodiesel Engines Technologies
http://dx.doi.org/10.5772/intechopen.97954
Edited by Freddie L. Inambao

Contributors
Semakula Maroa, Freddie L. Liswaniso Inambao, Anirudh Gautam, Ankita Singh, Yun Bai, Zhaoyang Chen, Wei Dou, Xiangdong Kong, Jing Yao, Chao Ai, Fugang Zhai, Jin Zhang, Liu Yang, Rizal Mahmud, Iis Rohmawati, Babu Dharmalingam, Santhoshkumar Annamalai, Deepakkumar Rajagopal, Ramakrishna Reddy Ramireddy, Malinee Sriariyanun, Venkata Ramana Katla, Omojola Awogbemi, Daramy Vandi Von Kallon, Josiah Pelemo, Vinoth Kannan Viswanathan, Pushparaj Thomai, Cesar Augusto Bautista Sterling, Gokul Raghavendra Srinivasan, Şafak Yildizhan, Lakshmanan Thangavelu, Ranjitha Jambulingam, Shalini Palani, Oleksandr Osetrov, Andriy Marchenko, Igor Parsadanov, Volodymyr Pylyov, Oleh Linkov, Serhii Kravchenko, Oleksandr Trynov, Denys Meshkov, Serhii Bilyk, Anatolii Savchenko, Rasul Arian, Inna Rykova, Emmanuel Idoko Onuh, Kayode Timothy Akindeji, Juma Sahar, Muhammad Farooq, Anita Ramli, Abdul Naeem

© The Editor(s) and the Author(s) 2022
The rights of the editor(s) and the author(s) have been asserted in accordance with the Copyright, Designs and Patents Act 1988. All rights to the book as a whole are reserved by INTECHOPEN LIMITED. The book as a whole (compilation) cannot be reproduced, distributed or used for commercial or non-commercial purposes without INTECHOPEN LIMITED's written permission. Enquiries concerning the use of the book should be directed to INTECHOPEN LIMITED rights and permissions department (permissions@intechopen.com).
Violations are liable to prosecution under the governing Copyright Law.

Individual chapters of this publication are distributed under the terms of the Creative Commons Attribution 3.0 Unported License which permits commercial use, distribution and reproduction of the individual chapters, provided the original author(s) and source publication are appropriately acknowledged. If so indicated, certain images may not be included under the Creative Commons license. In such cases users will need to obtain permission from the license holder to reproduce the material. More details and guidelines concerning content reuse and adaptation can be found at http://www.intechopen.com/copyright-policy.html.

Notice
Statements and opinions expressed in the chapters are these of the individual contributors and not necessarily those of the editors or publisher. No responsibility is accepted for the accuracy of information contained in the published chapters. The publisher assumes no responsibility for any damage or injury to persons or property arising out of the use of any materials, instructions, methods or ideas contained in the book.

First published in London, United Kingdom, 2022 by IntechOpen
IntechOpen is the global imprint of INTECHOPEN LIMITED, registered in England and Wales, registration number: 11086078, 5 Princes Gate Court, London, SW7 2QJ, United Kingdom

British Library Cataloguing-in-Publication Data
A catalogue record for this book is available from the British Library

Additional hard and PDF copies can be obtained from orders@intechopen.com

Diesel Engines and Biodiesel Engines Technologies
Edited by Freddie L. Inambao
p. cm.
Print ISBN 978-1-80355-786-1
Online ISBN 978-1-80355-787-8
eBook (PDF) ISBN 978-1-80355-788-5

# We are IntechOpen, the world's leading publisher of Open Access books

## Built by scientists, for scientists

**5,900+**
Open access books available

**145,000+**
International authors and editors

**180M+**
Downloads

**156**
Countries delivered to

Our authors are among the
**Top 1%**
most cited scientists

**12.2%**
Contributors from top 500 universities

WEB OF SCIENCE™

Selection of our books indexed in the Book Citation Index in Web of Science™ Core Collection (BKCI)

Interested in publishing with us?
Contact book.department@intechopen.com

Numbers displayed above are based on latest data collected.
For more information visit www.intechopen.com

For EU product safety concerns:
IN TECH d.o.o., Prolaz Marije Krucifikse Kozulić 3, 51000 Rijeka, Croatia,
info@intechopen.com or visit our website at intechopen.com.

# Meet the editor

Freddie Inambao obtained his MSc in 1981 and Ph.D. in 1985. He is a professor at the University of KwaZulu-Natal, South Africa, and an advisor to the university's Green Energy Solutions Research Group. His fields of expertise are sustainable energy, energy management, alternative energy systems, energy efficiency, fuels, nanomaterials, and biomaterials. Professor Inambao's research is quite diverse, focusing on, but not exclusively, the energy efficiency of commercial and industrial buildings, piezoelectric materials (PZT) power generation, low-temperature solar thermal energy conversion, combustion of fossil fuels and renewable fuels, and water desalination. Professor Inambao has authored sixteen books. He also has more than 210 research publications to his credit.

# Contents

# Preface

Diesel engines are some of the most common reciprocating engines for use in power generation applications worldwide. Recently, biodiesel has been used as a substitute for diesel fuel, positioning it as a key technology to help achieve cleaner air and lower greenhouse gas emissions for a sustainable environment. Over twelve chapters, this book provides a comprehensive overview of diesel engines and biodiesel technologies.

Chapter 1, "Research and Innovation to Improve the Efficiency of Modern Diesel Engines", summarizes the authors' experience in improving diesel engines, increasing specific volume power and reliability, and ensuring a low level of environmental pollution emissions. It presents results of research using Industry 4.0 technologies for systematization, choice of directions, and the search for rational ways to improve the efficiency of diesel engines. It also considers the application of an ergo-exergy method for analyzing the efficiency of the working process of the engine and its systems. Taking into consideration the operating conditions, the chapter proposes technical solutions to improve the reliability of the most heat-stressed parts of high-powered engines. The possibilities for a comprehensive assessment of fuel efficiency and environmental qualities of diesel engines have been expanded taking into account $CO_2$ emissions when using traditional, alternative, and hybrid diesel fuel.

Chapter 2, "The Influence of Exhaust Gas Recirculation on Performance and Emission Characteristics of a Diesel Engine Using Waste Plastic Pyrolysis Oil Blends and Conventional Diesel", focuses on finding the influence of exhaust gas recirculation (EGR) on waste plastic pyrolysis oil (WPPO) with diesel as a base comparison fuel. The chapter discusses engine performance and emission characteristics within the test fuel blends.

Chapter 3, "Effect of Injection Pressure on Local Temperature and Soot Emission Distribution of Flat-Wall Impinging Diesel Flame under Diesel Engine like-Condition", provides an investigation of near-wall temperature on heat transfer analysis, which has a crucial effect on the local heat flux, to understand the heat transfer phenomenon on the combustion chamber walls. The local temperature and KL factor (the factor for indicating the amount of combustion ) are investigated by using a high-speed video camera and a two-color method by using a volume vessel with a fix-impingement wall. Results show that the local temperature and KL factor distribution increase in low-injection pressure. The result had a dominant effect on local heat transfer.

Chapter 4, "A Comparative Evaluation of Biodiesel and Used Cooking Oil as Feedstock for HDRD Application: A Review", features an overview of a comparative evaluation of biodiesel and used cooking oil as feedstock for hydrogenation-derived renewable diesel (HDRD) applications. It discusses the major challenges that render biodiesel inefficient, including higher viscosity, lower energy content, higher nitrogen oxide ($NO_X$) emissions, lower engine speed and power, injector coking, engine compatibility, high cost, and greater engine wear. The novelty of this work is that it shows that biodiesel conversion to green diesel is possible using a biowaste heterogeneous catalyst to obtain high quality and yield of HDRD at a low cost. This renewable energy (HDRD) possesses properties that are directly compatible with compression ignition engines and transportation engines. This research reviews biodiesel and used cooking oil as feedstocks for the production of HDRD, including the cost-benefit of these feedstocks. Hydrogenation of biodiesel has the potential to overcome the drawbacks of conventional chemically catalyzed processes.

Chapter 5, “Replacement of Diesel Fuel by DME in Compression Ignition Engines: Case for India,” highlights decarbonizing of transport and industrial sectors of the economy as a necessity to stop or reverse global warming. Electrification of power trains is an effective way to decarbonize the transport sector. The use of batteries, fuel cells, hybrid topographies with smaller IC (Internal Combustion) engines, and the use of alternative fuels like methanol, ethanol, and DME (Dimethyl Ether or di-methyl ether ) in IC engines are some of the ways through which emission of greenhouse gases can be reduced or eliminated. DME is a single-molecule fuel-high cetane number that can be used as a drop-in fuel on the diesel engines albeit with retro-fitment of these engines with a new pressurized fuel system. DME with a chemical formula $CH3-O-CH3$ can be produced by different feedstocks such as coal, natural gas, biomass and bio-waste, and municipal solid waste. India has a large reserve of high-ash coal, which can be converted to DME without polluting the environment by the use of clean coal technologies.

Chapter 6, “Molecular Contribution of Fatty Acid Esters in Biodiesel Fueled CI Engines”, examines the contribution of fatty acid ester (FAE) molecules in deciding the performance, emission, and combustion characteristics of their biodiesel in compression ignition engines. The chapter also discusses engine characteristics, emissions characteristics, and fuel properties as per ASTM standards.

Chapter 7, “Feasibility of Biodiesel Production in Pakistan”, presents key solutions that address the country’s serious energy issues. Pakistan’s Alternative Energy Development Board (AEDB) has suggested introducing an energy mix to meet the increasing energy demand and fuel the economy. Renewable energy introduces unique environmentally friendly nature, constant supply, wider availability, and ease of integration into existing infrastructure. Biodiesel is considered the best and most easily accessible source of energy among all renewable energy resources. However, there is still substantial room for the development of renewable energies in Pakistan. This chapter examines the availability of biomass resources in Pakistan and their potential for meeting the country’s rapidly growing energy demand, boosting the country’s economy, and creating new employment in the future.

Chapter 8, “Zero Emission Hydrogen Fuelled Fuel Cell Vehicle and Advanced Strategy on Internal Combustion Engine: A Review”, provides a comprehensive review of hydrogen as an alternative fuel and advanced strategies for internal combustion engines.

Chapter 9, “Performance and Emission Characteristics of Hydrogenation Derived Renewable Diesel as Diesel Engine Fuel,” discusses ways of improving the performance of diesel engines as well as proposes mandates to reduce greenhouse gas emissions. The chapter describes HDRD as a sustainable, reliable, and cost-effective alternative to petroleum-based diesel (PBD) fuel for compression ignition (CI) engines. This may be because the physicochemical properties of HDRD are similar to that of PBD fuel. The chapter examines the performance and emission characteristics of HDRD in unmodified CI engines. Performance emissions characteristics such as power, torque, brake-specific fuel consumption, thermal efficiency, nitrogen oxides, carbon monoxide, carbon dioxide, particulate matter, and exhaust gas temperature are examined and compared with that of PBD fuel in a CI engine. The results of this comparison show that HDRD is better than biodiesel and a sustainable replacement for PDB fuel to achieve improved performance and reduced emissions of CI engines.

Chapter 10, "Characteristics Analysis of Performance as Well as Emission of Elaeocarpus Ganitrus Additive Based Pumpkin and Juliflora Mixed Biodiesel Blend in CI Engine", discusses the use of (pumpkin) Cucurbita pepo. L and Prosopis juliflora seed oil for the synthesis of mixed biodiesel with 5ml Elaeocarpus ganitrus (rudraksha) as an additive. Performance tests were conducted using a biodiesel blend in a water-cooled CI engine and the emissions were analyzed using a five-gas analyzer. The chapter also contains an analysis of smoke opacity.

Chapter 11, "Bio-Circular Engine: Simultaneous and Successive Use of BioDiesel as Bio-Lubricant and Bio-Fuel in Diesel Engines-(B100) New Bio-Lubricant for all Engines", is dedicated to the system named "Bio-Circular Engine," framed within the so-called circular economy, for using a single substance (B100) for two different functions (as bio-fuel and as bio-lubricant).

Chapter 12, "Pressure Fluctuation Characteristics of High-Pressure Common Rail Fuel Injection System", discusses the dynamic pressure fluctuation characteristics of the high-pressure common rail fuel injection system based on the injector inlet pressure. It examines the pressure fluctuation mechanism and its influence law and provides theoretical support for improving the control accuracy of multiple injection cycle fuel injection volume.

This book is a useful resource for all those involved with engines, including energy engineers, fuel specialists, researchers, consultants, analysts, policymakers, and professionals in the industry supply chain. I would like to thank all the chapter authors for their excellent contributions. I also wish to thank the staff at IntechOpen, particularly Author Service Manager Maja Bozicevic for all her efforts in seeing to the book's timely completion.

**Freddie L. Inambao**
Department of Mechanical Engineering,
University of KwaZulu-Natal,
Durban, South Africa

Chapter 1

# Research and Innovation to Improve the Efficiency of Modern Diesel Engines

*Andriy Marchenko, Igor Parsadanov, Volodymyr Pylyov, Oleksandr Osetrov, Linkov Oleh, Serhii Kravchenko, Oleksandr Trynov, Denys Meshkov, Serhii Bilyk, Anatolii Savchenko, Inna Rykova and Rasoul Aryan*

## Abstract

Modern diesel engines are one of the main mobile energy sources and are characterized by a high degree of workflow completeness, design, and manufacturing technology. The chapter summarizes the authors' experience in improving diesel engines, increasing specific volume power, and reliability, ensuring a low level of environmental pollution emissions. The results of research using industry 4.0 technologies for systematization, choice of directions, and the search for rational ways to improve the efficiency of diesel engines are presented. The application of anergo-exergy method for analyzing the efficiency of the working process of the engine and its systems is considered. Taking into consideration the operating conditions, technical solutions are proposed to improve the reliability of the most heat stressed parts of high-powered engines. The possibilities for a comprehensive assessment of the fuel efficiency and environmental qualities of diesel engines have been expanded taking into account $CO_2$ emissions when using traditional, alternative, and hybrid diesel fuel.

**Keywords:** fuel efficiency, toxic substances, power the reliability diesel engines alternative fuels, operation model, the anergo-exergy method, working processes, thermally stressed parts, the operating conditions

## 1. Introduction

Modern diesel engines are the main mobile energy sources, are widely used as stationary power plants, and are distinguished by a high degree of design, working, and technological processes.

The advantages of diesel engines are determined by the high level of fuel efficiency and reliability due to the high level of workflow, all systems, and components refining. But this does not mean that all reserves for further improvement of diesel engine performance have been exhausted.

 IntechOpen

The main disadvantages of diesel engines include the consumption of natural organic fuels and the contribution to environmental pollution.

Taking into account, the prospects for increasing energy potential in stationary, and especially in transport energy, limited natural resources, deterioration of the environment, today it seems relevant to solve the following problems.

- further improvement of the design of diesel engines in order to boost the liter capacity;
- reduction of operating fuel consumption;
- reduction of toxic emissions into the environment with exhaust gases;
- reduction of emissions ($CO_2$) into the environment with fuel combustion products.

The solution of these problems seems to be a much more rational direction, in comparison with the proposed solutions for the reduction and possible abandonment of the use of diesel engines in the future, which will invariably lead to the energy crisis, which may turn out to be much more painful for humanity in comparison with the ecological one.

In this chapter of the monograph, in order to systematize, select directions, and search for rational ways to improve the efficiency of diesel engines, the main results of fundamental and applied research carried out in recent years at the Department of Internal Combustion Engines of the National Technical University "KhPI" are considered. The experience of the authors in improving the quality of processes in cylinders of diesel engines, in increasing the reliability of the most loaded parts, in ensuring a reduction in the level of emissions of toxic substances and carbon dioxide, including the use of alternative fuels, is generalized.

It is proposed to evaluate the effectiveness of technical solutions to reduce the operating fuel consumption of transport diesel engines and emissions of toxic substances with exhaust gases, use of alternative and hybrid fuels, which includes green hydrogen, is proposed to be carried out using the fuel-ecological criterion. It is shown that further improvement of this criterion is associated with taking into account the impact of carbon dioxide emissions on the environment.

At the same time, the heat generated by the combustion of fuel in the engine cylinders cannot be completely converted into useful mechanical work. To study the efficiency of thermodynamic processes of diesel engines, it is proposed to use the anergy-energetic method of analysis, and the quality of heat conversion into work is estimated by the exergy efficiency, which makes it possible to identify the mechanisms of formation of internal and external losses and substantiate the ways to achieve optimal heat use.

Along with the improvement of economic and environmental indicators and the technical level, the improvement of diesel engines is associated with an increase in liter power, which requires ensuring reliability, first of all, the most heat-stressed engine parts and using modern industry 4.0 technologies.

## 2. Comprehensive assessment of the automotive diesel engines efficiency in terms of fuel consumption and exhaust gas toxicity

Improving the efficiency of power plants, preserving natural resources, and improving the quality of the environment are global problems of our time. Diesel engines are the main source of energy for transport, and at the same time, they are one of the main consumers of fuel oil and an active pollutant of the environment.

The level of excellence and technical level of modern diesel engines is largely determined by fuel consumption and exhaust gas emissions (EG). Diesel engines have higher fuel efficiency and lower mass emissions of toxic substances compared to gasoline and other heat engines. However, research data shows that, along with a high level of toxicity of nitrogen oxides emitted into the atmosphere together with EG, particulate matter (PM) poses a great danger to humans and the environment due to the adsorption of carcinogens.

At the same time, it is known that technical solutions aimed at reducing fuel consumption have an impact on the environmental performance of an engine, and fuel consumption can increase with an improvement in its environmental performance. Therefore, a compromise is needed. A targeted search for compromise technical solutions requires a comprehensive approach using a criterion that takes into account the level of fuel consumption indicators, EG emissions, and operating conditions. The solution to this problem is of paramount importance for automobile engines, since, they are used in crowded places in cities, suburbs, industrial areas and, therefore, pose the greatest danger to people and the environment.

## 2.1 Basics of calculating a comprehensive criterion

At the Department of Internal Combustion Engines of the National Technical University "Kharkiv Polytechnic Institute" (NTU "KhPI"), a dimensionless comprehensive criterion of fuel efficiency and EG toxicity for diesel engines has been developed [1].

This criterion is informative, simple, and user-friendly, takes into account the operating conditions, provides information on the degree of economic and environmental excellence, the effectiveness of the developed measures to improve the work process, engine design, and technology, the use of alternative and mixed fuels, exhaust gas neutralization systems for a specific diesel engine.

The initial data for the criterial comprehensive assessment of diesel engines are obtained with relatively simple, affordable, and minimal bench tests.

To determine the fuel-ecological criterion, it is necessary to know the average operating effective efficiency of the engine ($\eta_{e \cdot e \cdot}$), operating costs for fuel ($S_f$), and compensation for environmental damage from harmful emissions EG ($S_{ec}$) [2].

Then the comprehensive criterion can be represented as:

$$C_{f.ec} = \eta_{e.e} \cdot \beta \tag{1}$$

Here, $\beta$ is the ratio of relative operating environmental costs

$$\beta = \left(S_{f.ec} - S_{ec}\right)/S_{f.ec} \tag{2}$$

where $S_{f.ec} = S_f + S_{ec}$– are total costs fuel and compensation for environmental damage from harmful emissions.

Then, the unit costs for compensation of environmental damage from the harmful effects on the environment of the exhaust gases of a diesel engine during the combustion of 1 kg of fuel, referred to a unit of power for each representative fixed mode of operation of a diesel engine, are equal to:

$$S_{eci} = \left({}^{B_{hi}}/_{N_{ei}}\right) \cdot D_{eci} \text{ (EUR/kW} \cdot \text{hr),} \tag{3}$$

And the total unit costs for reimbursement of environmental damage from the harmful effects of toxic emissions of exhaust gases of a diesel engine for all representative fixed modes of the operating model.

$$S_{ec} = \frac{\sum_{i=1}^{z}(B_{hi} \cdot D_{eci} \cdot P_i)}{\sum_{i=1}^{z}(N_{ei} \cdot P_i)} \text{ (EUR/kW} \cdot \text{hr)} \tag{4}$$

In formulas (3) and (4): $B_{\text{hi}}$—the hourly fuel consumption for each engine operating mode (kW hr), $N_{\text{ei}}$—the effective power for each engine operating mode (kW); $D_{\text{eci}}$—value damage cost when burning one kg of fuel, in EUR/ kgf; $P_i$—partial operating time of the engine at each $i$-th fixed mode of the diesel engine operation model, $z$—the number of representative modes.

## 2.2 Automotive diesel engine operation model

As mentioned above, it is advisable to evaluate the indicators of fuel efficiency and toxicity of exhaust gases of diesel engines under operating conditions during bench tests on typical fixed operating modes, which are selected taking into account the type, purpose, and generalized data on engine operating time. Selected and justified fixed modes of operation, in which bench tests are carried out, represent a model of engine operation.

As a result of the analysis and processing of the operational test data, the authors proposed a generalized model of the operation of a diesel truck in the form of probabilistic distribution of the centers of the operating ranges (**Figure 1**).

Thus, it is possible, based on the results of bench tests of a diesel engine, to determine the level of fuel costs and compensation for environmental damage from the harmful effects of EG on the human body and the environment, as well as to apply a dimensionless fuel and environmental criterion for comparative assessment.

The disadvantages of the proposed model include the comparative complexity of the procedure for carrying out bench tests of a diesel engine, which provides for the determination of a large number of parameters at 28 modes. In this regard, based on summarizing the results of the research carried out for diesel engines of trucks, a 9 regime test cycle is proposed. The basis for the development of the 9th mode test cycle was ensuring the maximum possible compliance with the comprehensive criterion in comparison with tests for the 28th mode cycle. As a result, the developed cycle with a limited number of load modes and crankshaft speeds of a diesel engine makes it possible to determine a comprehensive criterion of fuel efficiency and exhaust gas toxicity without introducing additional errors.

The proposed 9th mode cycle of bench tests to determine the comprehensive criterion of fuel consumption and toxicity of exhaust gases during the operation of diesel engines of trucks is presented in **Table 1**. At the rated speed mode, the diesel engine is tested under loads corresponding to $P_n$ and 0.7 $P_{n.}$ Three load modes (1.1, 0.7 and 0.3$P_n$) correspond to crankshaft speeds of 0.8 and 0.6 $n_{n.}$ Another mode takes into account the share of fuel and environmental costs when the diesel engine is operating at the minimum idle crankshaft speed ($n_{x/x \text{ min}}$).

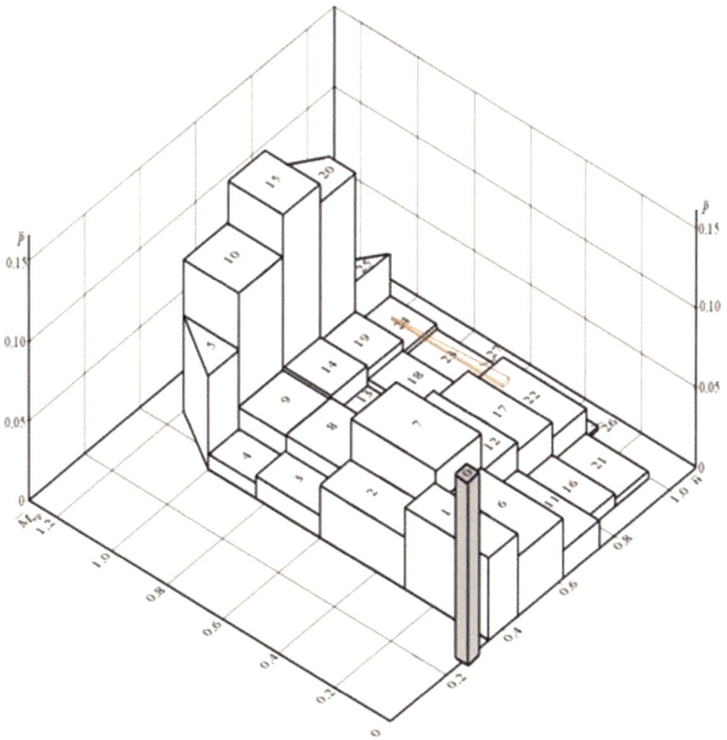


**Figure 1.**
*Probabilistic distribution of ranges of operating modes of a diesel engine of a truck during the aggregate movement in the city and on a suburban highway.*

| Modes Nos. | *n* | *P* | Significance coefficient, z |
|---|---|---|---|
| 1 | $n_n$ | $P_n$ | 0.05 |
| 2 | $n_n$ | 0.7 $P_n$ | 0.025 |
| 3 | 0.8 $n_n$ | 1.1 $P_n$ | 0.3 |
| 4 | 0.8 $n_n$ | 0.7 $P_n$ | 0.05 |
| 5 | 0.8 $n_n$ | 0.3 $P_n$ | 0.05 |
| 6 | 0.6 $n_n$ | 1.1 $P_n$ | 0.35 |
| 7 | 0.6 $n_n$ | 0.7 $P_n$ | 0.1 |
| 8 | 0.6 $n_n$ | 0.3 $P_n$ | 0.05 |
| 9 | $n_{x/x\ min}$ | 0 | 0.025 |

**Table 1.**
*Model of operation of a diesel engine of a truck with the combined movement of the city and suburban highway.*

For each mode, coefficients were selected that took into account the conditions for the distribution of fuel and environmental costs over the ranges of the diesel engine operating model of a fully loaded truck when driving in the city and on the highway.

Interestingly, if we compare the ESC cycles in accordance with the UNECE rules for diesel engines of trucks and the KhPI cycle, it can be noted that with a smaller number of test modes, the KhPI cycle covers almost equal ranges in terms of load and speed.

Consequently, for a balanced assessment of the environmental hazard of diesel engines for various purposes, it is necessary to take into account the real conditions of their operation and an objective approach to calculating the damage from the harmful effects of exhaust gases. Since the emissions of toxic components of the exhaust gases and the fuel efficiency of a diesel engine are directly related to the organization of mixture formation and combustion, an integrated approach to this problem is required.

The choice of the significant coefficient for each of the modes is based on generalizing the share of costs for fuel consumption and compensation for environmental damage from the harmful effects of exhaust gases in the total costs of the aggregate modes. The proposed dimensionless comprehensive criterion of fuel efficiency and toxicity allows a targeted search and assessment of the effectiveness of the developed measures aimed at reducing fuel consumption and toxicity of exhaust gas emissions under engine operating conditions.

The comprehensive criterion allows:

- to evaluate the efficiency of the internal combustion engine when operating conditions change;
- to determine the operating modes in which $S_f$, $S_e$, and $S_{\text{fe}}$ are the most significant;
- to develop measures aimed at increasing the efficiency of the internal combustion engine;
- to evaluate the efficiency of using alternative fuels or EG neutralization systems.

The dimensionless comprehensive criterion of fuel efficiency and toxicity of harmful EG emissions, taking into account the degree of diesel loading and the factor of operating time, makes it possible to evaluate the quality of a diesel engine when used on different vehicles or to assess the fuel and environmental efficiency of various engines when used on the same vehicle. The use of a comprehensive criterion, or, if necessary, the ratio of relative operating environmental costs, in turn, allows an analysis of a compromise situation when a decision is required on the permissible increase in fuel costs provided that the overall level of fuel and environmental costs decreases. In this case, it is necessary to additionally agree on the degree of complexity of the implementation of these solutions, taking into account the potential costs of a significant reconstruction of the diesel engine and the costs of using, for example, electronic control systems or neutralization of EG.

Further improvement of the integrated fuel and environmental criterion is associated with taking into account the compensation for damage caused by diesel engines by $CO_2$ emissions.

## 2.3 Efficiency of using alternative fuels in road transport

The most pressing for transport engines are fuel, energy, and environmental problems. These problems are directly related to the limitation of natural resources and environmental degradation. Currently, there are about 1 billion vehicles in the world that run on petroleum engine fuels and actively pollute the environment with hazardous toxic constituents of exhaust gases—carbon oxides (CO), hydrocarbons (CH), nitrogen oxides ($NO_x$), particulate matter (PM), and also contribute to the expansion of the greenhouse effect by emissions of carbon dioxide ($CO_2$).

In this regard, along with the further improvement of the power plants of vehicles, including those with diesel engines, the most important task is to expand the use of alternative fuels, as well as to reduce emissions of toxic components of exhaust gases and reduce the level of $CO_2$ emissions.

It should be noted that the share of the level of $CO_2$ emissions into the atmosphere by road transport and their average annual increase in relation to the total levels of emissions of $CO_2$ with fuel combustion products is ~23% and in relation to the technogenic $CO_2$ emission into the atmosphere ~ from 1 to 2%. These data give grounds to assert that vehicles with internal combustion engines, like all heat and power engineering, are not significant at the present stage in terms of the degree of accumulation of $CO_2$ in the atmosphere, and the corresponding warming of the climate, are not significant. But, on the other hand, vehicles with internal combustion engines negatively affect the change in the natural environment, as a component of the creation of transport systems, their operation and maintenance, including the search, production, transportation, processing of all natural resources, including the oil industry, which has a negative impact on the environment. Under the influence of the above-listed factors, the transformation and destruction of natural massifs, land desertification, pollution of the waters of the World Ocean occurs. All this leads to the degradation and destruction of the planet's photosynthetic systems, to a decrease in their natural biological productivity, a corresponding decrease in runoff levels of $CO_2$ and, as a consequence, an increase in the content of $CO_2$ in the atmosphere and the temperature of the surface air layer [3].

Currently, promising alternative fuels for diesel engines include:

- natural gas, which, in terms of reserves and cost, is currently considered one of the most acceptable energy carriers for vehicles, especially those operating in large cities;

- water-fuel emulsions, which are widely used in water transport, as well as for trucks with diesel engines [4, 5];

- biofuel, which can be used for road transport and especially for self-propelled agricultural vehicles.

"Green" hydrogen is currently being considered as an additional energy carrier for oil and alternative fuels for vehicles. The presence of additives of "green" hydrogen provides a decrease in the energy of ignition of fuels, an increase in the rate of its combustion, and reduces the level of formation of $NO_x$ and PM.

The Department of Internal combustion engines of NTU "KhPI" using a comprehensive fuel and environmental criterion (see Section 2.1 of this Chapter) has made a comparative analysis and a quantitative assessment of the effectiveness of the use of alternative fuels when operating a diesel engine of a truck in comparison with standard diesel fuel on a 6-cylinder diesel engine with a cylinder volume of 9.5 l.

The results of the tests and processing of experimental data are shown in **Table 2** and **Figure 2**. The Figure and the Table show the relative change in the integrated fuel and environmental criterion Cf.ec.

| Features of the piston design | Uncoated | | Coated |
|---|---|---|---|
| Engine boost level, kW/l | 25 | 29 | 29 |
| Temperature at a heavier stationary mode (section b), °C | 314 | 343 | 322 |
| Temperature at idle speed (section d), °C | 191 | 193 | 188 |
| Stress in a heavier stationary mode (section b), MPa | −36 | −42.5 | −7.6 |
| No-load stress (section d), MPa | −0.15 | 0 | −2.9 |
| Parameter of the physical reliability of the structure for the resource P | 0.552 | < −4 | 0.574 |

**Table 2.**
*Parameters of thermal tension of the piston combustion chamber edge.*

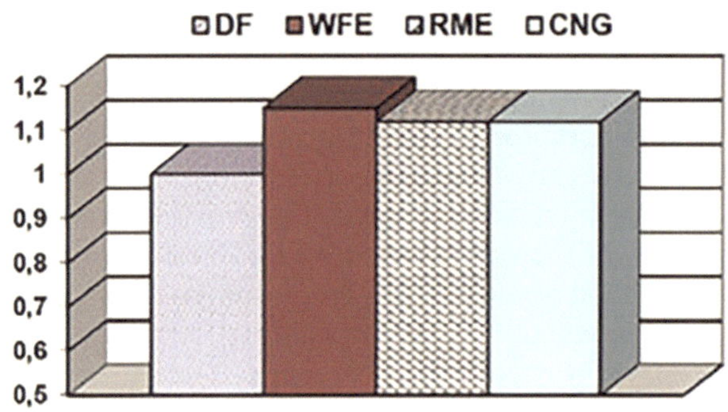


**Figure 2.**
*Relative change of Cf.ec., when using alternative types of fuel in a 6-cylinder automobile diesel engine with a cylinder capacity of 9.5 l.*

The criterion was determined based on the results of bench tests using a model of operation of a truck diesel engine. The engine was tested on diesel fuel, compressed natural gas (CNG) with 15% pilot diesel, rapeseed methyl ester (RME), and water-fuel emulsion (WFE) which contained diesel fuel and 10% water.

The price of natural gas and diesel fuel in the calculations was taken according to the averaged data of filling and gas filling stations in Ukraine.

The costs of water and the preparation of a water-fuel emulsion were not taken into account in the calculations.

The cost of 1 kg of rapeseed oil methyl ester, obtained in pilot plants and in small quantities, exceeds the cost of 1 kg of diesel fuel by 1.4–1.6 times. However, when calculating the comprehensive criterion, the price of rapeseed oil methyl ester was taken to be equal to the price of diesel fuel, taking into account its possible decrease with the expansion of the production of this fuel.

It follows from the above data that any of the investigated alternative fuels in an automotive diesel engine provides an increase in fuel and environmental efficiency. This is mainly due to a decrease in toxic emissions at engine operating modes at maximum load at reduced speeds.

The use of a gas-diesel cycle with CNG allows increasing the value of the comprehensive criterion of fuel efficiency and toxicity of exhaust gases of a truck diesel engine by 11.9%. It should be noted that in this case, as the proportion of partial modes increases, the ratio between the constant doses of ignition diesel fuel supplied to the cylinders and the amount of compressed natural gas increases. Accordingly, fuel costs increase and environmental efficiency from the use of gas fuel decreases.

The complex fuel and environmental criterion increase by almost the same amount when a truck diesel engine runs on rapeseed oil methyl ether. In this case, the deterioration in the average operating efficiency occurs to a large extent when the engine is running at partial conditions, in comparison with the engine running on diesel fuel.

When a truck diesel engine runs on a water-fuel emulsion, the comprehensive criterion of fuel efficiency and toxicity of exhaust gases increases most significantly—by 15%. This is due to the simultaneous reduction in environmental operating costs, and an increase in the average diesel engine operating efficiency.

It should be noted that the presented results were obtained without any changes in the diesel engine settings and without any changes in their design in order to adapt to a specific type of alternative fuel. Consequently, there are reserves for increasing fuel efficiency and improving the environmental performance of diesel engines when using each of the considered alternative fuels. These reserves include the use of alternative hybrid fuels, which contain green hydrogen.

## 3. Application of the anergo-exergy method for evaluating the efficiency of working processes in diesel engines

The heat generated by the combustion of fuel in the engine cylinders cannot be completely converted into useful mechanical work. In the thermodynamic cycle, the efficiency of converting heat into work is estimated by the thermal efficiency $\eta_t$, which is always less than one as a result of the transfer of part of the heat to a cold source. In a real engine, heat losses increase due to friction, heat transfer, incomplete combustion, and other reasons. In this regard, the effective efficiency of $\eta_t$ cycle is less important than the value of $\eta_t$.

Currently, there are two directions in the thermodynamics of investigating the efficiency of diesel processes. The traditional direction is that for the thermodynamic study of motors, a heat balance is used, based on the first principle of thermodynamics, when the criterion for the quality of converting heat into work is the effective efficiency ($\eta_e$). The capabilities of the traditional method are limited by the fact that the heat balance only records the final, qualitative result of energy transformations in the internal combustion engine cycle. To implement all the reserves for improving modern internal combustion engines, it is necessary to deeply study the quality of energy transformations in the engine. The anergo-exergetic method of analysis combines the first and second laws of thermodynamics, and the exergy efficiency ($\eta_{ex}$) acts as a criterion for the quality of converting heat into work. Only this method allows to identify the mechanisms of the formation of internal and external ICE losses, assess the possibilities of their reduction, and, therefore, substantiate the ways to achieve optimal heat use in diesel engines.

### 3.1 Theoretical foundations of the anergo-exergy method

It is known that heat and internal energy, as forms of energy, determined by the first law of thermodynamics, can only be partially converted into work. Accordingly, in a heat engine (**Figure 3**) it is possible to convert into work only a certain fraction of the energy supplied in the form of heat $Q$ [6].

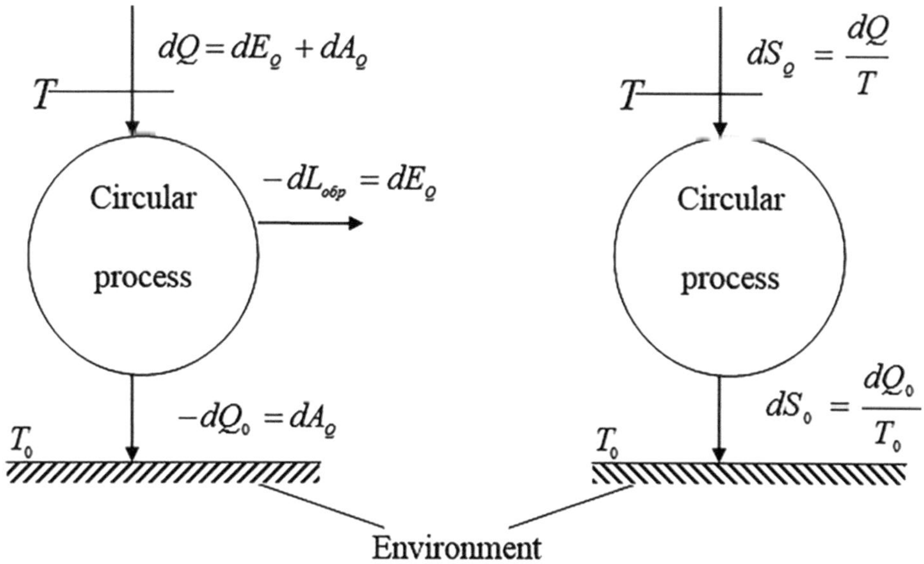


**Figure 3.**
*Diagram of the circular process of a heat power plant.*

In a generalized form, a consequence of the second law of thermodynamics is the statement that there are forms of energy that can be converted into any other form of energy. These forms of energy, covered by the general concept "exergy", are completely mutually convertible during reversible processes, and by reversible and irreversible processes they can be transformed into limited convertible forms of energy—internal energy and heat. At the same time, limited convertible forms of energy cannot be converted in any quantities into exergy. All forms of energy that are not transformed into exergy are summarized by the term "anergy".

"Exergy is the maximum possible work that the system can perform in the reversible transition from this state to a state of equilibrium with the environment; anergy is the energy that cannot be converted into exergy" [7].

For all forms of energy, the following general correlation is valid:

$$\text{Energy} = \text{Exergy} + \text{Anergy}.$$

According to the principle of irreversibility, all natural, actually occurring processes are irreversible. Thus, in these processes, the supply of exergy decreases due to its transformation into anergy. Part of the exergy that is converted into anergy during irreversible processes is the loss of exergy in the process.

To use the concept of exergy and anergy, it is necessary to know the proportions of these quantities for various forms of energy. When determining exergy, the heat supplied to the heat-power plant is considered, the working fluid of which performs a circular process. The exergy of heat appears here as useful work, and anergy as the unused heat of a circular process. However, the useful work of the circular process coincides with the exergy of the supplied heat under the following conditions:

- the circular process is reversible (otherwise it turns into anergy and useful work will be less than the applied exergy);

- heat removal is carried out at ambient temperature, so that the removed heat consists only of anergy and corresponds to the anergy of the supplied heat (**Figure 3**).

Heat supplied to the working fluid

$$dQ = dE_Q + dA_Q \tag{5}$$

As a result of the heat supply $dQ$, perceived at temperature $T$, the entropy of the working fluid will increase.

$dS_Q = dQ/T$. Since the entropy is not produced in a reversible process, the given heat $dQ_0$should be such that the entropy

$$dS_{Q_0} = dQ_0/T_0$$

transferred with it is equal to the perceived entropy $dS_Q$. From the balanced equation of entropy

$$dS_Q + dS_{Q_0} = \frac{dQ}{T} + \frac{dQ_0}{T_0} \tag{6}$$

for the given heat we get

$$-dQ_0 = \frac{T_0}{T} dQ \tag{7}$$

The heat removed to the environment consists only of anergy and represents the desired anergy of heat.

$$dA_Q = \frac{T_0}{T} dQ \tag{8}$$

The exergy of heat is manifested as the work of an imaginary reversible circular process

$$dE_Q = dQ - dA_Q = \left(1 - \frac{T_0}{T}\right) dQ \tag{9}$$

If heat is perceived or given off by the system in a certain temperature range, then the exergy of heat perceived or given off with heat $Q_{12}$ is determined by integrating:

$$E_{Q_{12}} = \int_1^2 \left(1 - \frac{T_0}{T}\right) dQ = Q_{12} - T_0 \int_1^2 \frac{dQ}{T} \tag{10}$$

In a similar way for the anergy of heat

$$A_{Q_{12}} = T_0 \int_1^2 \frac{dQ}{T}. \tag{11}$$

Here $T$ is the temperature of the energy carrier that gives or receives heat.

As well as $Q_{12}$, exergy of heat and anergy are characteristics of the process, and not parameters of state. The exergy and anergy of heat depend not only on $T_0$, but also on the temperature of the system that receives or gives off heat, which can be seen from expressions (10) and (11). This allows coming to the conclusion that in heat power plants, heat supply to the working fluid must be implemented at the maximum possible temperature for a given installation and to obtain the maximum possible work in the cycle.

### 3.2 Evaluation of the efficiency of the diesel engine working process based on the anergo-exergy method

According to the theory [8], any heat and enthalpy can be represented as components of exergy and anergy.

$$Q = E_Q + A_Q; I = E + A.$$

In addition, the anergy balance equations are valid for any ICE unit

$$\sum D_i = \sum A_{out} - \sum A_{in}$$

and balance of exergy

$$\sum D_i = \sum E_{in} - \sum E_{out}$$

where $\Sigma E_{in}$ and $\Sigma E_{out}$ may include work supplied to the assembly or taken away from it.

From these equations, it can be concluded that the exergy losses $\Sigma D_i$ arising due to the irreversibility of real processes increase the energy leaving the assembly and decrease the exergy entering the assembly. This allows representing the flows of anergy, exergy, and exergy losses in the form of anergo-exergy scheme of the thermodynamic unit. Based on the above equations and fundamental considerations, which are laid above in the basis for constructing an anergo-exergy scheme of a thermodynamic unit, will make it possible to build an anergo-exergy scheme of a diesel engine, shown in **Figure 4**.

The bifurcation of the ICE assemblies made it possible to reveal the corresponding losses of exergy: $D_{sc}$, $D_s$, $\Sigma D_i$, $D_w$, $D_{wp}$, $D_{ex}$, $D_t$, $D_{fp}$, $D_o$, $D_{op}$, $D_{МН}$. To determine the listed exergy losses, it is easier to use the following energy balances of the corresponding assemblies. $D_{sc} = A_{sc} - A_0$; $D_s = (A''_{os} - A'_{os}) - (A_{sc} - A_s)$; $A_m = A_{exh} - A_m$; $A_w = A''_w - A_w - A_{qw}$; $D_o = A''_o - A_o$; $D_{op} = A_o - A'_o$; $D_{fp} = A'_f - A_f$.

For a diesel engine cylinder, you can write

$$E_{Q_{ch}} = (Q_{ch} - Q_w) - A_Q,$$

where $A_Q = T_0 \int \frac{\delta Q_{ch}}{T} - T_0 \int \frac{\delta Q_w}{T} = A_{Q_{ch}} - A_{Q_w}$—heat anergy $Q = Q_{ch} - Q_w$.

At the same time

$$D_f = D_s + D_B = A_T - A_Q - A_s - A_f. \tag{12}$$

To find $A_Q$ and $D_f$, we divide the cycle in the engine cylinder into two parts: section (a–e)—compression, combustion, and expansion and section (e–a)—the gas exchange process.

In general, $dA = \delta A_Q + \delta A_f + \delta A_s - \delta A_T + \delta D_s + \delta D_B$, so for the process (a—f)

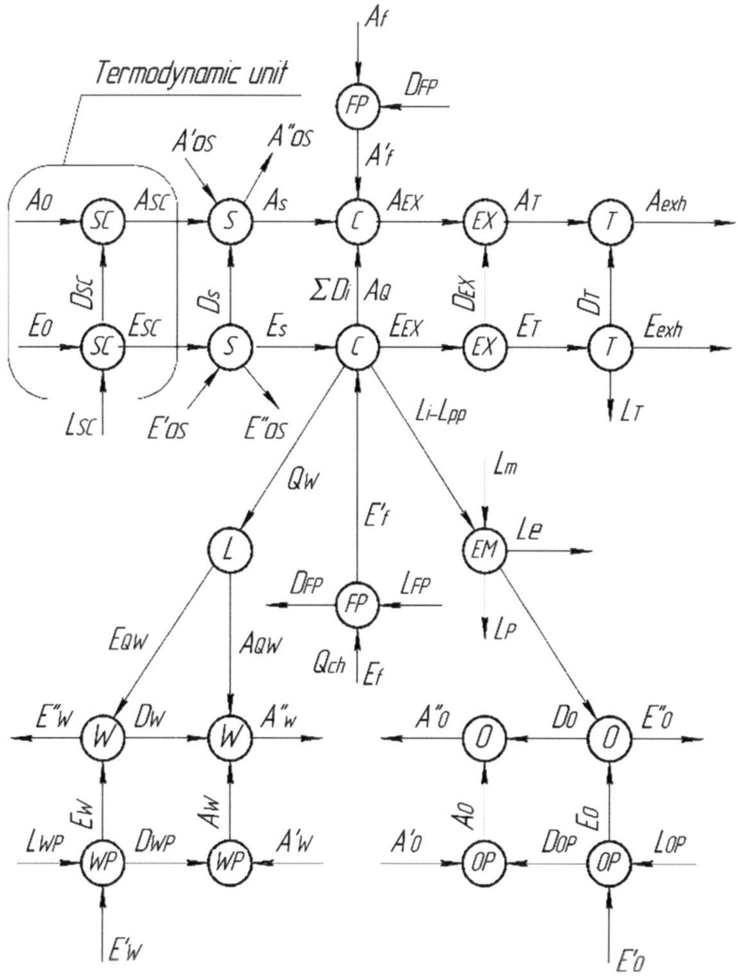


**Figure 4.**
*Anergo-exergy scheme of the internal combustion engine. SC—supercharger; S—air cooler; C—cylinders; EX—exhaust manifold; T—gas turbine; EM—engine mechanisms; FP—fuel pump; WP—water pump; OP—oil pump; W—water system unit that receives frictional heat transferred to water; O—oil system unit that receives frictional heat transferred to oil;* $A_i$, $E_i$ *respectively, are the flows of anergy and exergy of the working fluid; —respectively, with the cooling agent anergy and exergy;* $A'_{os}$, $a_{os}$, $a'_w$, $A_w$, $a_w$, $A'_o$, $A_o$, $A_o$, $A_f$, $a'_f$ *and* $E'_{os}$, $E_{os}$, $E'_w$, $E_w$, $E_w$ $E'_o$, $E_o$, $E_o$ $E_f$, $E'_f$ *respectively are the flows of anergy and exergy with the cooling agent, of water in the liquid cooling system of the engine, of oil in the engine lubrication system, of the fuel in the fuel supply system.*

$$A_e - A_a = A_{Q_{ch}} - A'_{Q_w} + A_f,$$

For process (e–a)

$$A_a - A_e = -A''_{Q_w} + A_s - A_T + D_f.$$

If the law of heat transfer in the gas exchange section is known, then

$$A''_{Q_w} = T_0 \int \frac{\delta Q_w}{T} = \frac{T_0}{T''_m} Q''_w,$$

where, $T''_m$ average process (e–a) temperature.

Having found $A''_{Q_w}$, we find the total losses of exergy during gas exchange $D_f$.

According to Eq. (12)), one can find $A_Q = A_T - A_s - A_f - D_f$.

If the law of heat transfer is known in the area of compression, combustion, expansion, then

$$A'_{Q_w} = T_0 \int \frac{\delta Q_w}{T} = \frac{T_0}{T'_m} Q'_w,$$

where $T'_m$ average process (a–f) temperature.

In this case $A_{Qch} = A_e - A_a - A_f + A'_{Q_w}$.

The balance of the exergy flows of the internal combustion engine can be obtained by considering the contour $E$:

$$E_0 + L_m + L_{op} + E'_o + A'_o + L_{fp} + Q_{ch} + E_f + A'_w + E'_w + L_{wp} + E'_{os} + L_{cs} =$$

$$= L_e + E''_o + A''_o + L_p + D_{fp} + E''_w + A''_w + E''_{os} + D_{sc} + D_s + A_Q + D_w + D_T + E_{exh} + L_{m.}$$

Let us take into account that

$$L_p = L_{op} + L_{wp} + L_{fp}; L_m + L_{sc} = L_{mi}; D_{\sum o} = D_o + D_{op} = A''_o - A'_o;$$

$$D_{\sum w} + A_{Qw} = D_w + D_{wp} + A_{Q_w} = A''_w - A'_w; D_{fp} + D_s + D_в = D_f.$$

With this in mind, we get

$$Q_{ch} = L_e + \left[\left(E''_o - E'_o\right) + \left(A''_o - A'_o\right)\right] + \left[\left(E''_w - E'_w\right) + \left(A''_w - A'_w\right)\right] + \left(E''_{os} - E'_{os}\right) +$$

$$+ \left(E_{exh} - E_0 - E_f\right) + D_{sc} + D_s + D_f + A_Q + D_T$$

or

$$Q_{ch} = L_e + \Delta E_o + \Delta E_w + \Delta E_{os} + \Delta E_{exh} + A_Q + D_{ch} + D_s + D_f + D_T + D_{\sum o} + D_{\sum w}.$$

Note that in this expression

$$\sum E_i + A_{Qch} + \sum D_i = Q_2 \text{ and } \sum E_i = \sum Q_2.$$

Since $Q_{ch} - A_{Q_{ch}} = E_{Q_{ch}}$, then

$$E_{Q_{ch}} = L_e + \sum E_i + \sum D_i.$$

This dependence is the equation of the energy balance of the engine. It can be seen from it that the exergy $E_{Q_{ch}}$, is supplied to the working fluid with the heat of the fuel $Q_{ch}$, is spent on the efficient operation of the engine $L_e$, covering the losses of exergy $\sum D_i$. Part of the fuel heat energy is removed to the environment with water $\Delta E_w$, with oil $\Delta E_o$, with intermediate air cooling $\Delta E_{os}$, and with exhaust gases $\Delta E_{exh}$. According to the well-known theory [9, 10], exergy supplied to the internal combustion engine

$$E_{\text{sup}} = E_{Q_{ch}} - \sum E_i.$$

Then

$$E_{\text{sup}} = L_e + \sum D_i.$$

The efficiency of converting exergy supplied to the internal combustion engine into useful work can be estimated by the exergy efficiency

$$\eta_e = \frac{L_e}{E_{\text{sup}}} = 1 - \frac{\sum D_i}{E_{\text{sup}}}.$$

Let us give an example of determining exergy losses using the proposed anergo-exergy method for a 6ChN12/14 tractor diesel engine with a power of 150 kW in one of its operating modes.

The performed calculation shows that in the diesel cycle, when the heat of the fuel is transferred to the working fluid, 21.3% of the anergy of this heat was formed. In addition, due to the irreversibility of real processes, losses of exergy amounted to 16.75%, that is, 38.05% of inoperable heat was also formed in the cycle. Part of the workable heat (20.95%) is carried away into the environment by heat carriers—oil, water, air, and exhaust gases. The rest of the heat turned into useful work (41.05%).

The largest amount of workable heat is carried away by waste gases (16.84%). The loss of performance in the lubrication and cooling systems is 2.16 and 7.84%, respectively. Attention is drawn to the noticeable total loss of performance during filling and when gases enter the exhaust manifold—1.34%. Noteworthy are $D_{SC}$ = 1.25% and $D_T$ = 3.83%—losses of exergy in the supercharger and gas turbine. In reducing the indicated losses of exergy, reserves for increasing the efficiency of a diesel engine are laid. **Figure 5** shows the items of the exergy balance.

In the diesel cycle, the exergy of the chemical heat of the fuel is supplied to the working fluid

$$E_{Q_x} = Q_{ch} - A_{Q_{ch}} = 3.19 kJ/cycle.$$

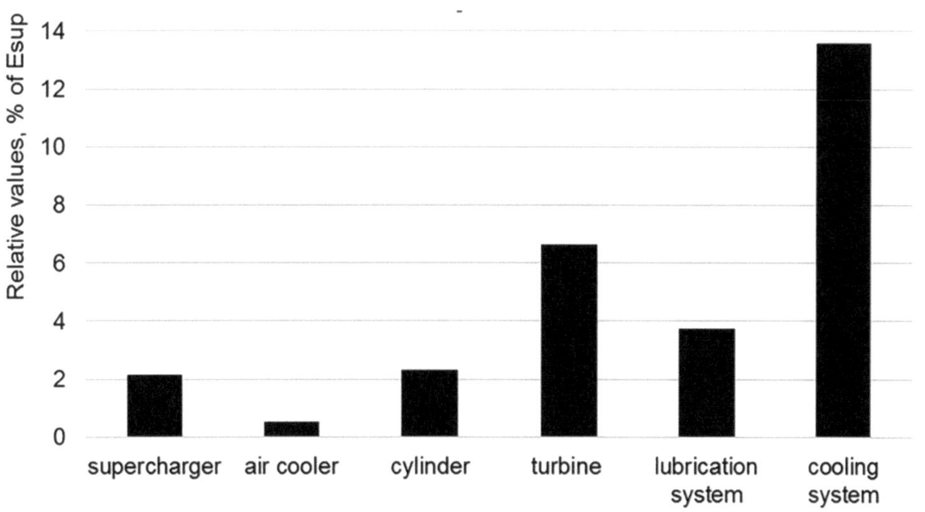


**Figure 5.**
*Exergy balance of a diesel engine.*

However, exergy took part in the process of converting this heat into work.

$$E_{\text{sup}} = E_{Q_{ch}} - \sum E_i - A_{Q_{ch}} = 2.34kJ/cycle.$$

Due to the irreversibility of real processes, 29% of $E_{\text{sup}}$ turned into loss of exergy, that is, a third of the exergy has irreversibly disappeared (turned into anergy).

The remaining 71% of exergy turned into effective work. The exergy losses in the exergy balance are "significant", since the exergy losses in water are 13.58%, that is, almost half of all exergy losses. Turbine losses are less and amount to about 6%. Both should be dealt with at the same time by the researcher in order to reduce them.

So, the use of anergo-exergy method of analysis makes it possible to identify the mechanisms of the formation of internal and external losses of diesel engines and their systems, and to substantiate the ways to achieve their maximum efficiency.

## 4. Improving the reliability of the most thermally stressed parts of highly accelerated engines, based on the operating conditions

Modernization of existing and creation of new engines of high specific power causes significant difficulties since it is necessary to minimize costs during the life cycle of a structure while ensuring a set of quality indicators during given service life. At the same time, for the most thermally stressed engine parts, the provision of their physical and parametric reliability must be taken into account. The practice of operating engines testifies to cases of failure of the combustion chamber parts due to their cracking during the declared resource and the appearance of chafes and scuffs on the lateral surface of the piston [11]. A substantiated increase in the reliability of heat-stressed parts of an internal combustion engine requires the use at the design stage of mathematical models that take into account a complex of factors affecting the physical and parametric reliability.

Let us consider the process of loss of structural reliability based on the model of material damage accumulation in time $d(\tau)$. At the initial moment of operation, it is usually assumed that there is no damage to the material, the reliability factor is $d(0) = 1$. Then the values $d(\tau^*) = 0$ correspond to the limiting state of the material. The dependence of the criterion $d(\tau)$ on the level of the engine-specific power is shown in **Figure 6**. Zone I corresponds to the absence of damage accumulation during the assigned resource P, zone II—to extensive accumulation of damage, zone III—to intensive accumulation of damage, and zone IV—to the loss of reliability during

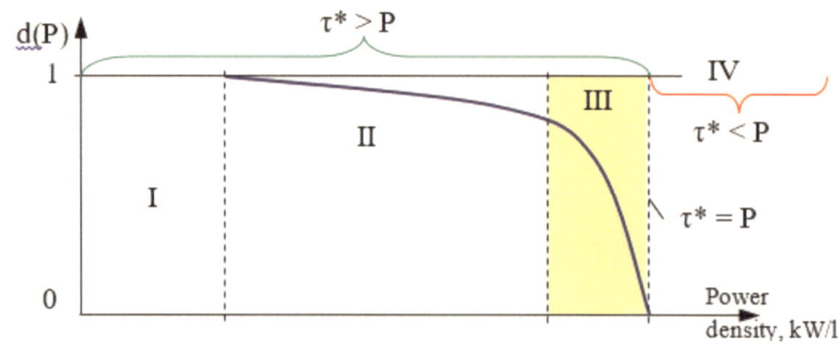


**Figure 6.**
*Typical zones of change in the reliability factor of the ICE heat-stressed part.*

operation, i.e. $d(P) < 0$. The main factors affecting the accumulation of damage are design features of the concerned part; properties of the part material(s); ICE operating conditions and features of the working process, cooling and lubrication systems. The influence of these factors is associated with the heat-stressed state of structures due to the material creep and fatigue over time.

**Figure 7a** shows a diagram of the part critical zone deformation for the case of possible instantaneous plastic deformations and creep deformations under the conditions of deformations structural limitation. Typical examples of such zones are the edges of the pistons combustion chamber and the cylinder head cross-sections between the valve orifice and the injector bore. Here, sections 1–2–3–4 denote the initial engine load, 4–5—work in a stationary heavy operating mode, 5–6–7–8—load reduction to a certain partial mode, 8–9—work in a stationary partial mode, 9–10—subsequent engine load to previous heavy-duty level. In this case, sections 1–2 and 5–6 correspond to the material elastic deformation, sections 2–3 and 6–7: creep deformation and stress relaxation, in sections 3–4 and 7–8 the creep process is accompanied by instant plastic deformations, and in Section 4–5 and 8–9 are characterized by stress relaxation.

In this case, the common condition for ensuring physical reliability during the work of the part material in such zones on the verge of strength are:

$$d_1(\Xi, P) = \varphi_{sf}(\Xi, t(\tau), \sigma(\tau)) = 1 - \sum_{k=1}^{N_P} \frac{1}{N_{sk}} - \sum_{k=1}^{N_P} \frac{1}{N_{fk}} > 0, \tag{13}$$

where $\Xi$ is the operating model of an engine for a specific purpose, $\Xi = \{\zeta_1, \zeta_2, \dots, \zeta_n\}; k = [1, N_p] \Xi = \{\xi_1, \xi_2, \dots, \xi_N\}$—single-engine load cycle; $N_p$—total number of engine load cycles during a given resource $P$; $t$—current temperature state of the part in the investigated area; σ—current tension value; $N_{sk}$—number of cycles until material failure due to creep under the conditions of the $k$-th cycle of loads; $N_{fk}$—number of cycles until material failure due to fatigue under the conditions of the $k$-th load cycle.

In practice, the choice of technical solutions to improve the physical reliability of the high specific power ICE heat-stressed parts corresponds to the solution of the

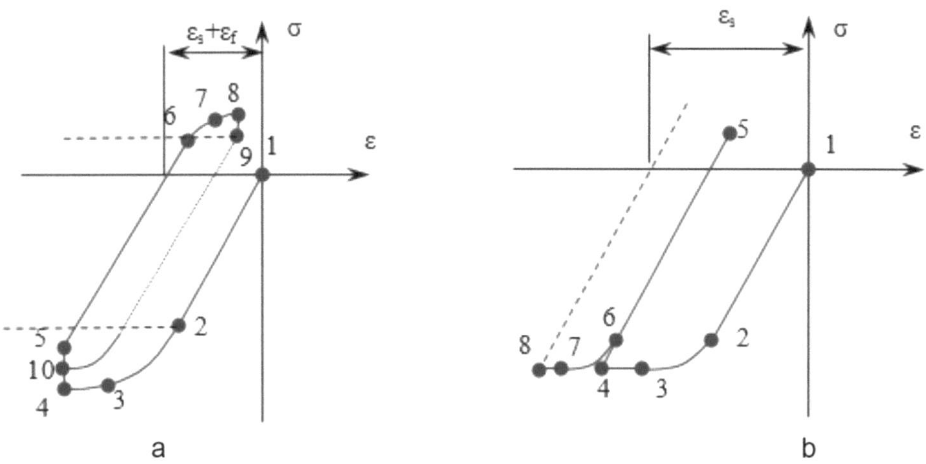


**Figure 7.**
*Typical variants of deformation of the critical zone of ICE heat-stressed part.*

problem of transition of the calculated result in Eq. (13), in accordance with **Figure 1**, from zone IV to zone III.

**Figure 7b** presents a variant of the part critical zone deformation with practically no limitation of creep deformations. Side surface of the piston is a typical example of such a zone. Here sections 1–2–3 denote the initial engine load, 3–4—work on the stationary heavy operation mode, 4–5—load reduction to a certain partial mode and work on the stationary partial mode, 5-6-7—subsequent engine load to the level of the previous heavy-duty and 7–8—subsequent work on the stationary mode. In this case, sections 1–2 and 5–6 correspond to the material elastic deformation, and 2–3–4 and 6–7–8 correspond to creep deformation $\varepsilon_s$, which increases with time.

Methods for determining the profile of the piston lateral surface are known. They consist in determining the clearance $\Delta R_{\text{set}}(h_i, \theta_i)$ when installing a piston in a cylinder that is variable in height $h_i$ in piston circumferential direction $\theta_i$. Subsequently, in the process of friction pair surfaces wear, the real gap increases to the permissible value $[\Delta R_{\text{set}}]$. Accordingly, the structure parametric reliability criterion is represented by the expression [12]:

$$0 < \Delta R_{set}(h_i, \theta_i, \tau) \le [\Delta R_{set}(P)] \tag{14}$$

With an increase in the level of engine boost due to the appearance of creep deformation, in accordance with **Figure 7b**, the size of the gap along with some coordinates $h_i$, $\theta_i$ may not increase, but decrease until chafing and scuffing occur, that is, the theoretical mutual penetration of the piston and cylinder materials, $\Delta R_{\text{set}}\ (h_i, \theta_i, \tau) < 0$. Therefore, to the condition for ensuring parametric reliability in Eq. (14), it is necessary to add the condition for parametric reliability not to exceed the material creep limit:

$$d_2(\Xi) = \begin{cases} 1, & \varphi_s(t,\sigma) \ge 1 \\ \varphi_s(t,\sigma), & \varphi_s(t,\sigma) < 1 \end{cases}, \tag{15}$$

where $\phi_s$ is the function of bringing the creep threshold to the criterion $d_2$.

Thus, the choice of technical solutions to improve the parametric reliability of the ICE piston side surface corresponds to the solution of the problem of transition of the calculated result from zone II to zone I (**Figure 6**).

To obtain a reliable result of the part guaranteed reliability, it is necessary to have input information about the non-stationary low-frequency and high-frequency temperature state of the structure in accordance with the adopted operating model:

$$t_k(\tau) = \bar{t}_k(\tau) + \tilde{t}_k(\tau) \tag{16}$$

$$\sigma_k(\tau) = \bar{\sigma}_k(\tau) + \tilde{\sigma}_k(\tau) \tag{17}$$

where the values $\bar{t}_k(\tau)$ and $\bar{\sigma}_k(\tau)$ correspond to the instantaneous averaged values of the low-frequency change in temperatures and thermal stresses in the investigated area of the part under the conditions of a single load cycle of the type $k$, and $\tilde{t}_k(\tau)$i$\tilde{\sigma}_k(\tau)$ correspond to the instantaneous deviation of temperatures and thermal stresses from the average value.

Formulation of the problem in the form Eqs. (16) and (17) with the subsequent use of model Eq. (13) significantly increases the design time. Therefore, a simplification of problem Eqs. (16) and (17) is proposed, which does not contradict the principle of guaranteed ensuring the strength of a part during design [13]. With a load surge and engine operation in a heavy stationary mode, we take:

$$t_k^{ab}(\tau) = \bar{t}_k(\tau) + \tilde{t}_k^{b\,\max} \tag{18}$$

$$\sigma_k^{ab}(\tau) = \bar{\sigma}_k(\tau) + \tilde{\sigma}_k^{b\,\min} \tag{19}$$

With a load drop and engine operation in a less heavy stationary mode, we take:

$$t_k^{cd}(\tau) = \bar{t}_k(\tau) + \tilde{t}_k^{d\,\max} \tag{20}$$

$$\sigma_k^{cd}(\tau) = \bar{\sigma}_k(\tau) + \tilde{\sigma}_k^{d\,\max} \tag{21}$$

A graphical explanation of the values used is shown in **Figure 8**.

On the basis of the proposed approaches, we determined the reliability criteria $d_1(\Xi, P)$ and $d_2(\Xi)$ relative to the piston of a tractor diesel engine of the SMD type (the engine cylinder diameter is 120 mm, the piston stroke is 140 mm, and the number of cylinders is 4), the boundary conditions of heat transfer for which are well known. The main piston material is the AK12M2MgN alloy. A piston of traditional design and with a heat-insulating layer on the surface of the combustion chamber is considered. Layer material—$Al_2O_3$. Layer thickness—0.25 mm. For a piston with a surface layer, the analysis was carried out according to the parameters under the thermal insulation layer. The calculation of the value $d_1(\Xi, P)$ was carried out for the edge of the piston combustion chamber with a liter diesel power of 25 and 29 kW/l using the RESURS program (Pylyov, V.O., Prokopenko, M.V., Shekhovtsov, A.F.: Resurs. UA Computer Software 5915, 16 July 2002). The value of $N_{sk}$ in Eq. (13) was determined by the energy criterion of Sosnin, the value of $N_{fk}$—on the basis of the generalized Neuber principle [14]. The calculated base is taken as $P$ = 10,000 hours. The engine operation model is adopted for an agricultural tractor [14]. The number of calculation cycles

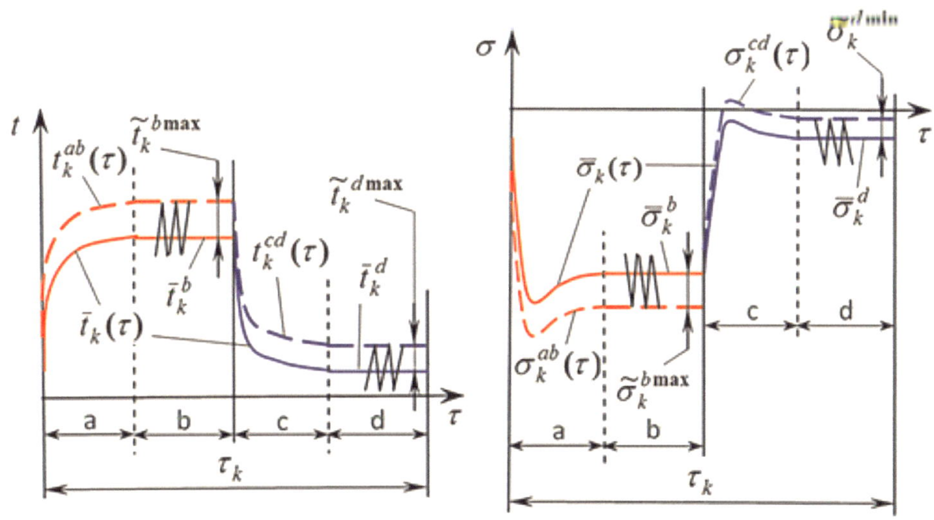


**Figure 8.**
*Local temperature (left) and thermal stress (right) in a single loading cycle of the studied zone of the piston: a—temperature and b—stress state of the studied area of the part: a—load surge; b—heavily loaded stationary mode; c—load drop; d—less loaded stationary mode.*

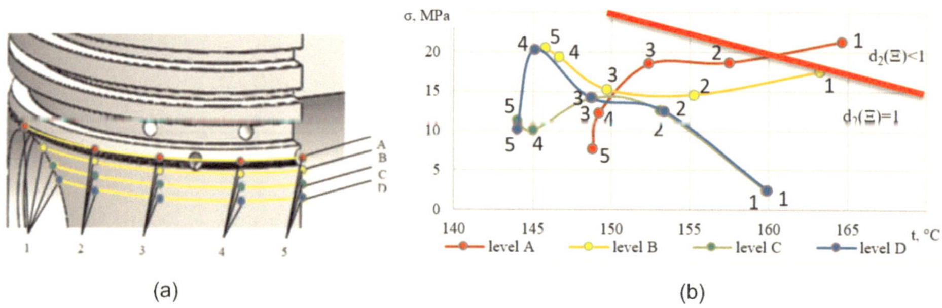


**Figure 9.**
*Control points on the piston: a—Location of control points on the piston for levels a—D; b—Change in the thermally stressed state for control points.*

$N$ = 80,800. The main calculation results are given in **Table 2**. It can be seen from the Table that for a given uncoated piston with a liter engine power of 25 kW/l, there is no cracking of the combustion chamber edge during a given resource, $d_1(\Xi, P) = 0.552$. But with a liter power of 29 kW/l, the resource is not guaranteed, $d_1(\Xi, P) < 0$. This means that at a power level of 25 kW/l, the piston is practically working on the verge of physical reliability. But with a thermal barrier layer with a liter engine power of 29 kW/l, the parameter of the physical reliability of the piston is 0.574, i.e. the reliability of the structure has been restored.

The calculated data on the structure thermal stress are also sufficient to determine the parametric reliability of the piston lateral surface $d_2(\Xi)$. The piston control points to be analyzed are shown in **Figure 9a**. The arrangement of the calculation array results for a piston with a thermal barrier layer at a liter engine power level of 29 kW/l is shown in **Figure 9b**. The data on the reduction of the creep threshold of the alloy to the criterion $d_2(\Xi)$ in Eq. (15) were taken on the basis of work [14], $\varphi_s = (2,5\sigma + t)/225$. The figure shows that the parametric reliability of the structure is ensured.

The proposed approach to the analysis of the reliability of structures of heat-stressed parts of highly accelerated engines takes into account the operating model and allows you to search for technical solutions while ensuring the operation of materials of structures on the verge of strength.

For monitoring and predicting the residual life of the most thermally stressed parts under engine operating conditions, the proposed methodology allows using modern 4.0 technologies.

## 5. Conclusions

The improvement of diesel engines, as the main source of modern energy, is associated with a further increase in fuel efficiency, liter capacity, and reliability, with a significant reduction in emissions of toxic substances and carbon dioxide into the environment.

The proposed systematic approach to a comprehensive assessment of fuel efficiency and emissions toxicity allowed proposing a dimensionless criterion that takes into account the operating conditions of the engine. Using this criterion, an assessment of the efficiency of using alternative fuels is given and the prospect of such an assessment is shown when using hybrid fuels that include green hydrogen.

To identify the mechanisms of the formation of internal and external losses and substantiate the ways to achieve optimal heat use, the use of the anergy-energetic method of analysis is justified.

In order to increase the reliability of the most thermally stressed parts of highly accelerated engines, taking into account the operating conditions, the approach has been proposed, taking into account the complexity of factors affecting the physical and parametric reliability.

The directions for improving diesel engines, considered in the chapter, are only a part of a set of tasks, the solution of which seems to be extremely relevant from the point of view of preventing an energy crisis and, at the same time, are only a part of the possible ones for practical implementation.

## Author details

Andriy Marchenko, Igor Parsadanov, Volodymyr Pylyov, Oleksandr Osetrov*, Linkov Oleh, Serhii Kravchenko, Oleksandr Trynov, Denys Meshkov, Serhii Bilyk, Anatolii Savchenko, Inna Rykova and Rasoul Aryan
National Technical University “Kharkiv Polytechnic Institute”, Kharkiv, Ukraine

*Address all correspondence to: oleksandr.osetrov@khpi.edu.ua

**IntechOpen**

© 2022 The Author(s). Licensee IntechOpen. This chapter is distributed under the terms of the Creative Commons Attribution License (http://creativecommons.org/licenses/by/3.0), which permits unrestricted use, distribution, and reproduction in any medium, provided the original work is properly cited.

## References

[1] Marchenko AP, Parsadanov IV, Tovazhnyansky LL, Shekhovtsov AF. Internal combustion engines: A series of textbooks. In: 6 volumes. V.5. The Environmentalization of ICE. 2nd ed. Kharkov: NTU "KPI" Publishing Center; 2014. p. 348. (in Ukrainian)

[2] Parsadanov I, Marchenko A, Tkachuk M, Kravchenko S. Complex Assessment of Fuel Efficiency and Diesel Exhaust Toxicity. SAE Technical Paper. 2020; 2020-01-2182. DOI: 10.4271/2020-01-2182

[3] Canillo P, Vnukova N, Turenko A, Gritsenko A. Global Energy and Climate Problems and the Urgency of their Solution. KhNADU: Kharkiv; 2020. p. 386. (in Ukrainian)

[4] Nadeem M, Rangkuti C, Anuar K, Haq MRU, et al. Diesel engine performance and emission evaluation using emulsified fuels stabilized by conventional and gemini surfactants. Fuel. 2006;**85**(14–15):2111-2119. DOI: 10.1016/j.fuel.2006.03.013

[5] Fahd M, Wenming Y, Lee PS, Chou SK, Yap CR. Experimental investigation of the performance and emission characteristics of direct injection diesel engine by water emulsion diesel under varying engine load condition Applied Energy. 2013; **102**:1042–1049. DOI: 10.1016/j.apenergy.2012.06.041

[6] Abramchuk FI, Marchenko AP, Razleytsev NF, Tretyak EI, Shekhovtsov AF,. Shokotov NK. Modern diesel engines: Improved fuel efficiency and durability. In: Shekhovtsov AF. editior. Kyiv: Tehnika; 1992. p. 272. ISBN: 5-335-01032-0. (in Russian)

[7] Energy and exergy. Translated from German by PhD (Sc) Kalinin NV, ed. By Dr. Tech. Sciences V.M. Brodyansky. M: Publishing House "Mir"; 1968. p. 192

[8] Baer GD. Technical Thermodynamics. Moscow: Publishing house "Mir"; 1977. p. 518

[9] Brodyansky VM, Fratscher V, Mikhalek K. Exergetic Method and its Applications M: Energoatomizdat; 1988. p. 288. ISBN 5-283-00152-0

[10] Marchenko AP, Kravchenko SS, Bekaryuk OM, Shelestov MS. Use of exergy method to assess the perfection of processes in the supercharging system of a diesel engine. Internal Combustion eEgines. 2021;**2**. DOI: 10.20998/0419-8719.2021.2.03 (in Ukrainian)

[11] DFC Diesel. Failure Analysis. Available from: https://www.dfcdiesel.com/warranty-info/failure-analysis. [Accessed 27 July 2021]

[12] Zabala B, Igartua A, Fernández X, Priestner C, Ofner H, Knaus O, et al. Friction and wear of a piston ring/cylinder liner at the top dead Centre: Experimental study and modelling. Tribology International. 2017;**106**:23-33. DOI: 10.1016/j.triboint.2016.10.005

[13] Marchenko A, Pylyov V, Linkov O. Estimation of Strength of the Combustion Chamber of the ICE Piston with a TBC Layer. In: Nechyporuk M, Pavlikov V, Kritskiy D. editiors. Integrated Computer Technologies in Mechanical Engineering—2020. ICTM 2020. Lecture Notes in Networks and Systems. 2021; 188: 415–426. DOI: 10.1007/978-3-030-66717-7_35

[14] Pylyov VO. Automated Designing of Pistons of High-Speed Diesels with a Given Level of Long-Term Strength. Kharkiv: Publishing center NTU "KhPI"; 2001. p. 332. (in Ukrainian)

Chapter 2

# The Influence of Exhaust Gas Recirculation on Performance and Emission Characteristics of a Diesel Engine Using Waste Plastic Pyrolysis Oil Blends and Conventional Diesel

*Semakula Maroa and Freddie L. Inambao*


## Abstract

Through an experimental study, this work focused on finding the influence of exhaust gas recirculation (EGR) on waste plastic pyrolysis oils (WPPOs) with diesel as a base comparison fuel. The results show the amount of carbon monoxide emissions seemed to decrease at low engine loads up to intermediate loads of (50%), thereafter continued to increase significantly but marginally. Among fuels tested, blend WPPOB100 reported the highest BSFC, at 0% EGR flow rate. The value was 0.4751g/kW.hr. compared with 0.7235 g/kW.hr. at 30% EGR flow rate. Increased blend ratio had a direct decrease in brake power linearly. At 30% engine load, CD, WPPOB10, WPPOB20, WPPOB30 and WPPOB40 recorded values of 2.125 kW, 2.15 kW, 2.05 kW, 1.98 kW, 1.86 kW and 1.75 kW, respectively. Exhaust gas temperature (EGT) at 30% EGR flow rate, blend WPPOB10 had the highest reduction in temperature compared with the any other WPPO blends at 320°C. Increased blend ratio and EGR percentage flow rate increased smoke emissions within the test fuels blends. At 15% EGR flow rate, the following data were recorded: 7.53%, 7.1%, 6.72%, 6.25%, 6.0% and 5.4% for CD, WWPO10, WPPO20, WPPO30, WPPO40 and WPPO100, respectively.


**Keywords:** biodiesel, $CO_2$, UHC, CO, conventional diesel, EGR flow rate, $NO_X$, smoke emissions, waste pyrolysis plastic oil

## 1. Introduction

Modern-day transport systems are important and critical, especially the transportation of goods, transport services and people. Internal combustion engines with diesel fuel as the primary source of energy form the bulk of commercial and personal transport. This is owing to their numerous advantages compared with other forms or types of propulsion in internal combustion engines.

Diesel engines are inherently lean burn engines, generating low carbon dioxide emissions compared with petrol-propelled internal combustion engines. Diesel engines have other merits such as high thermal efficiencies, durability and construction robustness [1]. This endears them to users, thus expanding application use as more countries move into urbanization and industrialization. However, there has been a formidable challenge to phase them out, based on environmental and human health issues due to the high levels of $NO_X$, smoke and PM emissions.

Therefore, there has been continued increase in stringent emission regulations enacted by global industrial powers, United States of America and the European Union environmental protection agencies the G-7 and G-20. The diesel engine has been accused as a pollutant, hence the search for alternative fuels in the interest of reducing energy consumption, environmental degradation and air pollution from $NO_X$ gases, which diesel engines emit, thus decelerating atmospheric carbon concentration globally. The road transport sector is an environmental concern due to its rapid expansion. This expansion has eroded all the technological developments and improvements achieved in the war against air pollution from diesel engines. Climatic change, erratic energy prices, uncertainty of future fossil fuel supplies, unending internal conflicts in major oil-producing countries create a compelling case for alternative fuels [2].

The alternative sources of fuel energy supply increase food insecurity as it makes the use of plant-based feedstocks for biodiesel fuel. This makes this option a far less viable option leading to high food prices and inflation [3]. Therefore, waste plastic from municipal solid waste management sites is increasingly becoming a popular alternative source of fuel and energy due to the widespread use of plastics in day-today activities. Despite the greater factor of environmental effects of plastic waste and disposal costs, plastics are still applied widely in daily economic and social activities. Plastic waste has created havoc to the environment due to challenges of proper disposal and non-biodegradable nature of plastics [4].

There are two types of plastics widely used today, namely PVC (poly-vinyl chloride) and HPDE (high-density polyethylene) also known as polyethylene high-density (PEHD) [5].

Globally plastic waste accounts for 8–12% of waste with a projected annual increase of 9–13% by 2025 [6, 7]. Back here at home in South Africa, 24,115,402 metric tonnes of general waste was produced, 6% of which is 1,446,924 metric tonnes of plastic waste with a national average waste production annual increment projection of 2–3%, since 2008 [8] as in **Figure 1**. This makes a sustainability case in managing waste into energy, using technology to degrade waste plastic mass into energy. Using techniques such as pyrolysis results in hydrocarbons similar in quality and characteristics to petroleum fuels due to its high yield achieved by pyrolysis [9, 10].

Originally, pyrolysis is a word coined from two Greek words pyro-'fire' and lysis-'decomposition' [11]. Pyrolysis is a chemical decomposition process of making fuel from plastic waste by heating [12]. Pyrolysis has been recommended as one of the solutions to ending the menace of plastic waste in the world. During pyrolysis, assorted waste plastic is introduced into a reactor and subjected to high temperatures of 400–600°C or sometimes 900°C at atmospheric pressure in the absence of oxygen for 3–4 hours to produce oil and other plastic waste by-products [13]. As a method of transforming waste plastic into biodiesel pyrolysis has been recommended by researchers and commercial entities. This is because of its cost-effectiveness and its high energy conversion rate besides the high yield compared with any other method of plastic waste extraction [14].

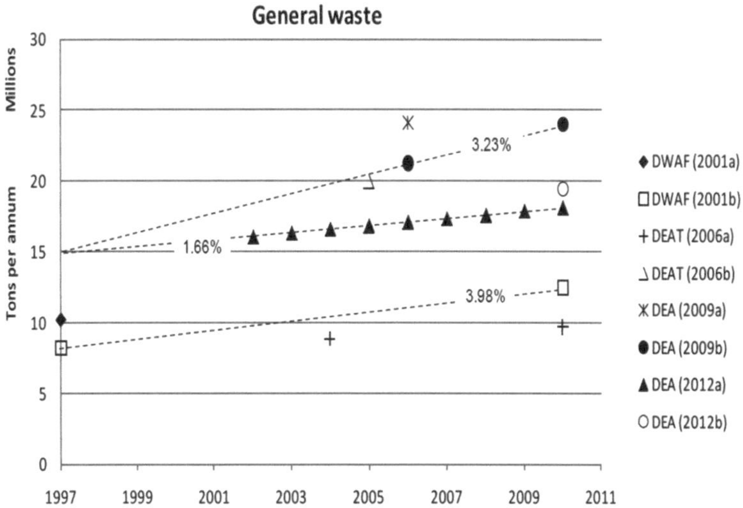


**Figure 1.**
*Waste data analysis (from municipalities) for South Africa [8].*

Catalysts are employed to maintain and sustain high temperatures during pyrolysis reaction [15]. These catalysts include calcium oxide (CaO), silica dioxide ($SiO_2$), aluminum tri-oxide ($Al_2O_3$) and zeolite ($NaAlSi_2O_6$-$H_2O$) [16]. Pyrolysis breaks down large molecules of plastic waste into minute molecules producing hydrocarbons with smaller molecular mass. For example, the addition of ethane enables fractional distillation to be applied and obtain fuels, chemicals and by-products from the process. The pyrolysis process gives yields with a weight factor of 75% of liquid hydrocarbons in mixtures of petrol, diesel and kerosene, in the proportion of 5–6% as residue coke while the remaining balance as liquidified petroleum gas (LPG) [17].

The use of biodiesel thus calls for $NO_X$ reduction techniques such as exhaust gas recirculation (EGR) due to the oxygen content inherent in most biodiesel fuels. This is the single most factor responsible for $NO_X$ formation as it reacts with high-temperature combustion mixture, thereby increasing the availability of $NO_X$ [7]. Diesel fuels and biodiesel fuels both require fuel additives to improve engine lubricity, better ignition qualities and better mixing. Oxygenates in biodiesels provide reduction in PM emissions since the $O_2$ content aids better combustion. It also lowers exhaust emissions with a clear-cut trade-off between PM and $NO_X$ as in the findings of [18–20]. Most of these researchers suggest modifications, for example, using thermal barrier coating [21]. Thermal coating improves efficiency, reduces NOx emissions and smoke density but minimally increases brake thermal efficiency with a decrease in fuel economy.

Saravanan, [22] Observed that with application of EGR percentage flow rate, a further reduction for both $NO_X$ and soot emission could be achieved with addition of n-pentanol. In an experiment conducted by [23], the authors reported a simultaneous reduction for both $NO_X$ and soot emissions using low-temperature combustion (LTC) strategy, with EGR % flow rates, late injection timing and n-pentanol blended diesel-biodiesel fuels. However, [24] reported a contrary finding with addition of n-pentanol to diesel-biodiesel resulting in increased BSFC and no decrease in BTE. This seems to confirm n-pentanol as a better fuel additive to waste plastic pyrolysis oil (WPPO) compared with n-butanol due to its high cetane number, better blend ratio stability and less hygroscopic nature [25].

In order to reduce combustion temperatures, ignition delay is suggested as it aids in the reduction of $NO_X$, which is temperature-dependent. The use of cetane improvers is also an alternative technique in reducing NOx as the poor cetane index of WPPO fuel blends leads to poor ignition quality. Particularly when biodiesel fuels are used such as glycol ether, which reduces PM, UHC and CO emissions in common rail direct injection diesel engines. These cetane improvers decrease cylinder pressure, ignition delay, heat release rate and engine knock or noise [26]. The inclusion of n-pentanol in diesel-biodiesel blends has been reported to shorten combustion duration and increases the HRR, while significantly reducing the $NO_X$, CO and UHC emissions [27].

In many experimental works, fuel additives have been utilized with diethyl ether as the most common. As an organic compound, diethyl ether has a high cetane number and capable of boosting ether cetane number [28]. When used as an additive, diethyl ether reduces ignition delay, cylinder peak pressure, heat release rate, CO, $CO_2$ and $NO_X$ with a trade-off in which the BTE increased [29]. Other researchers found that diethyl ether reduces the ignition delay period, UHC, NOx, whereas BTE seemed to increased, but [15] using WPPO fuel observed ignition delay and higher heat release rate with diethyl ether blends.

Diesel engines run stably on most medium blended ratios of waste plastic oil, although they produce high $NO_X$, UHC and CO emissions. However, to stabilize and improve performance for higher blend ratios, injection timing is a technique most recommended. This allows engine performance and stability without upgrading fuel, engine modification or fuel alteration through addition of additives [30]. Injection timing affects performance from WPPO and Jatropa blends of 20% tyre oil and 80%. This results into lower fuel consumption, CO, UHC and PM with increase in $NO_X$ emissions [31]. On the other hand, in a study by [32], the researchers observed increased BTE and $NO_X$ emissions. This was identical to the findings on emissions of NOx, with reduced fuel consumption, CO and UHC [31].

## 2. The experimental set-up and equipment

### 2.1 Introduction

As discussed in the introduction and literature review, one of the gray areas is the fewer literature works for plastic pyrolysis oil biodiesel on EGR. The few numbers of experimental works on the waste plastic pyrolysis oil biodiesel, the interaction and influences motivate this work. Considering this trend and the level of experimental work thus far achieved, it becomes important to mount an experiment to increase and deepen understanding. Therefore, using an experimental approach aims to achieve study aims and objectives as set forth to study the effects of exhaust gas recirculation (EGR) on the performance parameters of a diesel engine, using WPPO biodiesel. **Figure 2** shows engine experimental set-up schematic diagram.

### 2.2 Experimental apparatus and equipment

**Figure 3** shows the EGR schematic loop and definitions; 1. is the *direction of EGR gases*, 2. is the $x$ subscript representing the exhaust molar gas quantity, 3. is the direction of inlet gases fresh charge, 4. is the $z$ subscript representing remainder of

*DOI: http://dx.doi.org/10.5772/intechopen.105011*

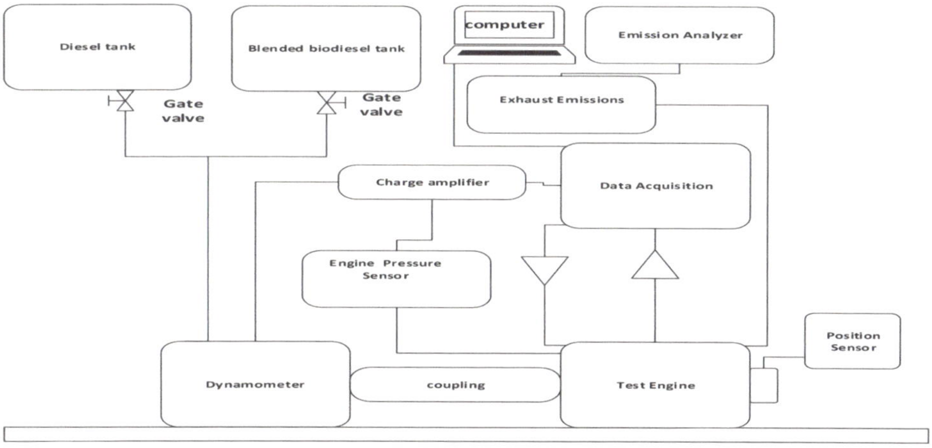


**Figure 2.**
*Experimental engine schematic diagram set-up.*

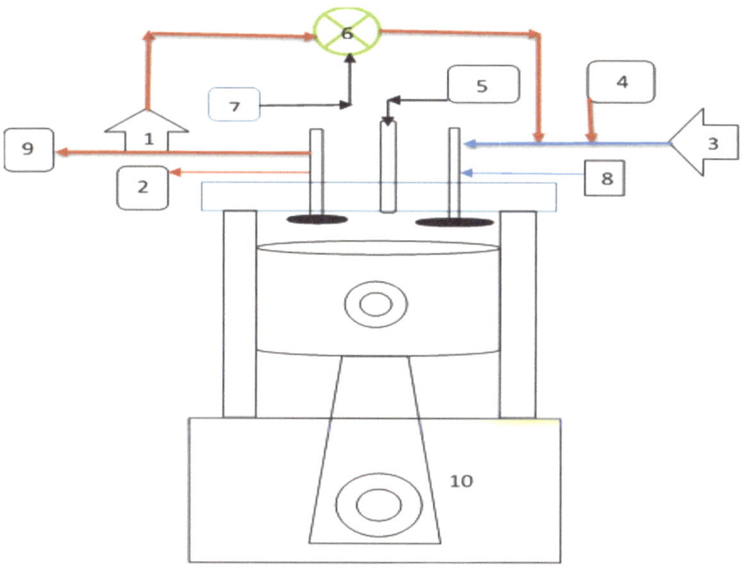


**Figure 3.**
*The EGR modification nomenclature schematic diagram.*

the intake charge, 5. is $n_f$, which is the fuel molar quantity, 6. is the EGR valve, 7. is the $R_x$ molar gas ratio, 8. is the subscript $y$ representing the inlet intake molar gas quantity, 9. is the direction of the exhaust gases exit, 10. is the engine unit.

## 2.3 Waste plastic preparation and the conversion process

a. The waste plastic materials are acquired from a municipal solid waste management site and taken to the sorting section of the pyrolysis plant. The dust and other fine wastes collected are filtered by the cyclone filter system and disposed through a vent with a particle size monitoring system.

b. After the sorting and removal of unwanted materials and dust. The waste plastics are taken through a conveyor machine for pressure cleaning and conveyed to the shredding machine, which reduces them to the required pyrolysis reactor size of 25.4–50 mm.

c. Loading into the pyrolysis reactor uses an automatic feeding machine for the waste material, and the reactor door is air tight locked to begin the next phase of pyrolysis process. As a caution, a manual loading system is provided in case of system power failure.

d. Using a power control panel by a machine operator, the system is started and operated. The preceding processes subsequently run automatically as the flow chart in **Figure 4** indicates. The first stage heats up the dry waste plastic materials, as the reactor's temperature increases to the required values of 400–500°C.

e. The heavy dense gas oil collects into the oil tank while the light oils go into the condenser where it is cooled and drops into the oil tank. However, the small quantities of liquefied gases, which failed to be converted into oil, are collected by the recycling system and burnt as fuel gas. The remaining gas is cleaned by the after-treatment system with the removal of sulphurets and black carbon, while the smoke and flue are emitted in the atmosphere.

f. After production of a batch, the system requires a cooling period of 4–5 hours through natural cooling. However, for faster cooling, nitrogen and carbon dioxide gases are utilized as cooling agents to shorten this process. This enables the removal of the carbon black compound without contamination or pollution to the environment.

g. Removal of steel and other metals is the final operation from the pyrolysis reactor plant, as this requires the opening of the reactor door in preparation for the next batch of the pyrolysis process.

## 2.4 Waste plastic pyrolysis oil (WPPO) physical properties

**Table 1** shows physical properties of the waste plastic pyrolysis oil obtained through the pyrolysis process using waste plastics from municipal solid waste (MSW) management sites. Compared with the properties of conventional diesel fuel oil, these properties are in optimized conditions. The oil is yellowish in color as shown by the picture shot in **Figure 5**. The liquid distillate, which is free of visible particulate sediments, has a flash point of 20°C and a gross calorific value (GCV) of 40.15 KJ/kg. This calorific value of the waste pyrolytic oil compares with the range of petroleum fuels including conventional diesel; thus, making it capable of giving the same comparative working performance results in diesel internal combustion engines.

The distillation report analysis shows that the waste plastic pyrolysis oil (WPPO) has an initial boiling point (IBP) of 119°C–353.5°C. This indicates a presence of other fuel oil components such as kerosene, gasoline and, to some extent, diesel oil in the tested samples. Therefore, it is possible for this oil to be a biofuel feedstock in future,

*DOI: http://dx.doi.org/10.5772/intechopen.105011*

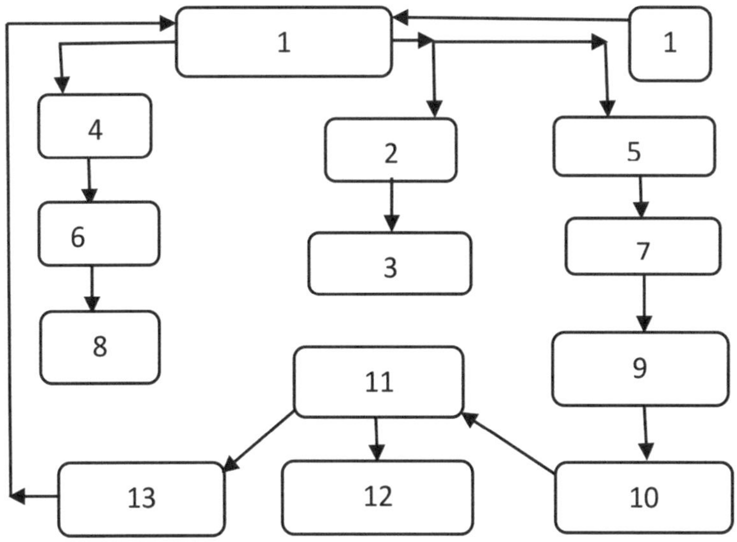


**Figure 4.**
*Waste plastic pyrolysis processing plant flow chart. Flow chart nomenclature 1. Pyrolysis reactor, 2. Carbon black discharge, 3. Carbon black deep processing, 4. Exhaust smoke discharge, 5. Gas separator, 6. Smoke scrubber to take out color and odor, 7. Condenser, 8. Chimney, 9. Oil tank, 10. Synchronized gas purification, 11. Synchronized gas-recycling system, 12. Extra gas burning, 13. Heating furnace during operation, 14. Loading of material.*

| Parameters | Position value |
|---|---|
| Ignition Type | 4 (Stroke) DICI |
| Number Of Cylinders | 1 |
| Model | TV 1 |
| Cooling Medium | Water |
| Manufacturer | Kirloskar |
| Revolutions Per Minute | 1500 |
| Brake Power | 3.5 Kw |
| Cylinder Bore | 87.5 mm |
| Piston Stroke | 110 mm |
| Compression Ratio | 18.5:1 |
| Connecting-Rod Length | 234 |
| Engine Capacity | 661 cc |
| Dynamometer Make | 234 |
| Injection Timing | 23.4 bTDC |
| Maximum Torque | 28 N-M@1500 |
| Injection Pressure | 250 Bar |

**Table 1.**
*The engine specifications and position value.*

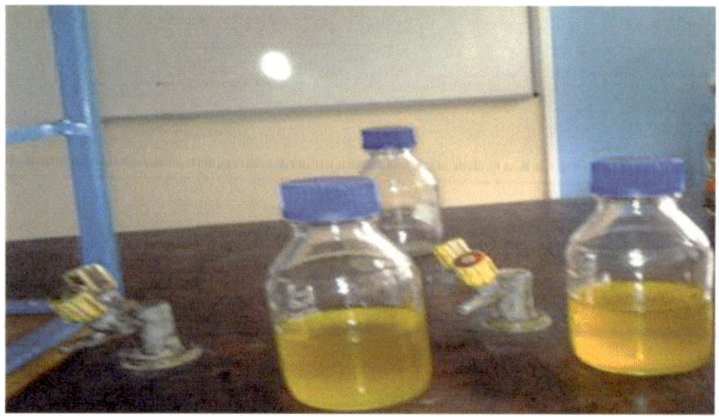

**Figure 5.**
*The liquid distillate samples from the waste plastic pyrolysis oil.*

| Property | Unit | CD | WPPO | STANDARD |
|---|---|---|---|---|
| Appearance | — | Clear/brown | Clear/amber | Visual |
| Density @20 | kg/M$^3$ | 838.8 | 788.9 | ASTM D1298 |
| Kinematic viscosity @40 °C | cSt | 2.32 | 2.17 | ASTM D445 |
| Flash point | $^0$C | 56.0 | 20.0 | ASTM D93 |
| Cetane index | — | 46 | 65 | ASTM D4737 |
| Hydrogen | % | 12.38 | 11.77 | ASTM D7171 |
| Cu corrosion | 3 hrs@100 °C | — | 1B | ASTM D130 |
| Carbon | % | 74.99 | 79.60 | ASTM D 7662 |
| Oxygen | % | 12.45 | 7.83 | ASTM D5622 |
| Sulfur content | % | <0.0124 | 0.15 | ASTM D4294 |
| IBP temperature | $^0$C | 160 | 119 | ASTM D86 |
| FBP temperature | $^0$C | 353.5 | 353.5 | ASTM D86 |
| Recovery | % | | 98 | — |
| Residue and loss | % | | 2.0 | — |
| Gross calorific value | kJ/kg | 44.84 | 40.15 | ASTM D4868 |

**Table 2.**
*Waste plastic pyrolysis oil test fuel properties, units of measurement, test standard methods, compared to conventional diesel.*

if upgraded into a lighter compound as diesel fuel or any liquid fuel in the near future (see **Table 2**).

## 2.5 Experimental procedure

a. The experimental engine is a Kirloskar variable compression engine, four-stroke single cylinder; water-cooled, developing 3.75 kW of power at 1500 rpm. The schematic diagram is provided in **Figure 1**.

b. The experimental engine's technical specifications are indicated in **Table 3**. Engine load was provided by a dynamometer during the experimentation.

| Molecular formula | Percentage composition |
|---|---|
| $C_{10}$ | 66.32 |
| $C_{10}$-$C_{15}$ | 4.38 |
| $C_{15}$-$C_{20}$ | 12.66 |
| $C_{20}$-$C_{25}$ | 8.22 |
| $C_{25}$-$C_{30}$ | 8.42 |

**Table 3.**
*The waste pyrolysis plastic oil chemical composition.*

Using a standard orifice mechanism, an air box was fitted to the engine intake manifold system, thus enabling engine intake airflow measurements.

c. Using a digital fuel gauge, the fuel flow rate to the experimental engine was measured and with aid of a stopwatch mounted to fuel outlet valve, time taken for the fuel consumed was measured.

d. Temperature thermocouples of k-2 type provide measurements for the exhaust the exhaust gas temperatures. This measurement for EGR temperature is done before it mixes with the fresh intake air charge and the combustion constituents.

e. In order to monitor cylinder combustion pressure in the engine cylinder head, a cylinder pressure transducer is mounted to collect data values through a system charge amplifier connected to a computer data acquisition machine.

f. The crank angle (crankshaft position) is monitored and measured through a mounted encoder near the crankshaft pulley area.

g. Experimental emissions were monitored through a five-channel gas exhaust analyzer, while for measurements for the smoke intensity, an AVL 437C smoke meter was utilized.

h. Since it is a variable compression engine which develops maximum power at 1500 rpm. The experiments were conducted based on nominal engine speed, at part load and full load. However, other data were obtained from different engine load as specified in the set-up. In this experiment, part engine load is 50% load, engine full load is engine running at 100% load, all with a fixed compression ratio of 18.5:1.

i. The modified experimental engine EGR system and data collection are shown in **Figure 3**. The exhaust gases were tapped from the exhaust pipe and joined to the intake manifold air intake system via the air flow meter box through a manually controlled gate valve, which makes it possible for the mixing of EGR gases and the fresh air intake.

j. The EGR percentage flow rate was divided into the following modes: 0%, 5%, 10%, 15%, 20%, 25% and 30% at intervals of 5%. The waste plastic pyrolysis oil fuel blends were prepared in the following percentages order and mixed with diesel fuel in 10%, 20%, 30%, 40% and 100%. For example, WPPO10 blend is 90% conventional diesel fuel and 10% waste plastic pyrolysis oil (WPPO) fuel in that order. Therefore, throughout this experiment, blends will be referred to as in this format WPPOB10 with digits denoting the percentage blend ratio of plastic oil by volume to conventional diesel supplied.

k. To avoid contamination, each test was conducted after a thorough evacuation procedure on the previous preceding experiment. The fuel lines and the fuel injection system mechanism were completely evacuated before a new set of experiment was initiated. This made it possible to collect good data with inputs from the test mode only, as there is no fear of contamination, poor results or error.

## 2.6 Percentage uncertainties and error analysis

In order to identify the accuracy and precision of the measuring tools and instruments used in this experiment work, this process error analysis was performed. Experimental errors occur due to conditions outside the experiment itself such as poor calibration of the instruments, observational errors, manufacturing errors, errors associated with experimental set-up and planning, besides environmental conditions existing during the experiment [33]. **Table 4** is a list of instruments, percentages of uncertainties of CO, $CO_2$, UHC, $NO_X$, exhaust gas temperature (EGT) and smoke opacity and percentage error analysis. These percentages of error analysis

| Instrument | Accuracy | Measuring Range | Percentage inaccuracies |
|---|---|---|---|
| AVL 437C (smoke meter) Smoke intensity | ±1% | 0–100% | ±1 |
| AVL pressure transducer GH14D | ±0.01 bar | 0–250 bar | ±0.01 |
| AVL 365C Angle encoder | $\pm 1^0$ | — | ±0.2 |
| AVL Digas 444 (Five Gas Analyzer) | | | |
| CO | ±0.03% to ±5% | 0–10% by vol | ±0.3 |
| $CO_2$ | ±0.5% to ±5% by vol | 0–20% vol | ±0.2 |
| $O_2$ | ±5% by vol | 0–22% by vol | ±0.3 |
| HC | ±0.1% to ±5% | 0-20,000 ppm by vol | ±0.2 |
| NOx | ±10% | 0–5000 ppm by vol | ±0.2 |
| K-2 Thermocouple | ±1 °C | 0-1250°C | ±0.2 |
| Digital Stop Watch | ±0.2 s | | ±0.2 |
| Digital Fuel Gauge | ±1 mm | | ±2 |
| Burette | ±0.2 cc | 1-30 cc | ±1.5 |

**Table 4.**
*Shows measuring instruments, range of measurement, percentages accuracies and inaccuracies, as calculated from Eq. (2).*

are derived from the root sum square method formula and expressed in equation form [1] as in Eq. (1).

$$R = \sqrt{\sum_{i=1}^{n} X_i^2} \tag{1}$$

Where R is the total uncertainty percentage, $X_i$ is the individual uncertainty of all the calculated operating parameters, $n$ is the total number of the parameters in the experiment and $i$ is the i[th] term of the computed parameters. The total percentage of the uncertainty is thus calculated based on Eq. (2) as follows:

$$R = \sqrt{X_1^2 + X_2^2 + X_3^2} + \ldots X_i^2 \tag{2}$$

## 3. Results and discussion

### 3.1 Brake-specific fuel consumption (BSFC)

**Figure 6** is the variation of brake-specific fuel consumption with full load under the effects of EGR percentage flow rate. Lower ratio blends of WPPOB10 and WPPOB20 show minimal reduction in the brake-specific fuel consumption (BSFC) at 0% EGR flow rate in **Figure 6** compared with the values of conventional diesel and WPPOB100, which are showing significantly high values of brake-specific fuel consumption (BSFC) at that mode. At 0% EGR, flow rate conventional diesel has a brake-specific fuel consumption value of 0.4 g/kW.hr., compared with WPPOB100 with a value of 0.4751 g/kW.hr. In other words, blend WPPOB100 has a higher brake-specific fuel consumption than diesel. The values for the other blends of WPPO are placed at 0.3225 g/kW.hr., 0.3615 g/kW.hr., 0.3645 g/kW.hr. and 0.3715 g/kW.hr., for WPPOB10, WPPOB20, WPPOB30 and WPPOB40 at this point, respectively.

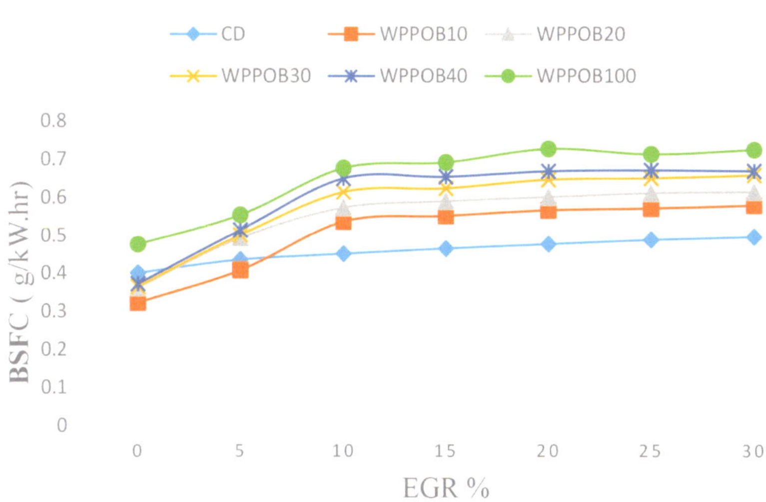


**Figure 6.**
*Brake-specific fuel consumption (BSFC) versus EGR percentage flow rate full load engine conditions.*

Similar trends are reported with application of EGR percentage flow rate such as at 20–25% EGR flow rate. At this point, the BSFC showed increasing tendencies, which is identical to the findings of [34]. This phenomenon is due to the effects of dilution of the fresh air intake as it mixes with exhaust gases. This mixture comes from the recirculated EGR system gases leading to incomplete combustion of the inducted mixture, hence a drop in power and engine torque. This scenario increases engine fuel consumption to maintain constant engine speed to meet increased load demand, which reflected increased BSFC. These findings are identical to the findings of [35].

The WPPO biodiesel blends showed better fuel economy with EGR percentage flow rate application. This is true especially for lower blend ratios of WPPOB10 and WPPOB20 compared with conventional diesel test fuels. However, increased EGR percentage flow rate increased the BSFC across all the test fuels used. For example, at 0% EGR, conventional diesel was 0.4 g/kW.hr. compared with 0.495 g/kW.hr. with application of 30% EGR flow rate. On the other hand, blend WPPOB10 was 0.3225 g/kW.hr. compared with 0.5780 g/kW.hr. with application of 30% EGR flow rate.

**Figure 6** shows that the highest BSFC among the blends of diesel and conventional diesel test fuel is from blend WPPOB100. This blend at 0% EGR flow rate had a value of 0.4751 g/kW.hr. compared with 0.7235 g/kW.hr. at 30% EGR flow rate. During experimentation with 10% EGR flow rate, the values for the BSFC across all the test fuel seemed to pick a lineal increment trend as in **Figure 6**. This was indicated by the flattening of the graph curves with close-packed value trends.

### 3.2 Brake thermal efficiency

The aim of brake thermal efficiency is to help us to understand the ability of the combustion system to utilize the fuel provided. Furthermore, it is a means of comparing and assessing how efficiently fuel conversion was carried by turning it into mechanical output [27, 28]. **Figure 7** is brake thermal efficiency (BTE) % variations, under different blends of WPPO and conventional diesel fuel, with EGR % flow rate.

**Figure 8** is variation of the brake thermal efficiency with EGR % flow rate using different blends of WPPO and CD. In this figure, a decrease in the brake thermal efficiency with all high blend ratio fuel such as WPPOB40 and WPPOB100 is indicated compared with conventional diesel fuel. However, blend WPPOB100 has the lowest decrease of brake thermal efficiency at 7.05%, with 10% EGR flow rate. The value drops further to 2.35% with application of 30% EGR flow rate.

**Figure 7.**
*Brake thermal efficiency (BTE) versus engine load percentage.*

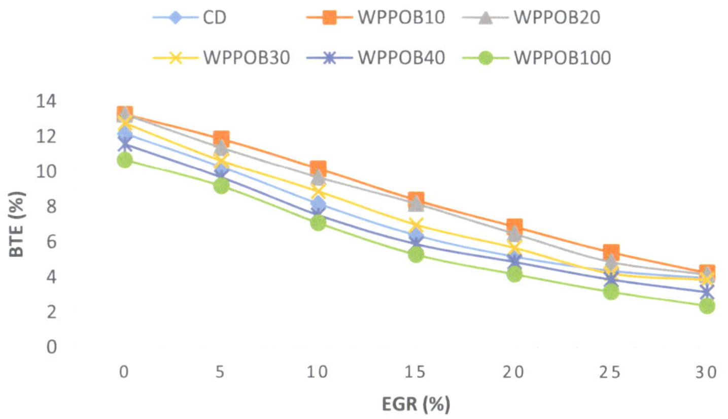


**Figure 8.**
*Brake thermal efficiency (BTE) % versus EGR % flow rate.*

In other words, there is a reduction in the brake thermal efficiency due to the application of EGR percentage flow rate as shown in **Figure 8**. For example, at 0% EGR flow rate, brake thermal efficiency for conventional diesel is 12.15% compared with WPPOB10 and WPPOB20 at 13.25% and 13.05%. The WPPOB100 blend has the lowest value for thermal efficiency for all EGR rate flow modes than any other test fuel as shown in **Figure 8**.

## 3.3 Brake power (BP)

**Figure 9** shows brake power variations with different blends of WPPO and conventional diesel fuel at full engine load. The results obtained show that there is a lineal increase in the brake power for all the test fuels applied with increase in engine load. Conventional diesel fuel has the highest increase in brake power values compared with the blends of WPPO. At 20% engine load, conventional diesel is at 1.45 kW while WPPOB10 has a value of 1.350 kW representing a difference of 6.8% in BTE when the two fuels are compared.

**Figure 9** also shows very close unitary increments with increase in engine load conditions. It also shows a decrease in brake power as the blend ratio increased for all

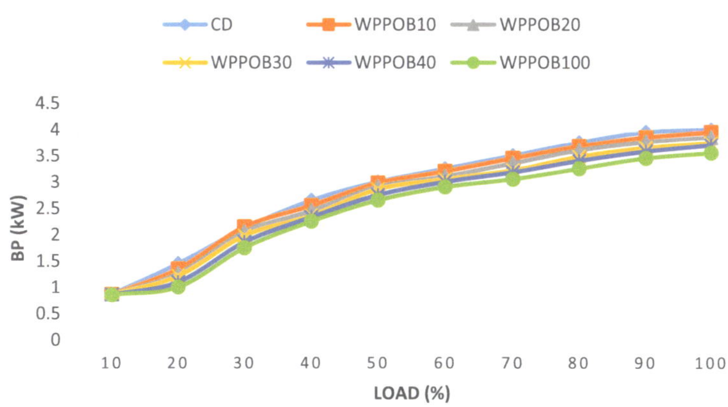


**Figure 9.**
*Engine brake power versus varying engine load percentage.*

the fuels tested. In other words, the increase in blend ratio showed a direct decrease in brake power linearly. For example, at 30% engine load CD, WPPOB10, WPPOB20, WPPOB30 and WPPOB40 reported values of 2.125 kW, 2.15 kW, 2.05 kW,1.98 kW, 1.86 kW and 1.75 kW, respectively, thus, showing a decrease in the value of the engine brake power throughout the experimentation and analysis period. Blend WPPOB100 showed the lowest values for the engine brake power compared with the blends of WPPOB10, WPPOB20, WPPOB30 and WPPOB40. These findings are identical with the findings of a research on WPPO blends [20].

The application of EGR percentage flow rate does not show significant changes in brake power. However, there is a negligible drop in the engine brake power application of EGR flow rate except for the blend WPPOB10. The blend has almost identical values to conventional diesel as the curve of the two fuels indicates in **Figure 9**; therefore, leading to a conclusion that the blends of WPPO have identical brake power values with conventional diesel.

### 3.4 Exhaust gas temperature (EGT)

Temperature is one of the key factors in determining the formation of engine exhaust emissions, besides providing or helping in the analysis and study of combustion processes in relation to fuel [36]. The result in **Figure 10** is showing a variation in exhaust gas temperature (EGT) with different fuel blends of WPPO and conventional diesel with the application of EGR percentage flow rate. The result indicates that EGT decreases with different blends of WPPO compared with conventional diesel test fuel.

The difference between conventional diesel and WPPO blends is the temperature increases in all the test conditions reported. However, it should be mentioned that as the blend ratio increased with EGR % flow rate application, the exhaust gas temperatures reduced significantly especially for WPPOB30 and WPPOB40 at 0% EGR flow rate, the highest temperature value obtained for conventional diesel is 456°C compared with WPPOB100 blend at 490°C. needless to mention at 30% EGR flow rate this blend has most reduction in temperature compared with the other WPPO blends with a temperature value of 320°C.

Applying increasing rates of EGR % flow rate modes as in **Figure 10** reduces exhaust gas temperature. For example, at EGR percentage flow rate of 5%, the

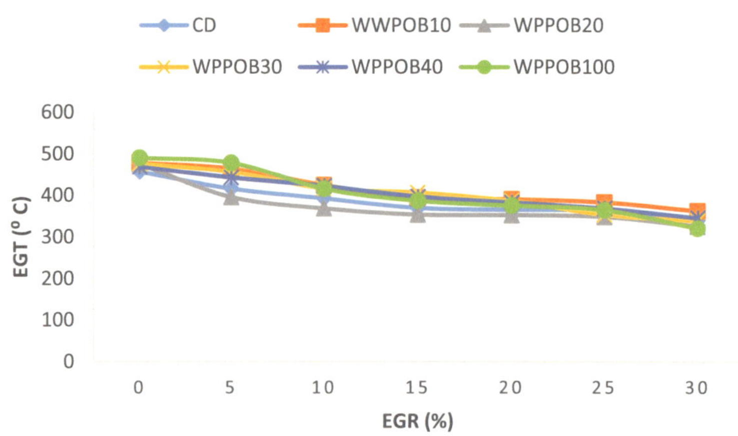


**Figure 10.**
*Exhaust gas temperature (EGT ° C) versus EGR percentage flow rate.*

highest value for conventional diesel test fuel obtained was 440°C. The minimum value was 340°C obtained at 30% EGR flow rate. This trend is repeatedly shown for other WPPO blends with the application of EGR percentage flow rate such as blend WPPOB10 with a high value of 467°C and the lowest being 362°C at 5% and 30%, respectively. On the other hand, WPPOB40 shows its highest value as 472°C and the lowest as 330°C when applying 5% and 30% EGR flow rate, respectively.

The reduction in exhaust gas temperature among the different blends of WPPO was due to low calorific value of the blends and the low exhaust loss. This result is identical to the findings of [37, 38]. According to results shown in **Table 2**, WPPO has a calorific value of 40.15 kJ/kg compared with the calorific value of conventional diesel at 44.84 kJ/kg. The third cause is the effects of dilution, chemical and thermal factors brought through exhaust gas recirculation rate flow [39, 40].

### 3.5 Hydrocarbon emissions

**Figure 11** is a variation of hydrocarbon emissions in parts per million under full engine load with the application EGR percentage flow rates, using different blends of WPPO and conventional diesel (CD). All the blends of WPPO tested indicated significantly higher hydrocarbon emissions, especially with higher engine load conditions as shown in **Figure 11**. However, conventional diesel still produced more and higher values of hydrocarbon emissions compared with all the blends of WPPO across all the engine loading conditions and operating modes.

For example, when the EGR percentage flow rate is 0%, in other words, no application effect, **Figure 11** shows there is less hydrocarbon emissions for all the test fuels. The following values were reported 22 ppm, 23 ppm, 21 ppm, 20 ppm, 19 ppm and 17 ppm for WPPOB10, WPPOB20, WPPOB30, WPPOB40 and WPPOB100 respectively compared with 20% EGR percentage flow rates with 77 ppm, 68 ppm, 52 ppm, 46 ppm, 44 ppm and 40 ppm, respectively.

The application of EGR percentage flow rate reduces the amount of hydrocarbon emissions emitted across board all test fuel blends. However, conventional diesel fuel produced more hydrocarbon emissions compared with all WPPO blends tested. For example, **Figure 10** shows that at EGR flow rates of 5%, 10%, 15%, 20%, 25% and 30%, conventional diesel had 43 ppm, 57 ppm, 70 ppm, 82 ppm and 85 ppm, respectively. On the other hand, the values for WPPOB10 were 23 ppm, 35 ppm, 40 ppm,

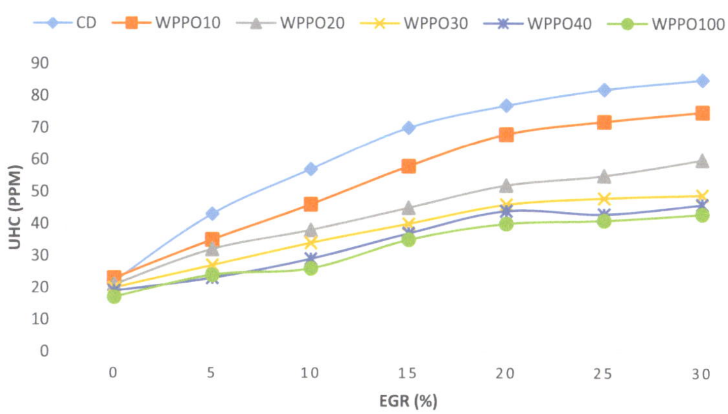


**Figure 11.**
*Unburnt hydrocarbons emissions versus EGR percentage flow rate.*

48 ppm, 50 ppm and 52 ppm, respectively. Therefore, the application of EGR percentage flow rate increased hydrocarbon emissions values as presented in **Figure 10**.

### 3.6 $NO_X$ emissions

The formation of $NO_X$ emission is dependent on cylinder temperature, the concentration of oxygen and the residence time spent in the combustion chamber by the fuel-air mixture during phase of pre-mixing [41]. All tested blends indicated a drop in $NO_X$ emissions with the application of EGR percentage flow rate, at all engine load conditions. This was due to the rise in the total heat capacity of the working gases as EGR % flow rate increased, which was identical with the studies and findings of [42–44]. **Figure 12** shows $NO_X$ emissions value for the conventional diesel was 920 ppm at full load without EGR percentage flow rate, compared with WPPOB100 at 1270 ppm. However, with application of EGR percentage flow rate of 30%, the values reduced to 401 ppm for CD and 432 ppm for WPPO100, respectively.

However, during study, engine part load values in **Figure 13** for $NO_X$ emissions reported lower values compared with the full load engine conditions for the same test fuels. The $NO_X$ emission for conventional diesel at 50% (engine part load) was 635 ppm compared with full load at 1100 ppm. On the other hand, the value for WPPOB100 at 50% (engine part load) was 850 ppm compared with 1250 ppm at full engine load. This result concurs that at 50% (engine part load), the values of $NO_X$ emissions emitted by tested blends of WPPO except WPPO100 were lower compared with full engine load conditions.

### 3.7 Carbon monoxide emission

**Figure 14** shows variations of carbon monoxide emission percentage with varying load under the effects of EGR percentage flow rate, with different fuel blends of WPPO and conventional diesel fuel. As a gas, carbon monoxide is toxic and requires control to acceptable levels. Carbon monoxide is a product of poor combustion of hydrocarbon fuels due to dependency on the air-fuel ratio relative to the stoichiometric proportions [42].

In the experiment conducted, the amount of carbon monoxide emissions decreased with engine loads up to part load (50%). For example, at 0% engine load,

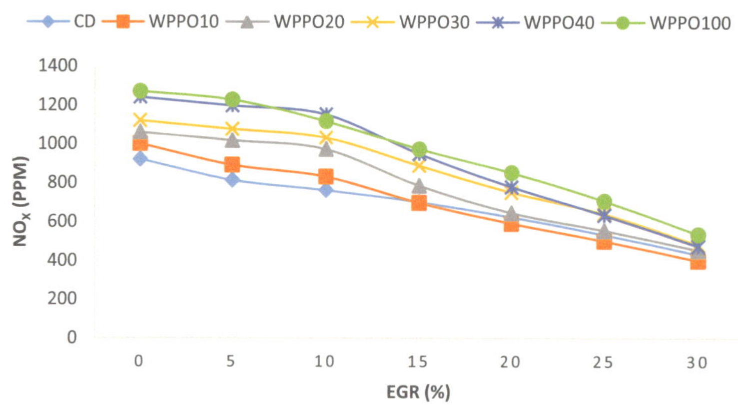


**Figure 12.**
*EGR percentage flow rate variations with $NO_X$ emissions.*

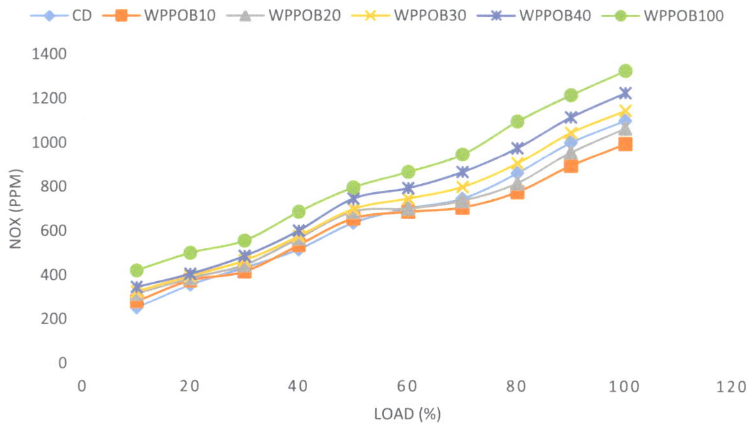


**Figure 13.**
*Variations of $NO_X$ emissions (ppm) versus varying engine load percentage without application of EGR flow rate.*

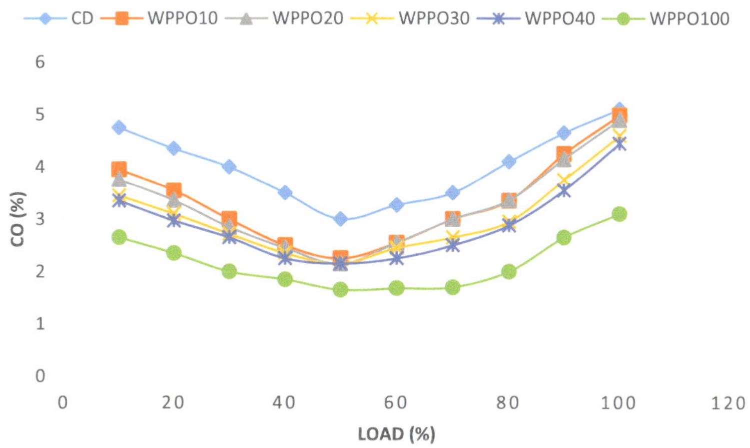


**Figure 14.**
*Carbon monoxide emissions percentage versus varying engine load.*

the value of conventional diesel is 0.051% compared with 50% engine load when the value dropped to 0.03% by volume. However, the CO emissions continued to increase significantly but marginally as in **Figure 14** as the load increased from this point. Increasing the engine load from 50% recorded continuous but marginal increases of carbon emissions by volume across all the test fuels irrespective of the EGR percentage flow rate. For example, at 80% engine load, the value for WPPOB100 is 0.02% up from 0.0165% by volume. The other WPPO biodiesel blends also show a similar trend and concurrency. WPPOB20 and WPPO30 test fuels at 50% engine load condition have values of 2.25% and 2.15% as compared with 3.36% and 2.95% respectively, at 80% engine load.

**Figure 15** is the variation of carbon monoxide with EGR percentage flow rate application under conventional diesel and different blends of WPPO. The WPPO blends produced continuous increase in smoke emissions almost doubling values with the application of EGR percentage flow rate. For example, at 10% EGR flow rate, the carbon monoxide emission values were 9.79%, 10.46%, 10.91%, 11.25% and 12.75% for WPPO10, WPPO20, WPPO30, WPPO40 and WPPO100, respectively.

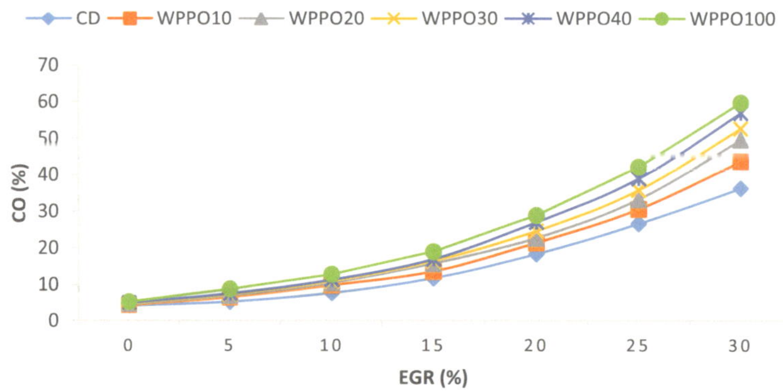


**Figure 15.**
*Carbon monoxide VS EGR percentage flow rate application.*

The application of same EGR percentage flow rate reports the lowest carbon emissions with a value of 7.65% for conventional diesel test fuel.

The blend ratio and EGR percentage flow rate have a correlation on the amount of CO emissions produced. In other words, increased blend ratio increased carbon monoxide emissions within the blends as the EGR percentage flow rate increased. For example, at 20% EGR flow rate, CO emission values recorded were 18.25%, 21.35%, 22.65%, 24.55%, 26.95% and 28.85%, respectively. These values are for CD, WPPO10, WPPO20, WPPO30, WPPO40 and WPPO100. However, blend WPPO30 reports values of 4.85%, 7.28%, 10.91%, 16.37%, 24.55%, 35.75% and 52.69% as the EGR flow rate increased to 30%, respectively. This is caused by dilution, thermal and chemical effects of EGR % flow rate application as some of the oxygen in the inlet charge is replaced with recirculated exhaust gas that causes incomplete combustion.

## 3.8 Carbon dioxide emissions

Carbon dioxide is the principal composition of the exhaust gas recirculation gases. However, carbon dioxide gas and the exhaust temperatures are indicators of combustion quality in the combustion chamber [6]. Carbon dioxide gas has a higher heat capacity making it a thermal heat sink during the combustion process. This helps in reducing peak cylinder temperatures, hence the reduction in the $NO_X$ emissions.

The value of $CO_2$ is considerably high without EGR percentage flow rate coupled with lower engine loads for all the fuel blends tested. For example, at 20% engine load blend WPPOB100 has 4.65% compared with all the other test fuels and is the highest carbon emissions value. The other blends reported are CD, WPPO10, WPPO20, WPPO30 and WPPO40 with 3%, 2.50%, 1.5% and 1.85% respectively, as shown in **Figure 16**.

Additionally, **Figure 16** shows that the amount of carbon dioxide increased with increased engine load. For example, as the engine load increases to 40%, the value of WPPOB40 is 2.75% compared with WPPOB30 at 3.25% while at 70% engine load, the values are 4.5% and 5.25%, respectively. The observation is that as the engine load is increased with increased blend ratios, lower-ratio blends are observed to emit more carbon dioxide emissions as compared with those blends with high ratios except blend WPPOB100 that releases more carbon emissions than any test fuel as aforementioned earlier. At full engine load, the value of carbon dioxide emissions is at the highest

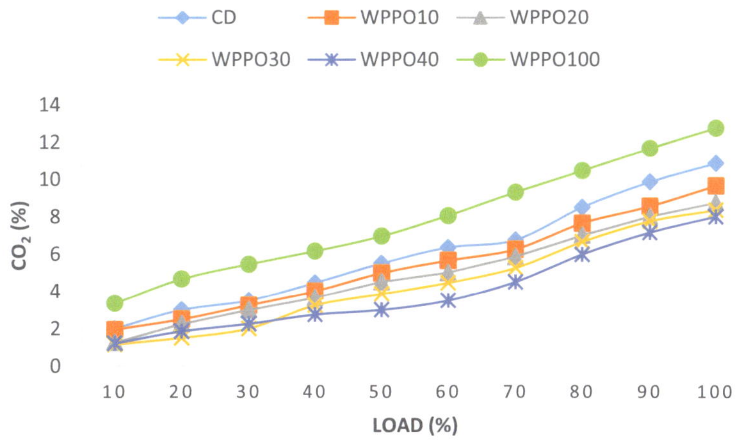


**Figure 16.**
*Variation of carbon dioxide percentage emissions versus engine load percentage, with different types of fuel blends of WPPO and conventional diesel.*

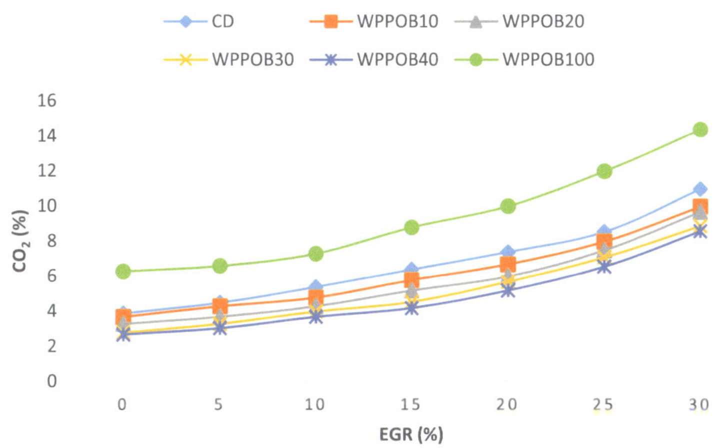


**Figure 17.**
*Variations of carbon dioxide percentage versus EGR percentage flow rate, with different types of WPPO fuel blends and conventional diesel.*

values as in **Figure 16** across all the test fuels. The values are 12.75%, 10.85%, 9.65%, 8.75%, 8.35% and 8% for WPPOB100, CD, WPPOB10, WPPOB20, WPPOB30 and WPPOB40, respectively.

The application of EGR percentage flow rate increases the carbon dioxide emissions exponentially by almost doubling the values as can be seen in **Figure 17**. For example, at 10% EGR flow rate, the value of conventional diesel is 5.35% compared with WPPOB100 at 7.25%, WPPOB10 at 4.75%, WPPOB20 at 4.25%, WPPOB30 at 3.95% and WPPOB40 at 3.65%, respectively. This result reinforces the idea that there exists a correlation between blending and the EGR percentage flow rate with carbon dioxide emissions. Hence, the conclusion that the higher the blend ratio, the higher the emissions values and vice versa with application of EGR percentage flow rate. For example, at 30% EGR flow rate, all test fuels show high carbon dioxide emission, such as conventional diesel at 10.95%, WPPOB10 at 9.95%, WPPOB20 at 9.65%, WPPOB30 at 8.85% and WPPOB100 at 14.35%, respectively.

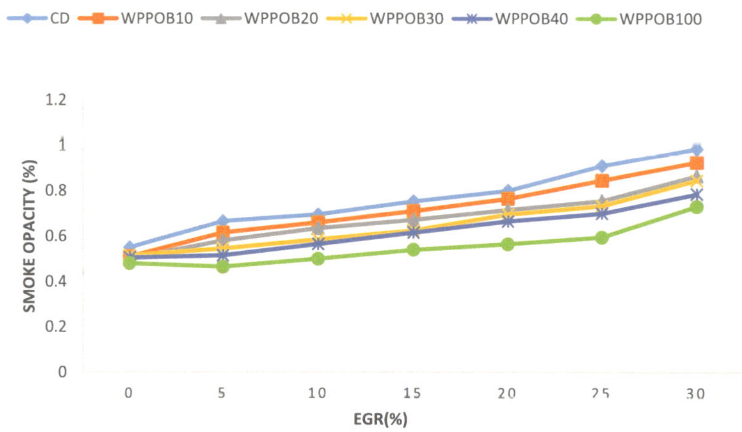


**Figure 18.**
*Variation of smoke emissions or opacity % versus EGR % flow rate, with different WPPO blends and conventional diesel.*

### 3.9 Opacity emissions

Smoke opacity emissions are defined as the solid hydrocarbon soot particles found in the exhaust system exit gases and linked to the formation of smoke emissions [45]. All tested blends of WPPO showed increased and aggravated levels of smoke emissions. However, they were of lower values compared to conventional diesel values.

**Figure 18** smoke emissions under the influence of EGR % flow rate using different WPPO blend and conventional diesel.

The incessant increase in smoke emissions in this figure is explained by high kinematic viscosity and the low volatility values of WPPO blends compared with conventional diesel test fuel. Furthermore, the poor injection and spray characteristics of WPPO blends of fuels compared with the spray and injection characteristics of conventional diesel fuel cause this phenomenon. WPPO blends are also associated with the high aromatic compounds compared with conventional diesel test fuel, hence the poor spray and injection quality.

The application of EGR percentage flow rate shows significant increases in the values of smoke emissions across all the test fuels. Blend WPPOB10 shows smoke emission of 7.2% lower compared with conventional diesel at 0% EGR flow rate. On the other hand, conventional diesel reports 11.5% higher emissions than WPPOB100 blend fuel when EGR flow rate is at 30%. This result is identical to the study findings of the following researchers [46].

The WPPOB10 blend emits the highest levels of smoke emissions for the blended fuels compared with the other WPPO fuel blends. In experimental analysis, blend WPPOB100 is the second highest emitter of smoke emissions. However, as the blend ratio and the EGR percentage flow rate increased, the smoke emissions increased incessantly across all the test fuels compared with EGR percentage flow rate of 0%.

## 4. Conclusion

- Increase in the percentage of the blend ratio of WPPO shows a marked decrease in the engine brake power compared with conventional diesel. This is due to lower percentage of the energy content presented in **Table 2** for WPPO test fuel compared with conventional diesel test fuel.

- WPPO biodiesel blends peak power produced did not match the peak power of conventional diesel. It ranged between 5% and 8% less compared with the peak power conventional diesel test fuel produced.

- Data presented in this work supports widespread use of WPPO as an alternative fuel for compression ignition engines as a feedstock. Through the experimental analysis conducted, this can be achieved with or without modifications to the engines, especially for blends WPPOB10 and WPPOB20.

- During part and intermediate engine load, it is observed that the engine operating under different blends of WPPO and conventional diesel fuel emits less $NO_X$ emissions as compared with the full load mode condition as demonstrated in the result observed in **Figure 12**.

- As the percentage ratio of the blended WPPO increased, there was a significant increase in the brake-specific fuel consumption, as in **Figure 5** for the blends of WPPO compared with the brake-specific fuel consumption (BSFC) values of conventional diesel test fuel.

- As the blend ratio is increased, there is a reduction in the percentage amount of CO emissions released by the test engine as shown in **Figure 13**. This is due to high oxygen content in WPPO blended fuels compared with conventional diesel, which aids combustion.

- The values of carbon monoxide emission obtained during experimentation for the two blends of WPPOB10 and WPPOB20 were close with minimal differences in terms of their volume percentage emissions produced by the test engine as in **Figure 13**.

- For all the test fuels, it was observed that at low engine load from 10 to 40% before it hits the 50% point, there is a decrease in emissions as shown in **Figure 14**. However, there is significant continuous and marginal increase in the percentage of carbon monoxide emissions by volume as the load is increased to 50% part load across all the test fuels irrespective of the EGR percentage flow rate.

- From **Figure 16**, it was observed during experimentation, there is an incessant increase in smoke emissions for all the blends of WPPO with or without EGR percentage flow with test fuel blend WPPOB10 producing the highest values of smoke emissions followed by WPPOB100 test fuel blend.

- Incessant increase in smoke emissions is observed due to WPPO blends of fuel having high kinematic of viscosity and low volatility compared with conventional diesel test fuel. Other possible causes could be the poor injection and spray characteristics associated with WPPO blends compared with conventional diesel fuel, with high spray qualities.

- $NO_X$ emission using conventional diesel at engine part load (50%) is 635 ppm compared at full load at 1100 ppm. The value for WPPOB100 at engine part load (50%) is 850 ppm compared with 1250 ppm at full engine load. This result data indicates a concurrence that at part engine load (50%), the values of $NO_X$ emissions by all the blends of WPPO are lower compared at full engine load condition except blend WPPO100.

## Abbreviations

| | |
|---|---|
| $R_X$ | Molar Gas Ratio |
| $n_f$ | Fuel Molar Quantity |
| $Al_2O_3$, | Aluminum Tri-Oxide |
| ASTM | American Society for Testing and Materials |
| BP | Brake Power |
| BSFC | Brake Specific Fuel Consumption |
| BTE | Brake Thermal Efficiency |
| CaO | Calcium Oxide |
| CD | Conventional Diesel |
| CO | Carbon Monoxide |
| $CO_2$ | Carbon Dioxide |
| Cu | Copper |
| DEA | Department of Environmental Affairs |
| EGR | Exhaust Gas Recirculation |
| EGT | Exhaust Gas Temperature |
| FBP | Final Boiling Point |
| G-20 | Group of 20 Highly Industrialized Countries |
| G-7 | Group of 7 Developed and Industrialized Countries |
| GC–MS | Gas Chromatography–Mass Spectrometry |
| GVC | Gross Calorific Value |
| HPDE | High-Density Polyethylene |
| HRR | Heat Release Rate |
| IBP | Indicated Brake Power |
| $SiO_2$ | Silica Dioxide |
| kW | Kilowatt |
| LPG | Liquid Petroleum Gas |
| LTC | Low Temperature Combustion |
| MSW | Municipal Solid Waste |
| $NaAlSi_2O_6$-$H_2O$ | Zeolite |
| $NO_X$ | Oxides Nitrogen |
| PEHD | Polyethylene High Density |
| PM | Particulate Matter |
| PVC | Poly Vinyl Chloride |
| UHC | Unburnt Hydrocarbons |
| WPPO | Waste Plastic Pyrolysis Oil |
| $x$ | Exhaust Molar Gas Quantity |
| $y$ | Inlet Intake Molar Gas Quantity |
| $z$ | Remainder of the Intake Charge |

## Author details

Semakula Maroa[1*] and Freddie L. Inambao[2]

1 University of Eastern Africa, Baraton, Kenya

2 University of KwaZulu-Natal, Durban, South Africa

*Address all correspondence to: ssemakulamaroa@gmail.com

IntechOpen

© 2022 The Author(s). Licensee IntechOpen. This chapter is distributed under the terms of the Creative Commons Attribution License (http://creativecommons.org/licenses/by/3.0), which permits unrestricted use, distribution, and reproduction in any medium, provided the original work is properly cited.

## References

[1] Kaimal VK, Vijayabalan P. A detailed study of combustion characteristics of a DI diesel engine using waste plastic oil and its blends. Energy Conversion and Management. 2015;**105**:951-956

[2] Reşitoğlu İA, Altinişik K, Keskin A. The pollutant emissions from diesel-engine vehicles and exhaust aftertreatment systems. Clean Technologies and Environmental Policy. 2015;**17**(1):15-27

[3] Kumar BR, Saravanan S. Use of higher alcohol biofuels in diesel engines: A review. Renewable and Sustainable Energy Reviews. 2016;**60**:84-115

[4] Damodharan D et al. Cleaner emissions from a DI diesel engine fueled with waste plastic oil derived from municipal solid waste under the influence of n-pentanol addition, cold EGR, and injection timing. Environmental Science and Pollution Research. 2018;**25**(14):13611-13625

[5] Kumar S et al. Performance and emission analysis of blends of waste plastic oil obtained by catalytic pyrolysis of waste HDPE with diesel in a CI engine. Energy Conversion and Management. 2013;**74**:323-331

[6] Muralidharan K, Vasudevan D, Sheeba K. Performance, emission and combustion characteristics of biodiesel fuelled variable compression ratio engine. Energy. 2011;**36**(8): 5385-5393

[7] Yamada H et al. Detailed analysis of diesel vehicle exhaust emissions: Nitrogen oxides, hydrocarbons and particulate size distributions. Proceedings of the Combustion Institute. 2011;**33**(2):2895-2902

[8] Department of Enviromental Affairs. National Waste Management Strategy. Pretoria: Department of Enviromental Affairs; 2012

[9] Azad A et al. Recent development of biodiesel combustion strategies and modelling for compression ignition engines. Renewable and Sustainable Energy Reviews. 2016;**56**:1068-1086

[10] Bowman CT. Control of combustion-generated nitrogen oxide emissions: Technology driven by regulation. In: Symposium (International) on Combustion. Elsevier. 1992;**24**(1):859-878

[11] Mangesh V, Thamotharan C. Evaluation of engine performance, emissions, of a twin cylinder diesel engine Fuelled with waste plastic pyrolysis oil. Ethanol and Diesel Blends with Cetane Additive AC2010A. Journal of Mechanical and Civil Engineering. 2015;**12**(2):10-15

[12] Geyer R, Jambeck JR, Law KL. Production, use, and fate of all plastics ever made. Science Advances. 2017;**3**(7):e1700782

[13] Kumar S, Panda AK, Singh R. A review on tertiary recycling of high-density polyethylene to fuel. Resources, Conservation and Recycling. 2011;**55**(11):893-910

[14] Sharuddin SDA et al. A review on pyrolysis of plastic wastes. Energy Conversion and Management. 2016;**115**:308-326

[15] Kalargaris I, Tian G, Gu S. Combustion, performance and emission analysis of a DI diesel engine using plastic pyrolysis oil. Fuel Processing Technology. 2017;**157**:108-115

[16] Wongkhorsub C, Chindaprasert N. A comparison of the use of pyrolysis oils in diesel engine. Energy and Power Engineering. 2013;**5**(04):350

[17] Kim D et al. Waste plastics as supplemental fuel in the blast furnace process: Improving combustion efficiencies. Journal of Hazardous Materials. 2002;**94**(3):213-222

[18] Güngör C et al. Engine performance and emission characteristics of plastic oil produced from waste polyethylene and its blends with diesel fuel. International Journal of Green Energy. 2015;**12**(1):98-105

[19] Mani M, Nagarajan G, Sampath S. Characterisation and effect of using waste plastic oil and diesel fuel blends in compression ignition engine. Energy. 2011;**36**(1):212-219

[20] Pratoomyod J, Laohalidanond K. Performance and emission evaluation of blends of diesel fuel with waste plastic oil in a diesel engine. Carbon. 2013;**79**:75-99

[21] Guruprakash V et al. Thermal barrier coating on IC engine cylinder liner. Archives of Materials Science. 2016;**38**:38

[22] Saravanan S. Effect of exhaust gas recirculation (EGR) on performance and emissions of a constant speed DI diesel engine fueled with pentanol/diesel blends. Fuel. 2015;**160**:217-226

[23] Kumar BR, Saravanan S. Effects of iso-butanol/diesel and n-pentanol/diesel blends on performance and emissions of a DI diesel engine under premixed LTC (low temperature combustion) mode. Fuel. 2016;**170**:49-59

[24] Wei L, Cheung C, Huang Z. Effect of n-pentanol addition on the combustion, performance and emission characteristics of a direct-injection diesel engine. Energy. 2014;**70**:172-180

[25] Calder J, Roy MM, Wang W. Performance and emissions of a diesel engine fueled by biodiesel-diesel blends with recycled expanded polystyrene and fuel stabilizing additive. Energy. 2018

[26] Gnanasekaran S, Saravanan N, Ilangkumaran M. Influence of injection timing on performance, emission and combustion characteristics of a DI diesel engine running on fish oil biodiesel. Energy. 2016;**116**:1218-1229

[27] Li L et al. Combustion and emission characteristics of diesel engine fueled with diesel/biodiesel/pentanol fuel blends. Fuel. 2015;**156**:211-218

[28] Devaraj J, Robinson Y, Ganapathi P. Experimental investigation of performance, emission and combustion characteristics of waste plastic pyrolysis oil blended with diethyl ether used as fuel for diesel engine. Energy. 2015;**85**:304-309

[29] Hariharan S, Murugan S, Nagarajan G. Effect of diethyl ether on Tyre pyrolysis oil fueled diesel engine. Fuel. 2013;**104**:109-115

[30] Kalargaris I, Tian G, Gu S. The utilisation of oils produced from plastic waste at different pyrolysis temperatures in a DI diesel engine. Energy. 2017;**131**:179-185

[31] Sharma A, Murugan S. Combustion, performance and emission characteristics of a DI diesel engine fuelled with non-petroleum fuel: A study on the role of fuel injection timing. Journal of the Energy Institute. 2015;**88**(4):364-375

[32] Wamankar AK, Murugan S. Effect of injection timing on a DI diesel

engine fuelled with a synthetic fuel blend. Journal of the Energy Institute. 2015;**88**(4):406-413

[33] Senthilkumar P, Sankaranarayanan G. Production of waste polyethylene bags in to oil and studies performance, emission and combustion characteristics in di diesel engine. International Journal of Humanities, Arts, Medicine and Science. 2015;**3**(10):149-158

[34] Hawi M, Kiplimo R, Ndiritu H. Effect of exhaust gas recirculation on performance and emission characteristics of a diesel-piloted biogas engine. Smart Grid and Renewable Energy. 2015;**6**(04):49

[35] Paul Daniel M et al. Performance and emission characteristics of diesel engine operated on plastic pyrolysis oil with exhaust gas recirculation. International Journal of Ambient Energy. 2017;**38**(3):295-299

[36] Jagadish D, Kumar PR, Madhu Murthy K. Performance characteristics of a diesel engine operated on biodiesel with exhaust gas recirculation. International Journal of Advanced Engineering Technolog. 2011;**2**(2):202-208

[37] Krishnan PN, Vasudevan D. Performance, combustion and emission characteristics of variable compression ratio engine fuelled with biodiesel. International Journal of ChemTech Research. 2015;**7**(1):234-245

[38] Yasin MHM et al. Study of a diesel engine performance with exhaust gas recirculation (EGR) system fuelled with palm biodiesel. Energy Procedia. 2017;**110**:26-31

[39] Sharma A, Murugan S. Potential for using a Tyre pyrolysis oil-biodiesel blend in a diesel engine at different compression ratios. Energy Conversion and Management. 2015;**93**:289-297

[40] Maroa S, Samwel, Inambao F. The effects of exhaust gas recirculation on the performance and emission characteristics of a diesel engine–a critical review. International Journal of Applied Engineering Research. 2017;**12**(23):13677-13689

[41] Heywood JB. Internal Combustion Engine Fundamentals. New Delhi: MacGraw-Hill Education India; 2012. p. 930

[42] Mani M, Nagarajan G, Sampath S. An experimental investigation on a DI diesel engine using waste plastic oil with exhaust gas recirculation. Fuel. 2010;**89**(8):1826-1832

[43] Ghazikhani M, Feyz ME, Joharchi A. Experimental investigation of the exhaust gas recirculation effects on irreversibility and brake specific fuel consumption of indirect injection diesel engines. Applied Thermal Engineering. 2010;**30**(13):1711-1718

[44] Abaas KI. Effect of exhaust gas recirculation (EGR) on the performance characteristics of a direct injection multi cylinders diesel engine. Tikrit Journal of Engineering Science (TJES). 2016;**23**(1):32-39

[45] Venkatesan H et al. Assessment of waste plastic oil blends on performance, combustion and emission parameters in direct injection compression ignition engine. International Journal of Ambient Energy. 2017;**40**(2):170-178

[46] Bridjesh P et al. MEA and DEE as additives on diesel engine using waste plastic oil diesel blends. Sustainable Environment Research. 2018

Chapter 3

# Effect of Injection Pressure on Local Temperature and Soot Emission Distribution of Flat-Wall Impinging Diesel Flame under Diesel Engine like-Condition

*Rizal Mahmud and Iis Rohmawati*


## Abstract

Increasing heat transfer and heat transfer coefficient as the combined effect of impingement distance and injection pressure has been explored in the previous report. However, local temperature distribution was limited to the discussion. Clearly, investigation of near-wall temperature in the heat transfer analysis is absolutely necessary. This has a crucial effect on the local heat flux to understand the heat transfer phenomenon on the combustion chamber walls. The local temperature and KL factor were investigated by using a high-speed video camera and a two-color method by using a volume vessel with a fix-impingement wall. We found that the local temperature and KL factor distribution increase in low injection pressure. This result had a dominant effect on local heat transfer.


**Keywords:** diesel engine, local temperature, local KL factor, soot emission, injection pressure

## 1. Introduction

Vehicle electrification is vigorously promoted to achieve net-zero $CO_2$ emissions by 2050, considering one of the significant contributors of global $CO_2$ emissions comes from road transport. However, even on the International Energy Agency (IEA's) most aggressive scenario toward future renewable society, around 40% of vehicles sold in 2030 worldwide are still predicted to be powered by internal combustion engines [1]. Therefore, in order to practically minimize the long-term $CO_2$ emission, not only vehicle electrification but also thermal efficiency improvement of internal combustion engines is absolutely necessary.

As an effective remedy to improve the thermal efficiency of diesel engines, reduction of heat loss through the engine combustion chamber wall is known to have significant potential. Numerous researches have been done in this field, particularly on improving the thermal efficiency of diesel engines. It was conducted using a single-cylinder diesel engine [2–5] as well as a wall insertion-type constant volume vessel (CVV) [6–8]. To enhance thermal efficiency in the design of future engines, complete knowledge of the heat loss pathway from combustion gas to cylinder wall is essential.

In order to elucidate the mechanism of the wall heat transfer during the diesel spray flame impingement, a series of parametric study of wall impinging diesel spray flame combining transient wall heat flux measurements and high-speed optical diagnostics was conducted detailed by authors [9–11]. The results indicate how various experimental conditions affect the spray/flame impingement behavior, with considerable heat loss resulting in some cases. Gas flame velocity, contact area, and temperature difference are important factors of affect substantial heat loss. Therefore, identifying local temperature distribution is most needed to clarify the temperature difference near the wall during the combustion period.

Regarding some experimental parameters study [12], combining higher injection pressure resulted in higher heat loss, which is naturally attributed to higher flame velocity impinging on the wall with increased heat transfer coefficient. Therefore, this study aims to investigate the local temperature distribution under different injection pressure conditions. Furthermore, it also attempts to investigate the soot emission (KL Factor) distribution to contribute to the realization of the carbon-neutral future society. We used the two-color method to observe the local temperature and soot emission distribution.

## 2. Experimental setup

### 2.1 High-pressure and high-temperature chamber vessel

The experimental research has been performed by the author [9] using high pressure and high-temperature chamber vessel with four side windows. The experimental setup was described in **Figure 1**. Two transparent quartz windows were placed for visualization purposes. The others were facing each other which are used for an injector and spray, respectively. Three K-type thermocouples were installed close to the wall with the gap of 5 mm. These thermocouples were placed among the nozzle

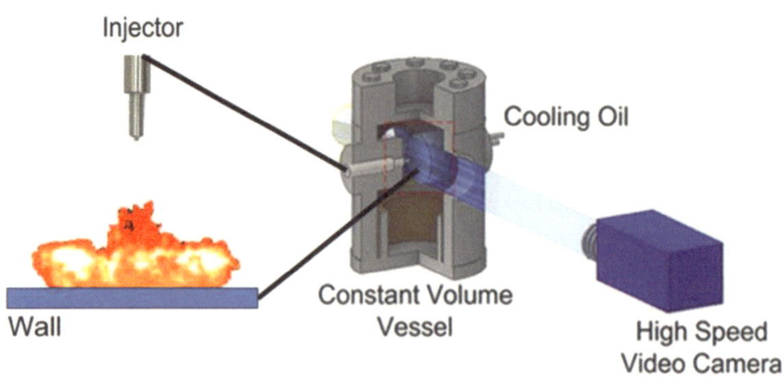


**Figure 1.**
*High-speed video camera setup.*

tip and wall for injection time purposes. If the ambient temperature was reached of set-point temperature, the diesel fuel will be injected to the chamber. Temperature gradient of these thermocouple was 5K. The injection pulse was defined as injection rate measurement.

**Figure 1** shows the high-speed camera setup and its arrangement. This high-speed video camera was installed to figure out the flame pattern at a frame rate of 20.000 fps (frames per second) with a resolution of 320 x 448 pixels. It indicates that each image has a resolution of 5 pixels per millimeter. Meanwhile, the two-color method was employed to investigate the relation between diesel flame and wall heat loss. Not only is the relation between diesel flame and wall heat loss, but also the distribution of soot and local temperature unable to observe by using the two-color method. The main principle of the two-color method is providing a number of assumptions by using different wavelengths of radiation intensity then the flame temperature and soot formation were found out. The soot concentration is defined from Hotel and Broughton's model which is represent the KL factor. More explanations about KL Factor have already been mentioned in the previous work [9, 10].

The two-color technique was calibrated using a standard light illuminant. Nac Image Technology's "Thermias" two-color pyrometry software was used to analyze the data. As a result, the flame temperature and KL factor were two-dimensional with line-of-sight information.

### 2.2 Test conditions

Combusting spray conditions were tested, which is containing the amount of air at high temperature to figure out the local temperature and KL factor under injection pressure. Uniform gas density is set to 16 kg/m$^3$ with keeping the ambient pressure at 4.1 MPa. Measurement conditions on the small-size diesel engines were determined in actual operation to reach the optimum results. Temperature and pressure were adjusted to compression TDC (top dead center) in low load operation for self-ignite

| Fuel | | Diesel fuel | |
|---|---|---|---|
| Ambient gas (MPa) | | 4.1 | |
| Ambient temperature (K) | | 873 | |
| Ambient density (kg/m$^3$) | | 16 | |
| Injector type | | Piezo actuator | |
| Number of nozzle holes | | 1 | |
| Injection quantity (mm$^3$) | | 5 | |
| Injection pressure (MPa) | 80 | 120 | 180 |
| Injection duration (ms) | 1.4 | 1.2 | 0.9 |
| Nozzle hole diameter (mm) | | 0.133 | |
| Impingement distance (mm) | | 40 | |
| Wall temperature (K) | | 460±10 | |
| Cooling method | | Oil cooling | |

**Table 1.**
*Experimental conditions.*

the fuel purposes. The impinging distance of 40 mm was decided between the nozzle tip and the wall. Meanwhile, the test conditions were performed by using nozzle with a hole diameter 0.133 mm at three types of injection pressure 80, 120, and 180 MPa. Experimental conditions in detail can be seen in **Table 1**.

## 3. Result and discussion

### 3.1 Combustion characteristics

In this section, the combustion characteristics of impinging flame at three types of injection pressure are discussed. At 40 mm of impingement distance, three types of injection pressure were employed i.e. 80, 120, and 180 MPa.

**Figure 2** shows the impinging flame at three types of injection pressure. Regarding injection rate graphs in previous work [10], luminous flames occur just before the end of injection. Furthermore, comparing three different injection pressure, the luminous flame appears earlier with higher injection pressure. It means increasing injection pressure leads to shorter ignition delay due to generate higher premixing of fuel and air. The flame luminosity was figured out at 0.9 ms after start of injection (ASOI) only during injection pressure of 120 and 180 MPa. However, it was unseen at injection pressure 80 MPa at 0.9 ms ASOI when increasing the time to 1.2 ms ASOI, the flame luminosity appeared. During the time of 1.2 ms ASOI, most brightness of flame luminosity was captured at injection pressure of 180 MPa. In this case, the luminous flame is starting to decrease then disappear. Similarly, at timing of 1.8 ms ASOI, the injection pressure of 120 MPa shows luminous flame start to decrease. However, in this timing the luminous flame continuous to develop at injection pressure of 80 MPa. It evident from **Figure 2**. that injection duration was longer with lower injection pressure than combustion duration showed finish later.

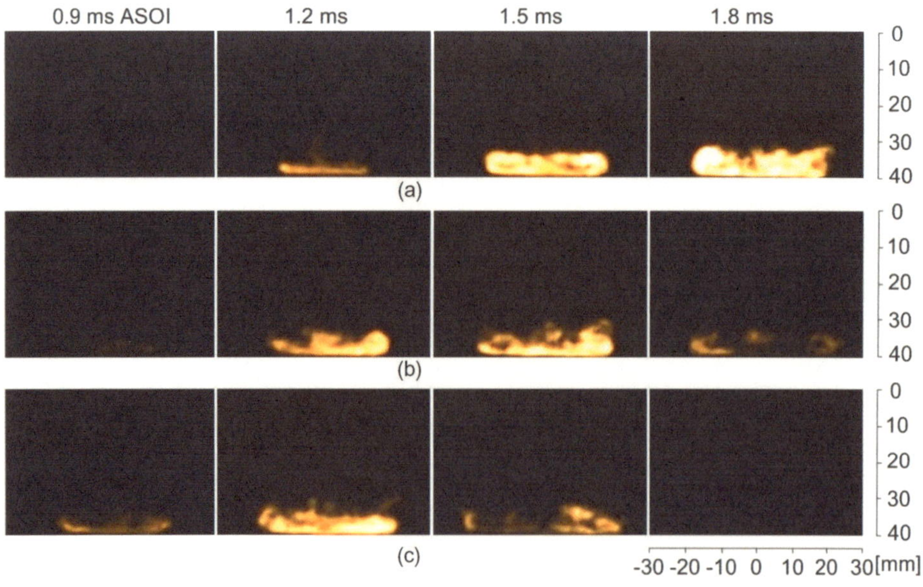


**Figure 2.**
*Impinging flame at three types of injection pressure. (a) $P_{inj}$ = 80 MPa. (b). $P_{inj}$ = 120 MPa. (c) $P_{inj}$ = 180 MPa.*

**Figure 3(a–c)** shows the temperature and KL factor distribution of impinging flame at different injection pressure. These figures were extracted from flame natural luminosity images by using a two-color method analysis. Brightness color in its distribution indicates temperature and KL factor. Flame luminosity was obtained to investigate the temperature and KL factor distribution, therefore both of them seem similar shapes.

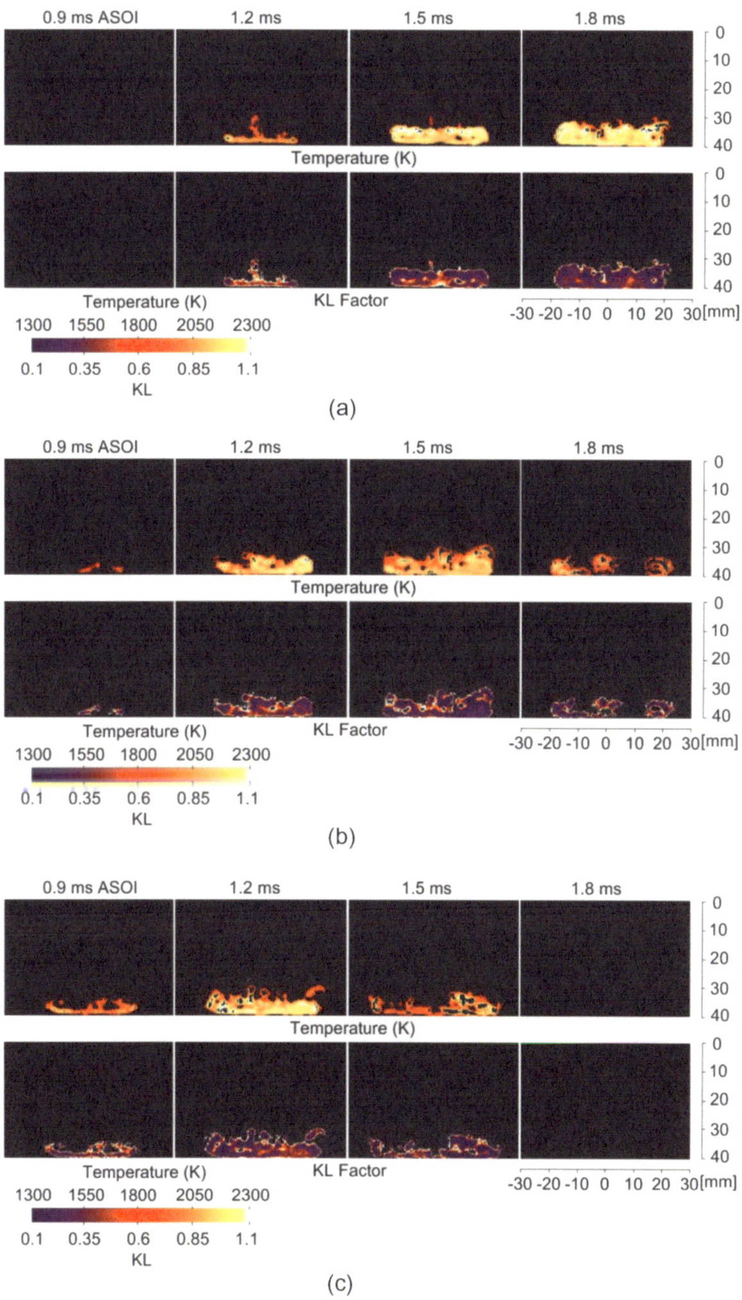


**Figure 3.**
*Flame temperature distribution (Top) and KL Factor (Bottom) at different injection pressures. (a) $D^{imp}$ = 40mm $P^{inj}$ = 80 MPa. (b) $D^{imp}$ = 40mm $P^{inj}$ = 120 MPa. (c) $D^{imp}$ = 40mm $P^{inj}$ = 180 MPa.*

At high injection pressure, complicated distribution of temperature was spread to wider area. However, the phenomenon was contrarily at lower injection pressure. The temperature was contributed to heat transfer on the wall [9]. The lowest injection pressure in this study is 80 MPa where it has the highest temperature compared to other injection pressure variations. This is probably due to more flame natural luminosity was captured at lower injection pressure as shown in **Figure 2**.

**Figure 3(a–c)** describes the KL factor distribution at injection pressure variations. Based on liquid length data in evaporating conditions, the soot formation region was formed [10]. This liquid length existed before impingement during the injection period. A shorter period with higher injection pressure has been found through the soot formation that occurred around the center of the impingement wall as shown in **Figure 3**. Mixing the fuel and air better is an important decisive factor in the reduction of soot formation.

A similar trend of integrated flame luminosity, luminous flame area, mean temperature, and integrated KL factor was shown in **Figure 4(a–d)**. It starts increasing and reaches the maximum point before decreasing at time variation in each injection pressure. The integrated luminosity is shown in **Figure 4(a)**. The flame luminosity

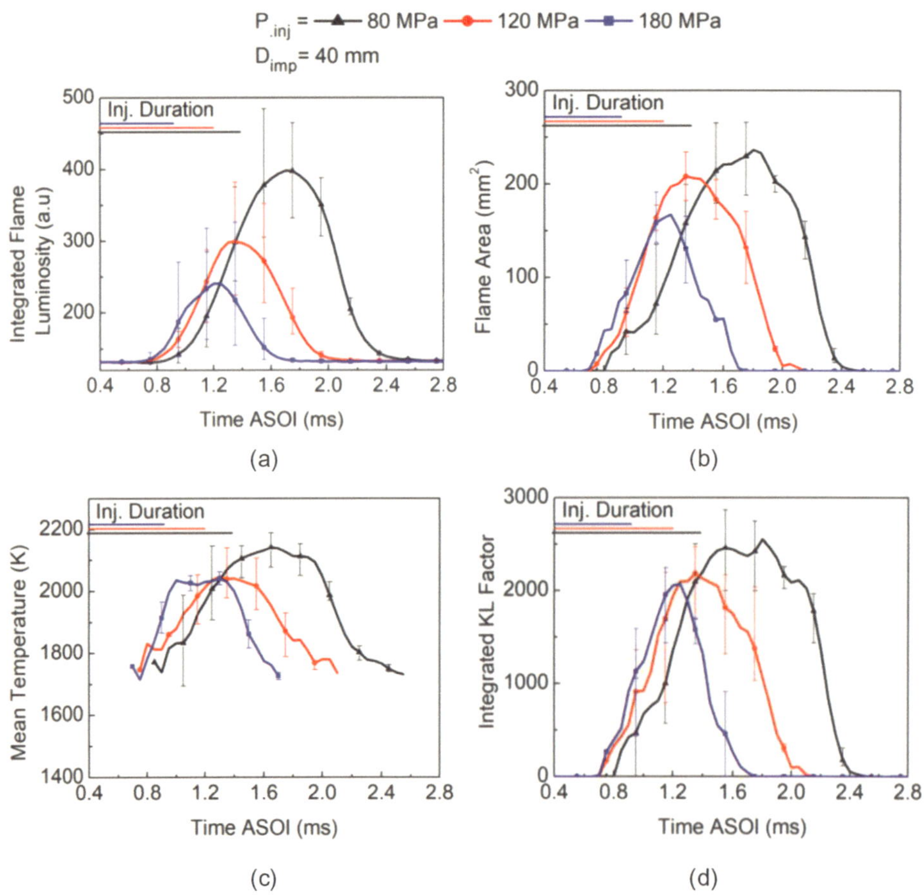


**Figure 4.**
*Integrated Flame luminosity, luminous flame, mean temperature, and KL factor under injection pressures. (a) Integrated flame luminosity. (b) Luminous flame area. (c) Mean temperature. (d) Integrated KL factor.*

indicates soot combustion in case of insufficient oxygen by a rich mixture. Opposite relation was found between injection pressure and luminosity where higher injection pressure will have smaller and shorter luminosity. Increasing injection pressure means increasing the velocity. Therefore, air entertainment will improve spray atomization and premixing of fuel and air. **Figure 4(b)** shows the shorter flame area at higher injection as a result of premixing fuel and air.

**Figure 4(c)** describes the mean temperature of three types of injection pressure where it has a lower temperature compared to the temperature at injection pressure of 80 MPa. The differences in temperature among them were approximately 100K. We can see that flame temperature distribution at three types of injection pressure were 1.2, 1.5, and 1.8 ms ASOI as shown in **Figure 4(c)**. At the time 1.5 and 1.8 ms ASOI, the temperature distribution was almost uniform where the mean temperature will be high as shown in **Figure 4(c)**. On other hand, the reduction of soot formation may affect premixed combustion at high injection pressure with a shorter injection duration.

**Figure 4(d)** shows the integrated KL factor which is consist of the soot formation. This formation was decreasing when the injection pressure increased. Meanwhile, in **Figure 4(a–c)**, the integrated KL factor increases to peak value at maximum flame natural luminosity and flame area. As mentioned before, opposite relation between injection pressure and KL factor was found which is due to less air entertainment than

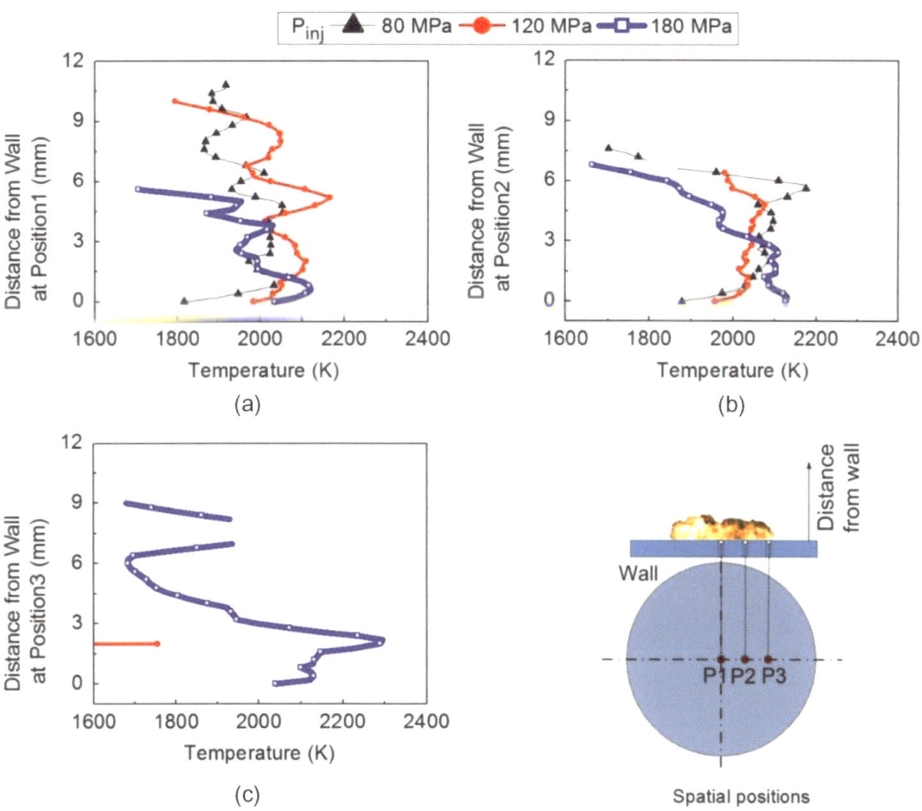


**Figure 5.**
*Distribution of temperature from wall surface at 1.2 ms ASOI under injection pressures and positions. (a) Position1 at 1.2 ms ASOI. (b) Position2 at 1.2 ms ASOI. (c) Position3 at 1.2 ms ASOI.*

spray atomization and fuel-air mixing decreasing. Next, it affects a high equivalent ratio. For soot production purposes, the correlation between high temperature and fuel-air mixing is interesting to discuss.

### 3.2 Local temperature and KL factor distribution

In this study, we evaluate the axial temperature distribution from the wall surface with position variations. **Figure 5** shows the time 1.2 ms ASOI in three types of positions where it reaches the maximum at injection pressure 180 MPa. These temperature distributions shown in **Figure 4** were obtained from **Figure 3**. On other hand, the increasing temperature has a reverse effect from some injection pressure. It means that 180 MPa of injection pressure contributed to the highest heat loss. In this case, the temperature gradient has an effect on the local heat flux and the heat transfer. In **Figure 5(c)** at position 3, the dominant temperature distribution at 180 MPa affect the wider area of flame distribution.

One of the necessary parts of heat loss on the wall during flame impinging is the spatial distribution of local temperatures. **Figure 6** shows the mean temperature near the wall at 0.8 mm for all variations. The result shows that the injection pressure of 180 MPa has the maximum temperature at all surface areas. It indicates that the flame temperature along the near wall is uniform. This made the local temperature gradient was higher.

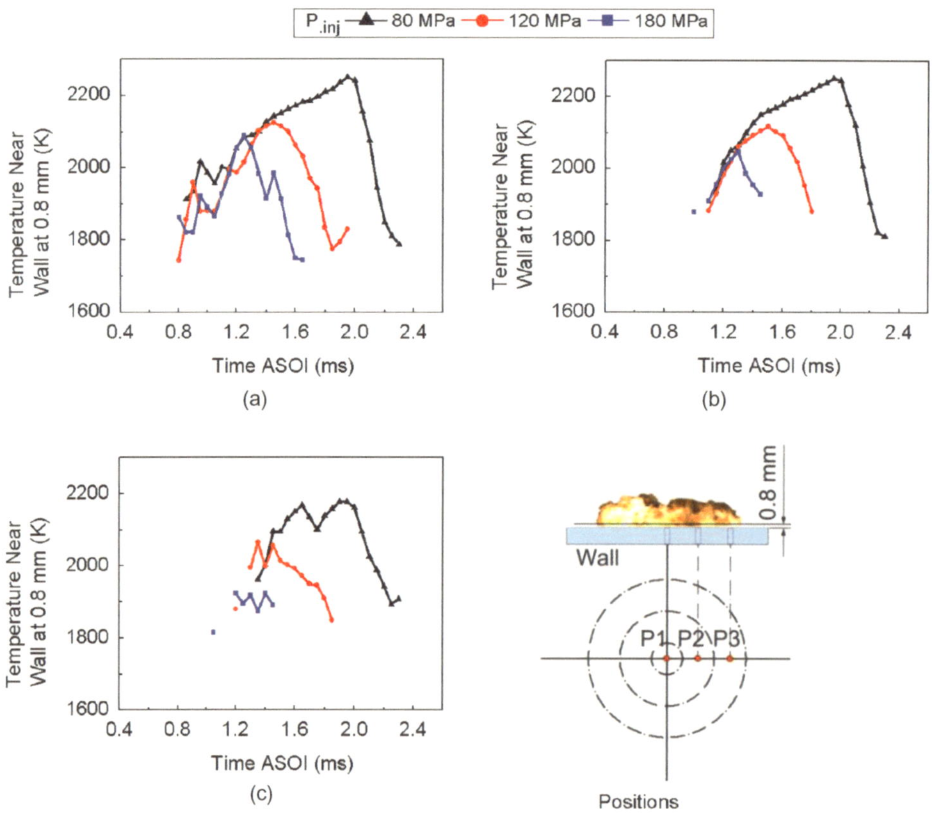


**Figure 6.**
*Temperature near wall at 0.8 mm from wall during injection pressure. (a) Position 1. (b) Position 2. (c) Position 3.*

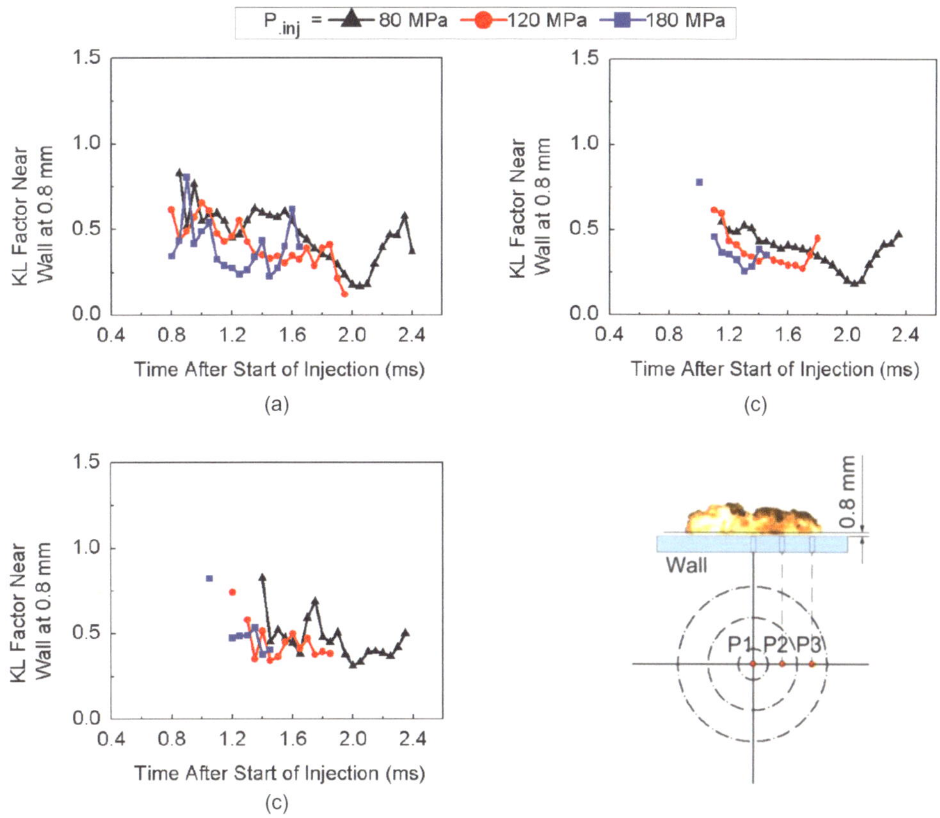


**Figure 7.**
*KL Factor near wall at 0.8 mm from wall under injection pressures. (a) Position 1. (b) Position 2. (c) Position 3.*

The spatial distribution of the local KL Factor near a wall at 0.8 mm is shown in **Figure 7(a–c)**. The figure shows that the distribution of KL Factor varies at three areas namely Position1, Position2, and Position3 under different Injection Pressures. KL Factor decreases with increasing time after the start of injection at all injection pressures and Positions. It can be seen from **Figure 7(a–c)** that the higher injection has a shorter KL Factor period where it will have an impact on the total amount of KL factor. This means that the injection of 180 MPa can be considered to have contributed to the reduction of carbon neutrality for a future society.

## 4. Conclusion

The effects of injection pressure on local temperature and soot emission distribution were investigated in the flat-wall impinging diesel flame under diesel like-condition. General conclusions obtained from the study are summarized as follows:

1. Injection pressure of 80 MPa reached the most brightness in flame luminosity, which is attributed to the higher mean flame temperature and soot emission.

2. The axial temperature distribution from the wall surface was found higher local temperatures at 180 MPa compared with other pressures under 1.2 ms ASOI at all positions.

3. Injection pressure of 80 MPa has the maximum local temperature near a wall (0.8 mm) at all of the surface areas. This indicates that flame temperature is uniform along a near-wall which is lead large the local temperature difference between wall and flame.

4. Injection pressure of 180 MPa has a shorter KL Factor period which identifies a small amount of the total KL factor.

## Acknowledgements

The authors thank to Prof. Keiya Nishida at Mechanical Power and Motor System Laboratory, University of Hiroshima for their support with measurement in this study and we gratefully acknowledge Prof. Tetsuya Aizawa-Meiji University for his valuable suggestions and discussions.

## Appendices and nomenclature

| | |
|---|---|
| $P_{inj}$ | Injection Pressure |
| ASOI | After Start of Injection |
| TDC | Top Dead Center |
| Fps | Frame per Second |
| CVV | Constant Volume Vessel |

## Author details

Rizal Mahmud[1*] and Iis Rohmawati[2]

1 Adhi Tama Institute of Technology, Surabaya, Indonesia

2 Hiroshima University, Higashihiroshima, Japan

*Address all correspondence to: rizal@itats.ac.id

IntechOpen

© 2022 The Author(s). Licensee IntechOpen. This chapter is distributed under the terms of the Creative Commons Attribution License (http://creativecommons.org/licenses/by/3.0), which permits unrestricted use, distribution, and reproduction in any medium, provided the original work is properly cited.

## References

[1] IEA. Electric Vehicles. Paris: IEA; 2021

[2] Kogo T, Hamamura Y, Nakatani K, et al. High efficiency diesel engine with low heat loss combustion concept: Toyota' s inline 4-cylinder. SAE paper. 2016

[3] Wakisaka Y, Inayoshi M, Fukui K, et al. Reduction of heat loss and improvement of thermal efficiency by application of "temperature swing" insulation to direct injection diesel engines. SAE paper. 2016

[4] Enya K, Uchida N. Improvement in thermal efficiency of a diesel engine by homogenized flame distribution. SAE Technical Paper. 2019

[5] Horibe N, Bao Z, Taguchi T, Egoshi K, et al. Improvement of thermal efficiency in a diesel engine with high-pressure split main injection. SAE Technical Paper. 2018. DOI: 10.4271/2018-01-1791

[6] Tatsumi T, Maeda S, Miyata S, et al. A study on the wall heat loss in diesel spray flame (First report: Effect of injection pressure on the heat flux). JSAE Annual Congress (Spring) in Pasifico Yokohama, Japan; May 25-27, 2016 (in Japanese)

[7] Tatsumi T, Maeda S, Nakata M, et al. A study on the wall heat loss in diesel spray flame (third report: effect of fuel heating on the heat flux). In: Proceeding of the 27th Internal Combustion Engine Symposium. Tokyo, Japan; 2016. p. 25

[8] Nakata M, Arai N, Maeda S, et al. A study on the wall heat loss in diesel spray flame (Sixth report: effects of impingement distance and inclination angel of the wall on the heat flux). In: Proceeding of 2016 JSAE Annual Congress, Yokohama, 2016. (in Japanese)

[9] Mahmud R, Kurisu T, Nishida K, Ogata Y, Kanzaki J, Tadokoro T. Experimental study on flat-wall impinging spray flame and its heat flux on wall under diesel engine–like condition: First report—Effect of impingement distance. Proceedings of the Institution of Mechanical Engineers, Part D. 2019;**233**(8):2187-2202. DOI: 10.1177/0954407018778153

[10] Mahmud R, Kurisu T, Nishida K, Ogata Y, Kanzaki J, and Akgol O, Effects of injection pressure and impingement distance on flat-wall impinging spray flame and its heat flux under diesel engine-like condition. Advances in Mechanical Engineering. July 2019;**11**(7):168781401986291. DOI: 10.1177/1687814019862910

[11] Mahmud R, Kurisu T, Ilminnafik N, Nishida K, et al. Wall heat flux on impinging diesel spray flame: effect of hole size and rail pressure at similar injection rate condition. SAE Technical Paper. 2020. DOI: 10.4271/2020-32-2313

[12] Mahmud R, Kurisu T, Akgol O, Nishida K, et al. Characteristics of flat-wall impinging spray flame and its heat transfer under diesel engine-like condition: Effects of injection pressure, nozzle hole diameter and impingement distance. SAE International Journal of Advances and Current Practices in Mobility. 2020;**2**(1):319-329. DOI: 10.4271/2019-01-2183

Chapter 4

# A Comparative Evaluation of Biodiesel and Used Cooking Oil as Feedstock for HDRD Application: A Review

*Josiah Pelemo, Kayode Timothy Akindeji, Freddie L. Inambao, Omojola Awogbemi and Emmanuel Idoko Onuh*

## Abstract

The search for clean energy for transportation fuel across the globe has grown in intensity. The use of biodiesel as a fuel for compression ignition (CI) engines has shown some deficiencies, e.g., poor storage, and poor pour point. The carbon chain of biodiesel is one of the factors to be considered; the longer carbon chain length leads to decreased ignition delay, which leads to the formation of OH during the premixed combustion phase. The major challenges that render biodiesel inefficient are discussed, like higher viscosity, lower energy content, higher nitrogen oxide ($NO_X$) emissions, lower engine speed and power, injector coking, engine compatibility, high cost, and higher engine wear. The novelty of this work is that it shows that biodiesel conversion to green diesel is possible using a biowaste heterogeneous catalyst to obtain quality and high yield of HDRD with lower cost. This renewable energy (HDRD) possesses properties that are directly compatible with CI engines and transportation engines. This research reviewed biodiesel and UCO as feedstocks for the production of HDRD, including the cost–benefit of these feedstocks. Hydrogenation of biodiesel has the potential to overcome the drawbacks of conventional chemically catalyzed processes.

**Keywords:** hydrogenation, biodiesel, HDRD, hydrogen, UCO

## 1. Introduction

The global population growth and the resulting development of commercial and industrial activities, especially in the transportation sector, have stimulated scholars in various institutions to search for sustainable renewable energy. With the continual depletion of conventional primary energy, the need for renewable alternative energy sources becomes more and more important for energy utilization in social and industrial activities. Biodiesel has as a renewable energy has received much attention and research over the year. Biodiesel in neat form or mixed with conventional diesel can be used in compression ignition (CI) engines and stationary engines [1, 2].

The production of biodiesel can be from vegetable oils or animal fats by means of a transesterification reaction, which uses alcohols in the presence of a catalyst [3]. A catalyst is used to chemically convert triglyceride molecules into alkyl esters, generally known as biodiesel fuels [4, 5]. Methanol and ethanol are the most commonly used alcohols for transesterification due to their low cost and high activity [6]. A free fatty acid (FFA) content of higher than 0.5% in vegetable oil renders it a low-grade feedstock due to saponification under alkali catalyzed reaction. The production of conventional biodiesel comprises two stages: the first stage is acidic catalytic esterification and the second stage is the transesterification method using an alkaline or base catalyst. The production process is a time-consuming and tedious procedure because of certain mandatory stages, i.e., time taken for the water to settle at the base and the transesterification process in the presence of an alkaline catalyst in the second stage. The volume of wastewater generated during esterification is high. The drying process of the esterified mixture and further application of transesterification is also a highly taxing process. The catalytic conversion process of low-grade vegetable oil with a high percentage of FFA or high moisture content into biodiesel requires a high-temperature reaction as reported in the literature [7, 8]. The high content of oxygen in biodiesel results in deficiencies such as low oxidative stability, high viscosity, low cloud point, and high pour point in the cold region [9]. Biodiesel also shows lower stability during storage and it attacks metals like copper, zinc, tin, and lead, which can lead to corrosion of some parts of the engines. Another shortcoming of biodiesel is the low energy content and nitrogen oxide ($NO_X$) [10] which reduces thermal and break power efficiency [11]. The deficiencies mentioned above limit the application of biodiesel in CI engines.

The focus of this study is the processing of biodiesel as a feedstock to obtain a green diesel product which can offer sufficient properties without any adverse effects on the CI engine and environment. The following vegetable oils have been discovered and have been used as a feedstock for the production of biofuel for decades, viz., rapeseed, palm, cottonseed, sunflower, peanut oil, soybean oil [12, 13] and animal fats like butter, fish oil, and tallow. These renewable energy sources from biomass sources are the major feedstock sources of biofuel production. However, vegetable oil cannot be used directly to fuel CI engines because it is not compatible due to its high viscosity. The triglycerides and fatty acids present in vegetable oil are the promising components of vegetable oil feedstocks for the production of sustainable biofuel. These feedstocks produce diesel and gasoline type of hydrocarbons via hydroprocessing that can be used in CI engines [14]. HDRD is produced via hydroprocessing of triglycerides contained in edible oil such as used cooking oils and vegetable oils (e.g. rapeseed, soybean, cottonseed, palm, corn, sunflower, coconut, peanut, camelina, carinata, and jatropha oils), fats, and micro-algal oils [15]. These vegetable oils cannot be applied directly in the modern CI engine due to their high viscosity but can be used as a fuel source after some modifications in the fuel properties [16]. Feedstock sourced from edible oil for the production of biofuel has become a problem because of the threat to food security. The need for large land space for farming, the cost, and the resulting threat of deforestation is a major challenge for edible oil, therefore, UCO as a feedstock for HDRD application has recently been adopted. Sunflower oil constitutes about 40–50% of vegetable oil produced in Europe, Russian Ukraine, Turkey, and Argentina. It was reported in literature that sunflower and rapeseed oil are the major sources of feedstock for renewable energy in Europe [17]. The percentage production of the main vegetable oils across the globe are sunflower (10%), rapeseed (55%), cottonseed (10%), and soybean (55%) [18]. Palm oil has been discovered as a potential feedstock for biofuel production in Malaysia [19, 20]. This novel research study focuses on the potential of biodiesel fuel as a better feedstock for production of green diesel.

The main feedstock for the production of biodiesel is vegetable oil [21]. Biodiesel is an alternative fuel that has similar properties to conventional or 'fossil' diesel. Conventional homogeneously catalyzed processes for fatty acid methyl esters (FAME) biodiesel production can be used to convert waste vegetable oils but is limited to oils with a relatively low FFA content. Some of the shortcomings of biodiesel that makes it necessary to convert it first to green diesel are: variation in the quality of biodiesel, food shortages, clogging in engine, not suitable for use in low temperatures, water shortages, slight increase in nitrogen oxide emissions. Biodiesel can be hydrotreated to obtain a quality green diesel fuel.

Da Rocha Filho et al. reported that more than 600,000 tons of used cooking oil are generated in South Africa per/annum [22, 23]. HDRD could be produced annually from this waste; given a yield rate of 80%, this will provide 205 million liters. However, this would cater for 50% supply targeted for renewable fuel in the biofuel policy of the government of the South Africa government. The current price of UCO is R3/liter and diesel is R14/liter. A value-added industry generating R2.04 billion can be created producing premium diesel (HDRD) along with the creation of thousands of jobs. Other potential secondary sources of feedstock are cellulose from pulp and paper industries plus a diverse range of agricultural waste with a far greater capacity than UCO [24]. UCO is an oil generated from vegetable oils after frying. UCO is readily available and abundant from food industries, restaurants, households, and fast food outlets using vegetable oils for cooking and frying. The demand for vegetable oil is on the increase in the continent. The yearly consumption of edible vegetable oils in China is approaching 22 million tons, and the country produces more than 4.5 million tons of used oil and grease per year [25]. Vegetable oil used for cooking undergoes various form of chemical and physical changes. Some unwanted compounds like FFAs and some polymerized triglycerides are formed during frying which causes a rise in the molecular mass and condenses the volatility of the oil. Used cooking oils are renewable and do not contain any aromatics, metal, or sulfur contaminants. Reuse of UCO can exacerbate environmental problems, health challenges including hypertension, diabetes, vascular inflammation, and other health effects [26]. Vegetable oil is an

| Fuel properties | Used cooking oil | Biodiesel | Commercial |
|---|---|---|---|
| Kinetics viscosity ($mm^2/s$, @313 k) | 36.4 | 5.3 | 1.9–4.1 |
| Flash points (k) | 485 | 469 | 340–358 |
| Cetane Number | 49 | 54 | 40–46 |
| Density (kg/l, @ 288 k) | 0.924 | 0.897 | 0.75–0.840 |
| Pour points(k) | 284 | 262 | 254–260 |
| Sulfur contents (%) | 0.09 | 0.06 | 0.35–0.55 |
| Water contents (%) | 0.42 | 0.04 | 0.02–0.05 |
| Ash contents (%) | 0.06 | 0.004 | 0.008–0.010 |
| Free fatty acid (mgkoH/g oil) | 1.32 | 0.10 | — |
| High heating value (MJ/kg) | 41.40 | 42.65 | 45.62–46.48 |

**Table 1.**
*Comparison of properties of UCO, biodiesel, and commercial diesel fuel [27].*

| S/N | Name of industry | Feedstock/ Technology | Targeted product/ton | Properties of product | References |
|---|---|---|---|---|---|
| 1 | UOP/Ecofining | Triglyceride, hydro-processing of UCO | Green diesel | Similar to the properties of fossil fuels, Good cool flow | [28–31] |
| 2 | Tyson Foods Inc. Gesmar LA, USA | Hydrotreating of non-edible and animal fats | Green diesel, 75 Mil.P/ annum | Good storage stability, High cetane numbers, properties similar to petroleum fuel | [32] |
| 3 | Haldor Topsoe | New hydrotreating raw tall | Green diesel | Similar to fossil fuel properties | [33] |
| 4 | ConcocoPhillips | Hydrogenation of Vegetable oil | Green diesel 365,000 barrel/ annum | Sulful free fuel content, emission of less NOx | [33] |
| 5 | Valero Energy Corporation, St. Charles | Hydrogenation of UCO/Animal fats | Renewable diesel | Similar to fossil fuel properties | [34] |

**Table 2.**
*Commercialization of green diesel by selected industries.*

oil used for cooking a various type of food items which include, chicken, beef, yam/ potatoes. Currently, the major factor that hinder the commercialization of renewable fuel the high cost of feedstock compared to fossil fuel. It has been reported in literature that about 70–85% cost of production of HDRD arises from the raw materials. However, the use of UCO as a feedstock for production of HDRD will enhance the commercialization of green diesel due to the availability at a low price. The HDRD production process involves the conversion of fatty acids in triglycerides into normal and/or iso-paraffin which can be obtained by hydrodeoxygenation, decarbonylation, decarboxylation, isomerization and hydrocracking or a combination of two or more thereof (**Tables 1** and **2**).

The high acid value of UCO is due to the high content of FFAs [23]. In recent years, several petroleum companies have directed their resources into the production of renewable green fuels from hydro-processing of vegetable oils feedstock, with considerable commercial success.

The Neste Oil Co. developed a technology to convert vegetable oil and animal fat into high-quality hydrocarbons. The plant start operation by Neste Oil in Singapore in 2012 using NExBTL technology targeted production of over 800,000 tons renewable diesel per annum from feedstocks [35]. The experimental analysis of the samples by Neste Oil shows a high cetane value of between 84 and 99, a low cloud point value (as low as minus 30°C), and can withstand storage for extended periods. These properties enhance its performance in both car and truck engines [36].

The commercialization of bio hydroformed diesel (BHD) by the joint effort of Toyota Motor Corporation (TMC), Hino Motors, the Tokyo Metropolitan Government, and Nippon Oil Corporation (NOC) have commenced operation in recent time, a second-generation renewable diesel fuel produced by hydrogenating a vegetable oil feedstock. Nippon Oil and Toyota have worked jointly on the development of BHD technology since 2005. The use of refinery-based for hydro processing of vegetable oil for the production of a synthetic, second-generation biofuel depends on several issues, including the properties and effects of first-generation FAME (storage, oxidation, possible effect on fuel handling systems). In its studies, Nippon Oil explored reaction temperatures ranging from 240–360°C, with reaction pressures of 6 MPa and 10 MPa, and used a common hydrodesulfurization catalyst. The resulting fuel is claimed to be aromatics- and sulfur-free, with a cetane number of 101 [33].

The daily consumption of vegetable oil has witnessed a tremendous increase globally due to the increasing population and modernization. The global total primary energy consumption(GTPEC) has recorded over 150,000,000 GW h and this is expected to increase by 57% in the year 2050 [37]. This significant growth of energy consumption will eventually result in more environmental problems [38]. Currently, over 80% of the total energy used across the globe is sourced from fossil fuels, leading to their high contribution to environmental and health challenges [39]. UCO, which is considered waste, is collected before disposal. The annual collection of UCOs is evidence of the high consumption rate of UCO in some countries. The Energy Information Administration (EIA) of the United States reported an estimate of 100 million gallons of UCO produced per day in the USA [40]. About 135,000 and 140,000 tons of UCO are generated per annum in Canada [41, 42]. In South Africa, 0.6 million tons of UCO are collected annually from bakeries, takeaway outlets, and restaurants [43, 44]. The UK and the European Union countries produced 0.7 million tons to1.0 million tons and 0.2 million tons of UCO per annum, respectively [45]. The generation of a large amount of used cooking oil is a panacea to food security and fuel sustainability if properly harnessed for hydrogenation purposes. UCO is readily available, sustainable and cost–effective. Reports show that 17% of UCO offer a yield of 11.92 million tons while about 9% of the feedstock for the production of 26.62 million tons of biofuel is obtained globally in 2015 [46]. UCO was investigated, and the outcome of the chemical properties show that oleic acid has the highest value of 43.67%, followed by palmitic acid with 38.35%, and linoleic acid, 11.39% [47]. These properties of used cooking oils make it viable as a l feedstock for conversion into hydrocarbon. **Table 3** shows the properties of UCO samples of sunflower oil, palm oil and sunfoil.

The feedstock is one of the key resources in determining the production costs of biofuel. The adoption of biodiesel as feedstock will reduce the total cost of production and this will support the profitability and commercialization of an HDRD product.

| **Properties** | **UCO Samples** | | |
|---|---|---|---|
| | **Sunflower oil** | **Palm oil** | **Sunfoil** |
| Density (Kg/m3) | 920.4 | 913.4 | 923.2 |
| pH value | 5.34 | 6.19 | 6.61 |
| Viscosity ($mm^2/s$) | 31.381 | 38.407 | 35.236 |
| Acid value | 2.29 | 1.13 | 1.44 |
| Congealing temperature °C | –5.15 | 14.7 | –3.4 |
| Molecular weight (g/mol) | 51.94 | 586.05 | 395.28 |
| Iodine value (cg/g) | 111.1 | 54.2 | 54.2 |

**Table 3.**
*Properties of feedstock compared to other vegetable oil [46].*

Biodiesel can be hydrotreated to combat the challenges of storage stability, cetane number to obtain superior HDRD as known as renewable fuel at a lesser cost and labor. UCO oil makes up approximately 80% of the total production expenses [29]. Biodiesel oil is readily available, does not affect food security, is cheap, requires not much effort to source, and offers a good yield when used. Green diesel is oxygen-free, hence oxidation stability is high, and has a high cetane number (CN) similar to fossil fuel. Hydrogenation derived renewable diesel(HDRD) possess high pour point better than biodiesel, It reduces $NO_X$ emissions, and has a high heating value which is a significant property of diesel fuel because it gives the energy content of the fuel and aid the performance of CI engines. Furthermore, green diesel produced by the hydro-processing of triglycerides has propane as a by-product which is a gaseous fuel of good market value. This property makes HDRD production more feasible in economic terms when compared to the production of FAME [48]. The composition of biodiesel products can be improved by hydro-processing techniques to obtain HDRD.

## 2. Parameters and metrics of study

Biodiesel fuel has been investigated by researchers with varying outcomes. **Table 4** shows some problems and possible solutions related to biodiesel fuel. Some important properties of biodiesel are cloud and pour point, storage stability, viscosity, acid value, cetane number. These are the properties of biofuel that must meet the set ASTM standard of biofuel. These properties are deficient when considering using biodiesel in CI engines, particularly in a cold, temperate, regions. The cloud points of ethyl ester produced from use cooking oil, linseed oil, canola, sunflower, and rapeseed oil are −1°C, −2°C, −1°C, −1°C, and −2°C, respectively. Lang et al. (2001) [49] reported that the cloud point of ethyl esters of linseed oil, canola, sunflower, and rapeseed oil were −2°C, −1°C, −1°C, and −2°C, respectively,whereas the corresponding methyl esters had cloud points of 0°C, 1°C, 1°C, and 0°C. Currently, internal combustion engine are controlled by compression ignition(CI) engine and the spark ignition (SI) engine. The operations of Spark engine (SI) is done by premixed charge near stoichiometric air-fuel ratio (Ø ~ 1), the flame is spread with the aid of a spark plug, the throttling effect of the charges into the cylinder leads to low thermal efficiency at partial load. Moreover, with the high temperature at peak load, cause generation of massive $NO_X$, this have advert effect on the environment, is form a major challenge of the SI engine applications. The compression ratio of CI engine is between 12 and 24, in addition CI engine, is characterized by turbulent flame, diffused flame, auto-ignition via elevated pressure and temperature around the top dead center. The CI engine has a higher thermal efficiency compared to a SI engine. The major part of its operation is ignition delay (this is the time difference between the start of injection and the self-ignition). The ignition delay mechanism is controlled by physical and s chemical kinetic processes. The physical process proceeds sequentially through droplet formation, collision, break-ups, evaporation, and vapor diffusion. The chemical kinetic process involves species and radical formation proceeding through low-temperature reaction (LTR), negative temperature coefficient (NTC), and high-temperature reaction (HTR). The challenge is that the time scales for the physical processes are often larger than those of the chemical processes, hence ignition often commences before the physical processes are completed. This inevitably leads to a complex system of charge, flame, and thermal stratification that produces both high $NO_X$ in some regions and high soot precursors and unburnt hydrocarbon (UHC) in others. Research focus, therefore, has been to hydroprocess biodiesel fuel into HDRD to mitigate $NO_X$, UHC, and soot

| Problem | Causes | Possible solutions |
|---|---|---|
| Cold weather starting | High viscosity, low cetane, and low flash point of vegetable oils | Preheat fuel before fuel injection. Chemically alter fuel to an ester |
| Plugging and gumming of filters lines and injectors | Natural gums (phosphatides) in vegetable oil. Other ash | Partially refine the oil to remove gums. Filter to 4-microns |
| Excessive engine wear | The high viscosity of vegetable oil, incomplete combustion of fuel. Poor combustion at part load with vegetable lubricating oil due to blow-by of vegetable oil | Heat fuel before injection. Switch engine to diesel fuel when operating at part load. Chemically alter the vegetable oil to an ester. Increase motor oil changes. Motor oil additives to inhibit oxidation |
| Engine knocking | Very low cetane number of some oils. Improper injection timing | Adjust injection timing. Use higher compression engines. Preheat fuel before injection. Chemically alter fuel to an ester |
| Coking of injectors on piston and head of engine | The high viscosity of vegetable oil, incomplete combustion of fuel. Poor combustion at part load with vegetable oils | Heat fuel before injection. Switch engine to diesel fuel when operating at part load. Chemically alter the vegetable oil to an ester |
| Failure of engine lubricating oil due to polymerization | Collection of polyunsaturated vegetable oil blow-by in the crank case to the point where polymerization occurs | Heat fuel before injection. Switch engine to diesel fuel when operating at part load. Chemically alter the vegetable oil to an ester. Increase motor oil changes. Motor oil additives to inhibit oxidation |

**Table 4.**
*Problems, causes, and possible solutions regarding biodiesel products.*

formation (as is the case with the SI engine). From the foregoing, it is clear that there are many challenges facing the efforts to utilize biodiesel as a fuel in transportation.

## 3. Potential feedstock for biodiesel production

The rapid urbanization, abundant land resources, and vegetable oil in Africa offer adequate opportunity for biodiesel production on a large scale. FAME product obtained from this resource provide insight into the new option of sourcing for renewable energy. Traditionally, biodiesel had been produced from vegetable oil; about 93% of biodiesel is produced from edible oil, and the feedstock sourced from agricultural sector [50]. However, the edible vegetable oil is unsustainable due to the food security threat [51]. Biodiesel production uses around 4.4 million hectares of arable land in the European Union [52]. The consequence of deforestation is greenhouse gas (GHG) which great effect on man, plant, animal. The alternative option to these is the use of non-edible oils, hence the focus on biodiesel and used cooking oil as feedstock. Other vegetable oils can be propagated easily in drought-prone areas and are highly adaptable in tropical areas [53]. Biodiesel fuel converted to HDRD is set to play an important role in the future around the world. Africa's population is projected to reach 1.5 billion by 2030, with 53.5% of this population living in developed countries [54]. This trend is set to accelerate and a greater supply of green diesel will be necessary in order to meet the growing demand for HDRD. The only economically beneficial and sustainable ways to achieve this goal is to embark on aggressive production of HDRD using biodiesel as a feedstock.

## 4. Production and feedstock value chain

The technology for petroleum-based fuel production has been in existence for many years. The same catalyst, reactor type, and distillation facilities used for the production of fossil-based fuel are also applicable for hydro-processing of vegetable oil-derived feedstocks to obtain high-quality hydrocarbon. Thus, massive savings are achievable since the same production plant facilities can be used for the purpose. Conversion of used vegetable oil via hydro processing using stand-alone units can be achieved by optimizing and controlling the facilities to obtain high yield green diesel. The design and construction of this facility as an attachment to existing plants and hydrogen in the refinery can be streamlined as recycled gas. The only shortcoming of a stand-alone unit as a production facility is the high cost of construction. In recent times the feedstock used for the production of HDRD has been plant-derived oils such as rapeseed, soybeans, and palm [55–58] which are edible oils, with the non-edible oils like Jatropha, algal oils and waste cooking oil products being the most popular feedstock ion recent times [57, 59]. Many researchers have confirmed UCO as being the most viable, cost-effective, and available feedstock. Researchers are still investigating the most suitable technology for the production of green diesel that is cost-effective and CI engine compatible. No research has been done yet on the viability of biodiesel oil as a potential feedstock for HDRD production. When this is explored it will boost the supply of renewable fuel in energy sector. Green diesel obtained from biomass can be processed through four technologies: (i) Pyrolysis and upgrading of bio-oil, (ii) Hydro-processing, (iii) Catalytic upgrading of sugars, starches, and alcohols. (iv) Biomass to liquid (BTL) thermochemical processes. FT green diesel is produced by the Fischer-Tropsch [60, 61]. Hydrocracking and hydrotreating are the two major step in hydroprocessing technique [62]. The following yield was obtained via hydro processed used vegetable oil; biofuel (85%), non-condensable gases (10%), and water(5%) [63]. Green diesel is a biofuel product that comprises of branched saturated hydrocarbons and straight chain which contain carbon atoms of C15–C18. The properties of green diesel is similar to fossil fuel making it compatible with CI engines [64, 65]. Therefore, green diesel has superior fuel properties like high cetane number, oxidation stability, and cold flow and cloud point compared to FAME and petroleum based diesel [66]. Othman et al. investigated the value of cetane number, the outcome shows that green diesel has value of cetane number between 80 to 90 which is higher than the petrol based diesel standard [67]. The density and the net heating value of green diesel are in the range 0.77 g/ml to 0.83 g/ml and 42 MJ/kg and 44 MJ/kg respectfully, which also meets the biodiesel and petrol diesel standard [68].

## 5. Sourcing of UCO for production of FAME

The global population and consumption of vegetable oils have a direct relationship. The increase in population has triggered the consumption of vegetable oil that gives rise to massive volumes of UCO. The largest percentage of these vegetable oils is used in households, restaurants, and fast-food outlets for cooking and frying. **Table 5** shows the estimated amount of UCO collected by some countries. Canada is reported to generate between 0.120 and 0.135 million tons of UCO per year [42, 69], while the United States of America produced 0.6 million tons of yellow grease in 2011. The United Kingdom and the European Union countries generate ~0.7 million tons to 1.0 million tons and 0.2 million tons of UCO per year, respectively [45, 70].

| Country | UCO collection (million ton/annum) |
|---|---|
| Canada | 0.14 |
| China | 0.15 |
| Malaysia | 0.5 |
| South Africa | 0.6 |
| United Kingdom | 0.2 |

**Table 5.**
*Annual collection of UCOs in some countries.*

In South Africa 0.6 million tons of UCO are collected per year, while more than an estimated 0.2 million tons of UCO is produced but not collected from households, bakeries, takeaway outlets, and restaurants per year [44, 70], which contributes to soil and water contamination, sewage blockages, and damage to aquatic life [71]. China, Malaysia, and Japan generated 0.6 million tons, 0.5 million tons, and 0.6 million tons of UCO, respectively, annually. It is reported that more than 60% UCO generated globally is indiscriminately disposed of [24].

UCO is produced when vegetable oils sourced from palm, soybean, sunflower, cottonseed, olive, palm kernel, and rapeseed or animal fats like butter, fish oil, and tallow, are used for cooking or frying food [72]. With feedstock accounting for between 80% and 85% of the production cost of biodiesel, the use of UCO can result in a substantial reduction in production costs, thereby significantly reducing the cost of biodiesel fuel. This makes biodiesel more viable as a feedstock for green diesel production as a substitute fuel for internal combustion engines, particularly unmodified CI engines. The focus of this research work has been on the conversion of biodiesel product into HDRD via the hydrogenation technique.

## 6. Evaluation of properties of UCO and biodiesel

The pH value of UCO varies between 5.13 and 6.61, indicating a weak acid for a biodiesel feedstock. The pH value reported in the literature were not uniform but depended on the type and degree of usage. Food such as sausage triggered higher pH values than fish etc. This is a result of fats from fish being more acidic than those of beef [39]. The value of acidity is reduced in used palm oil after repeated frying due to the effects of thermal degradation and contamination from the food items. UCO which is normally subjected to heating witnesses a reduction in pH as a result of usage. Generally, due to repeated and high cooking temperatures, the acid value of the oil tends to reduce. However, the transformation and the mechanism for the generation of cyclic and noncyclic hydrocarbon in vegetable oil during high-temperature repeated cooking can be difficult to predict as a result of the many reactions that produce many unstable intermediate hydrocarbons. In contrast, the results obtained as reported in the literature show that biodiesel is a feasible feedstock for the production of HDRD which is also known as green diesel. From the previous report research analysis, EN 14214 and other standard test methods were used for the analysis of biodiesel. Density at 15°C and kinematic viscosity at 40°C of the fuel were measured under EN ISO 3675:1998 and EN ISO 3104:1994,with 926 $kg/cm^3$ and 37.3 1$mm^2/s$ values being reported respectively [73]. The acid value and iodine value analysis were done by titration under EN 14104:2003 and EN 14111:2003, with 0.63 mgKOH/g and 109.98 $mgI_2$/100 g being reported respectively [74]. The flashpoint was measured using the Pensky-Martens method ISO 2719:2002 [75], and the sulfur content was quantified under EN ISO 13032:2012 [76]. The aforementioned properties prove the viability of biodiesel as a potential feedstock for HDRD production. The application of HDRD derived from biodiesel fuel offers high yield, excellent storage stability, high cetane number, and flashpoint.

## 7. Conclusions

The continuous search for sustainable renewable energy has stimulated researchers to develop solutions to the global environmental crises and demand for renewable biofuel. HDRD has received significant attention across the global economy due to its higher stability, higher heating value, and emission of harmless pollutants. The use of biodiesel as a feedstock is more viable as it reduces the production cost and has the potential to settle the concerns related to the edible oil market. This critical review has identified biodiesel fuel as a potential feedstock for hydrogenation into HDRD. The main conclusions are:

- Commercialization of HDRD production requires government attention without delay. Implementation of government policies and regulations that directly address the challenge of demand, production, and trade of this HDRD must be spearheaded by ministries and governmental bodies.

- The outcome of this research shows that biodiesel fuel can be harnessed as a potential feedstock for the commercial production of HDRD.

- HDRD possesses superior fuel qualities such as CN, low pour point, and excellent oxidation stability which offers a great benefit for the CI engine and the environment.

- The goal of producing cost-efficient HDRD is still far from being reached. The economic feasibility of the production of HDRD using biodiesel is sustainable.

## Acknowledgements

The authors are grateful to the leadership of Green Energy Solutions, Discipline of Mechanical Engineering, Howard College, University of KwaZulu-Natal, Durban for their contributions towards the success of this work.

## Conflict of interest

No conflict of interest.

## Notes/thanks/other declarations

The Authors hereby acknowledge the funding support from TETFUND.

## Author details

Josiah Pelemo[1,2]*, Kayode Timothy Akindeji[1,3], Freddie L. Inambao[2], Omojola Awogbemi[4] and Emmanuel Idoko Onuh[5]

1 College of Agriculture Science and Engineering, University of KwaZulu-Natal, Durban, South Africa

2 Discipline of Mechanical Engineering, Green Energy Group, University of KwaZulu-Natal, Durban, South Africa

3 Department of Electrical power Engineering, Durban University of Technology, South Africa

4 Ekiti State University, Ekiti, Nigeria

5 Federal Polytechnic Bauchi, Nigeria

*Address all correspondence to: josiahpelemo27@gmail.com

IntechOpen

© 2022 The Author(s). Licensee IntechOpen. This chapter is distributed under the terms of the Creative Commons Attribution License (http://creativecommons.org/licenses/by/3.0), which permits unrestricted use, distribution, and reproduction in any medium, provided the original work is properly cited.

## References

[1] Devan P, Mahalakshmi NJF. Performance, emission and combustion characteristics of poon oil and its diesel blends in a DI diesel engine. Fuel. 2009;**88**(5):861-867

[2] Sahoo P, Das L. Combustion analysis of Jatropha, Karanja and Polanga based biodiesel as fuel in a diesel engine. Fuel. 2009;**88**(6):994-999

[3] Demirbas A. Comparison of transesterification methods for production of biodiesel from vegetable oils and fats. Energy Conversion and Management. 2008;**49**(1):125-130

[4] Madhu D, Singh B, Sharma YC. Studies on application of fish waste for synthesis of high quality biodiesel. RSC Advances. 2014;**4**(59):31462-31468

[5] Akbar E, Binitha N, Yaakob Z, Kamarudin SK, Salimon J. Preparation of Na doped SiO 2 solid catalysts by the sol-gel method for the production of biodiesel from jatropha oil. Green Chemistry. 2009;**11**(11):1862-1866

[6] Chen C-L, Huang C-C, Tran D-T, Chang J-S. Biodiesel synthesis via heterogeneous catalysis using modified strontium oxides as the catalysts. Bioresource Technology. 2012;**113**:8-13

[7] Sarma AK, Kumar P, Aslam M, Chouhan APS. Preparation and characterization of Musa balbisiana colla underground stem nano-material for biodiesel production under elevated conditions. Catalysis Letters. 2014;**144**(7):1344-1353

[8] Wakil M, Kalam M, Masjuki H, Fattah IR, Masum B. Evaluation of rice bran, sesame and moringa oils as feasible sources of biodiesel and the effect of blending on their physicochemical properties. RSC Advances. 2014;**4**(100): 56984-56991

[9] Sonthalia A, Kumar N. Hydroprocessed vegetable oil as a fuel for transportation sector: A review. Journal of the Energy Institute. 2017;**92**(1):1-17

[10] Ertunc Tat M, Van Gerpen JH. Fuel Property Effects on Biodiesel. American Society of Agricultural and Biological Engineers. St. Joseph, MI. 2003. Available from: https://www.asabe.org/

[11] Al-Dawody MF, Bhatti SK. Optimization strategies to reduce the biodiesel NOx effect in diesel engine with experimental verification. Energy Conversion and Management. 2013;**68**:96-104. DOI: 1016/j.enconman.2012.12.025

[12] Pryor RW, Hanna MA, Schinstock JL, Bashford LL. Soybean oil fuel in a small diesel engine. Transactions of ASAE. 1983;**26**(2):333-0337

[13] Moser BR. Camelina (Camelina sativa L.) oil as a biofuels feedstock: Golden opportunity or false hope? Lipid Technology. 2010;**22**(12):270-273

[14] Abdulkareem-Alsultan G, Asikin-Mijan N, Lee HV, Rashid U, Islam A, Taufiq-Yap YH. A review on thermal conversion of plant oil (edible and inedible) into green fuel using carbon-based nanocatalyst. Catalysts. 2019;**9**(4):350. Available from: http://www.mdpi.com/2073-4344/9/4/350

[15] Anuar MR, Abdullah AZ. Challenges in biodiesel industry with regards to feedstock, environmental, social and sustainability issues: A critical review. Renewable and Sustainable Energy Reviews. 2016;**58**:208-223

[16] Calero J et al. An overview on glycerol-free processes for the production of renewable liquid biofuels, applicable in diesel engines. Renewable and Sustainable Energy Reviews. 2015;**42**:1437-1452

[17] Yaakob Z, Sukarman ISB, Narayanan B, Abdullah SRS, Ismail M. Utilization of palm empty fruit bunch for the production of biodiesel from Jatropha curcas oil. Bioresource Technology. 2012;**104**:695-700

[18] Qian J, Shi H, Yun Z. Preparation of biodiesel from Jatropha curcas L. oil produced by two-phase solvent extraction. Bioresource Technology. 2010;**101**(18):7025-7031

[19] Irwan S, Yaakob Z, Kumar MS, Primandari S, Kamarudin SK. Biodiesel progress in Malaysia. Energy Sources, Part A: Recovery, Utilization, and Environmental Effects. 2012;**34**(23): 2139-2146

[20] Aziz HA, Aroua MK, Yusoff R, Abas NA, Idris Z, Hassan HA. Production of palm-based esteramine through heterogeneous catalysis. Journal of Surfactants and Detergents. 2016;**19**(1):11-18

[21] Mazubert A, Aubin J, Elgue S, Poux M. Intensification of waste cooking oil transformation by transesterification and esterification reactions in oscillatory baffled and microstructured reactors for biodiesel production. Green Processing and Synthesis. 2014;**3**(6):419-429

[22] Da Rocha FG, Brodzki D, Djega-Mariadassou G. Formation of alkanes, alkylcycloalkanes and alkylbenzenes during the catalytic hydrocracking of vegetable oils. Fuel. 1993;**72**(4):543-549

[23] Hazrat MA, Rasul MG, Khan MMK, Ashwath N, Rufford TE. Emission characteristics of waste tallow and waste cooking oil based ternary biodiesel fuels. Energy Procedia. 2019;**160**:842-847. DOI: 10.1016/j.egypro.2019.02.149

[24] Onuh E, Inambao F, Awogbemi O. Green diesel, renewable energy and waste cooking oil: The potential for synergy. International Journal of Applied Engineering Research. 2018;**13**(10):8714-8727

[25] Meng X, Chen G, Wang Y. Biodiesel production from waste cooking oil via alkali catalyst and its engine test. Fuel Processing Technology. 2008;**89**(9):851-857

[26] Falade AO, Oboh G, Okoh AI. Potential health implications of the consumption of thermally-oxidized cooking oils–a review. Polish Journal of Food and Nutrition Sciences. 2017;**67**(2):95-106

[27] Pelemo J, Inambao F, Onuh E. Potential of used cooking oil as feedstock for hydroprocessing into hydrogenation derived renewable diesel: A review. International Journal of Engineering Research & Technology. 2020;**13**:500-519.

[28] Kalnes T, Marker T, Shonnard DR. Green diesel: A second generation biofuel. International Journal of Chemical Reactor Engineering. 2007;**5**:1

[29] Petri JA, Marker TL. Production of Diesel Fuel from Biorenewable Feedstocks. Google Patents. 2009

[30] Kalnes TN, Koers KP, Marker T, Shonnard DR. A technoeconomic and environmental life cycle comparison of green diesel to biodiesel and syndiesel. Environmental Progress & Sustainable Energy. 2009;**28**(1):111-120

[31] Krár M et al. Fuel purpose hydrotreating of vegetable oil on

NiMo/γ-Al2O3 catalyst. Hungarian Journal of Industry and Chemistry. 2009;**37**(2):2009

[32] Leader E. Tyson Foods, Syntroleum Partner to Turn Grease into Fuel. 2012. Available from: https://www.environmentalleader.com/2010/11/tysonfoods-syntroleum-partner-to-turn-grease-into-fuel/

[33] Sotelo-Boyás R, Trejo-Zárraga F, Hernández-Loyo FJ. Hydroconversion of triglycerides into green liquid fuels. Hydrogenation. 2012;**338**:338

[34] New Orleans Net. New Renewable Diesel Plant Headed to Norco with Federal Backing. 2012

[35] Karamé I. Hydrogenation. Janeza Trdine, Rijeka, Croatia: BoD–Books on Demand; 2012

[36] Neste Oil. NExBTL Diesel. Available from: http://www.nesteoil.com/default.asp? path= 1,41,11991,12243 12335. Last accessed April 4th. 2012

[37] Hajjari M, Tabatabaei M, Aghbashlo M, Ghanavati H. A review on the prospects of sustainable biodiesel production: A global scenario with an emphasis on waste-oil biodiesel utilization. Renewable and Sustainable Energy Reviews. 2017;**72**:445-464

[38] Hosseini SS, Aghbashlo M, Tabatabaei M, Najafpour G, Younesi H. Thermodynamic evaluation of a photobioreactor for hydrogen production from syngas via a locally isolated Rhodopseudomonas palustris PT. International Journal of Hydrogen Energy. 2016;**40**(41):14246-14256

[39] Sharma MP. Selection of potential oils for biodiesel production. Renewable and Sustainable Energy Reviews. 2015;**56**:1129-1138

[40] Radich A. Biodiesel performance, costs, and use. US Energy Information Administration; 2006

[41] Hachinski V et al. National Institute of Neurological Disorders and Stroke–Canadian stroke network vascular cognitive impairment harmonization standards. Stroke. 2006;**37**(9):2220-2241

[42] Issariyakul T, Kulkarni MG, Dalai AK, Bakhshi NN. Production of biodiesel from waste fryer grease using mixed methanol/ethanol system. Fuel Processing Technology. 2007;**88**(5):429-436

[43] Mbohwa C, Mudiwakure A. The status of used vegetable oil (UVO) biodiesel production in South Africa. In: Proceedings of the World Congress on Engineering. 2013

[44] Hamze H, Akia M, Yazdani F. Optimization of biodiesel production from the waste cooking oil using response surface methodology. Process Safety and Environment Protection. 2015;**94**:1-10

[45] Martinez-Guerra E, Gude VG. Transesterification of waste vegetable oil under pulse sonication using ethanol, methanol and ethanol–methanol mixtures. Waste Management. 2014;**34**(12):2611-2620

[46] Awogbemi O, Onuh EI, Inambao FL. Comparative study of properties and fatty acid composition of some neat vegetable oils and waste cooking oils. International Journal of Low Carbon Technologies. 2019;**14**(3):417-425

[47] Asli H, Ahmadinia E, Zargar M, Karim MR. Investigation on physical properties of waste cooking oil–rejuvenated bitumen binder. Construction and Building Materials. 2012;**37**:398-405

[48] Vassilev SV, Vassileva CG, Vassilev VS. Advantages and disadvantages of composition and properties of biomass in comparison with coal: An overview. Fuel. 2015;**158**:330-350

[49] Lang X, Dalai AK, Bakhshi NN, Reaney MJ, Hertz P. Preparation and characterization of bio-diesels from various bio-oils. Bioresource Technology. 2001;**80**(1):53-62

[50] Gui MM, Lee K, Bhatia S. Feasibility of edible oil vs. non-edible oil vs. waste edible oil as biodiesel feedstock. Energy. 2008;**33**(11):1646-1653

[51] Refaat A. Different techniques for the production of biodiesel from waste vegetable oil. International Journal of Environmental Science and Technology. 2010;**7**(1):183-213

[52] Balat M. Potential alternatives to edible oils for biodiesel production–a review of current work. Energy Conversion and Management. 2011;**52**(2):1479-10492

[53] Divakara B, Upadhyaya H, Wani S, Gowda CL. Biology and genetic improvement of Jatropha curcas L.: A review. Applied Energy. 2010;**87**(3):732-742

[54] Cohen B. Urbanization in developing countries: Current trends, future projections, and key challenges for sustainability. Technology in Society. 2006;**28**:64-80

[55] Tat ME, Wang PS, Van Gerpen JH, Clemente TE. Exhaust emissions from an engine fueled with biodiesel from high-oleic soybeans. Journal of the American Oil Chemists' Society. 2007;**84**(9):865-869

[56] Begum H, Siwar C, Alam AF, Choy EA, Ishak S, Alam L. Enhancing sustainability amongst oil palm smallholders in Malaysia. International Journal of Agricultural Resources, Governance and Ecology. 2018;**14**(1):62-79

[57] Onay O, Koçkar OM. Fixed-bed pyrolysis of rapeseed (Brassica napus L.). Biomass and Bioenergy. 2004;**26**(3):289-299

[58] Venter R, Booysen J, Marx S, Schabort C. Evaluation of Bio-char based Products as Hydrotreating Catalysts for the Production of Renewable Fuel. 25th European Biomass Conference and Exhibition. 2017

[59] Duman G, Okutucu C, Ucar S, Stahl R, Yanik J. The slow and fast pyrolysis of cherry seed. Bioresource Technology. 2011;**102**(2):1869-1878

[60] Srifa A, Faungnawakij K, Itthibenchapong V, Assabumrungrat S. Roles of monometallic catalysts in hydrodeoxygenation of palm oil to green diesel. Chemical Engineering Journal. 2015;**278**:249-258

[61] Gousi M et al. Green diesel production over nickel-alumina co-precipitated catalysts. Applied Catalysis A: General. 2017;**536**:45-56

[62] Perez-Cisneros ES, Sales-Cruz M, Lobo-Oehmichen R, Viveros-García T. A reactive distillation process for co-hydrotreating of non-edible vegetable oils and petro-diesel blends to produce green diesel fuel. Computers and Chemical Engineering. 2017;**105**:105-122

[63] Kumar V, Sindhu RK, Kumar S. Comparative analysis of green diesel versus petro-diesel in compression ignition engine. Bioscience Biotechnology Research Communications. 2018;**11**:128-135

[64] Orozco LM, Echeverri DA, Sánchez L, Rios LA. Second-generation green diesel from castor oil: Development of a new and efficient continuous-production process. Chemical Engineering Journal. 2017;**322**:149-156

[65] Kordouli E, Pawelec B, Bourikas K, Kordulis C, Fierro JLG, Lycourghiotis A. Mo promoted Ni-Al2O3 co-precipitated catalysts for green diesel production. Applied Catalysis B: Environmental. 2018;**229**:139-154

[66] Kalnes T, Marker T, Shonnard D, Koers K. Green diesel and biodiesel: A technoeconomic and life cycle comparision. In: 1st Alternative Fuels Technology Conference. Vol. 18. Prague, Czechoslovakia; 2008

[67] Othman MF, Adam A, Najafi G, Mamat R. Green fuel as alternative fuel for diesel engine: A review. Renewable and Sustainable Energy Reviews. 2017;**80**:694-709

[68] Mwangi JK, Lee W-J, Chang Y-C, Chen C-Y, Wang L-C. An overview: Energy saving and pollution reduction by using green fuel blends in diesel engines. Applied Energy. 2015;**159**:214-236

[69] Rolfson DB, Majumdar SR, Tsuyuki RT, Tahir A, Rockwood K. Validity and reliability of the Edmonton frail scale. Age and Ageing. 2006;**35**(5):526-529

[70] Mbohwa C, Mudiwakure A. The status of used vegetable oil (UVO) biodiesel production in South Africa. Proceedings of the World Congress on Engineering. 2013;**1**:3-5

[71] Abdulkareem-Alsultan G, Asikin-Mijan N, Lee H, Rashid U, Islam A, Taufiq-Yap Y. A review on thermal conversion of plant oil (edible and inedible) into green fuel using carbon-based nanocatalyst. Catalysts. 2019;**9**(4):350

[72] Said N, Ani F, Said M. Review of the production of biodiesel from waste cooking oil using solid catalysts. Journal of Mechanical Engineering Science. 2015;**8**:1302-1311

[73] International Organization for Standardization. ISO. 3675: Crude Petroleum and Liquid Petroleum Products. Laboratory Determination of Density. Hydrometer Method. Geneva, Switzerland: International Organization for Standardization; 1998

[74] British Standards Institution. Fat and Oil Derivatives—Fatty Acid Methyl Esters (FAME)—Determination of Oxidation Stability (accelerated oxidation test). 2003

[75] International Organization for Standardization. Determination of Flash Point—Pensky-Martens Closed Cup Method. 2002

[76] Harabi M, Neji Bouguerra S, Marrakchi F, Chrysikou LP, Bezergianni S, Bouaziz M. Biodiesel and crude glycerol from waste frying oil: Production, characterization and evaluation of biodiesel oxidative stability with diesel blends. Sustainability. 2019;**11**(7):1937

Chapter 5

# Replacement of Diesel Fuel by DME in Compression Ignition Engines: Case for India

*Anirudh Gautam and Ankita Singh*

## Abstract

Decarbonising of transport, industrial and all sectors of economy is a necessity to stop or reverse global warming. Use of batteries, fuel-cells, hybrid topographies with smaller IC engines and use of alternative fuels like methanol, ethanol, DME in the IC engines are some of the ways through which emission of green-house gases can reduced/eliminated. Diesel engines are highly efficient due to higher compression ratios and are used in the heavy-duty transportation vehicles. DME is a single molecule fuel having high cetane number and which can be used as a drop-in fuel on the diesel engines albeit with retro-fitment of these engines with a new pressurized fuel system. DME with a chemical formula CH3-O-CH3 can be produced by different feedstocks such as coal, natural gas, biomass and bio-waste and municipal solid waste. India has a large reserve of high ash coal and generates high quantities of biomass and MSW, all of which can be converted to DME by use of clean production technologies. India's transport and industrial sectors consume about 100 billion liters of diesel fuel per year produced entirely from imported petroleum. This amount of diesel can be replaced by indigenously produced DME from locally available coal, biomass and MSW.

**Keywords:** diesel, petroleum, DME, engine, fuel injection equipment, coal, biomass, gasification, decarbonising

## 1. Introduction

Decarbonizing the transportation sector is the highest priority to reduce global warming, air pollution and associated health hazards. There is a need to develop ultra-low carbon or carbon-negative fuels which can be produced from renewable sources like waste, organic substances, plants and trees etc. DME is a low cost, low carbon and zero soot and PM chemical stock that can be used on CI engines. DME can be produced from various feedstocks such as wood, methanol, wastes, biomass etc. Its use as a substitute for diesel fuel is due to high cetane number of 55–60 and superior combustion characteristics. It is a colorless, non-toxic, mildly narcotic and easily inflammable gas at normal temperature and pressure similar to LPG.



IntechOpen

Many CI engine manufacturers like Isuzu, Nissan, Mitsubishi, Volvo have developed DME fueled CI engine powered vehicles. DME vehicle's reliability have been validated in the field with running tests of 100,000 Km or more. DME fueled CI engines exhibit higher combustion efficiency and soot free combustion in comparison to corresponding diesel engines.

Diesel engines are used widely in power generation industry. Expansion of the power generation sector is fueling the growth of diesel engine production. Global diesel power engine market is estimated to grow at a CAGR of more than 3.5% during the period from 2021 to 2026. Demand for reliable electricity due to the industrial expansion, development of commercial infrastructure, electrification of human habitats, and uninterrupted power supply is the major cause for the growth of diesel engine power market [1]. Therefore, there is a valid case for developing these engines to operate on DME.

## 2. Energy consumption trends and future of diesel engines and DME

**Figure 1** presents the global energy consumption from 1971 to 2019 in Exajoules ($10^{18}$ J). World energy consumption has been increasing steadily from 1971 (176.8 EJ) to 2019 (418 EJ), an increase of 136% [2]. Largest increase has been in oil and electricity consumption. It is expected that the increase in these two sources will continue in the future. With the development of new battery technologies and solar PVs, increase in rate of electricity consumption including that generated from renewables will increase further.

**Figure 2** gives the world energy consumption history and projections. **Figure 2(a)** shows that increase in industrial energy consumption is expected to be the largest followed by the transportation sector although growth in residential and commercial sectors is also expected at a relatively moderate rate. **Figure 2(b)** gives the world energy consumption on the basis of fuel. Use of petroleum and other liquids are anticipated to increase till 2050 and will be about 250 quadrillion Btu. Renewables, during the same period, are expected to rise to 40 quadrillion Btu. Therefore, the share of petroleum and other liquids will be around six times that of renewables. In the heavy-duty sector diesel-based compression ignition engines are the preferred power source. Therefore, there is a sound future for the CI engines in all sectors, i.e., power generation, transportation, industry etc. Conversion of CI engines to use DME will lead to reduction in fossil fuel consumption and prevent degradation of environment.

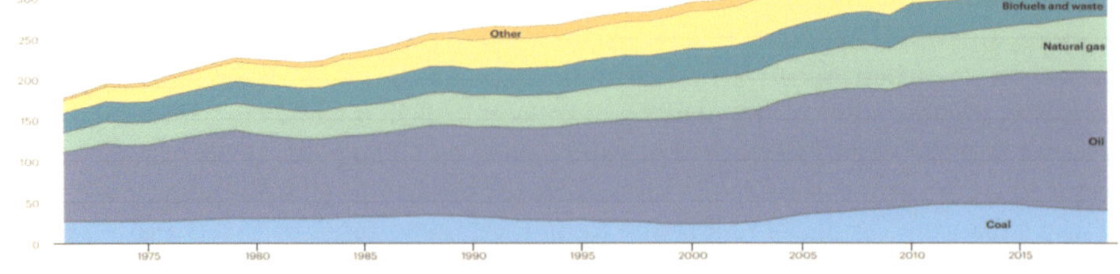


**Figure 1.**
*World total energy consumption by source (1971–2019) [2].*

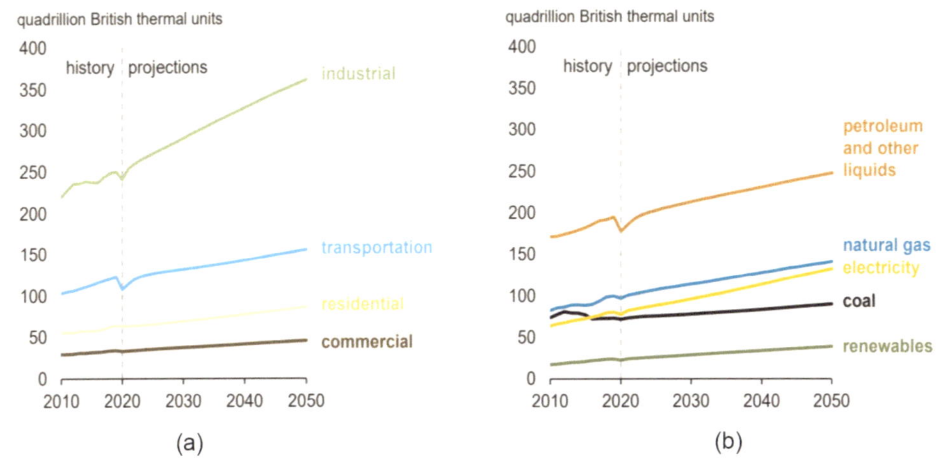


**Figure 2.**
*World energy consumption, history and projections [3]. a) World end-use consumption by sector. b) World end-use energy consumption by fuel.*

## 2.1 Consumption of diesel by end use

Among the petroleum fuels, diesel is a critical fuel for transportation especially heavy-duty vehicles, agricultural vehicles, construction and earth moving equipment. Power generation is another important area where diesel is used. **Figure 3** shows the end-use of diesel fuel in the USA, Japan and India, countries which are the one of the largest consumers of diesel fuel. In all three countries transport sector is the major consumer of mineral diesel. Conversion of trucks and busses to DME will lead to major reduction of fossil fuel and associated emissions.

## 2.2 Categorization of diesel engine market

IMARC group has categorized the global diesel engine market as shown in **Figure 4**.

Power rating of the diesel engines varies from 0.5 MW to more than 5 MW used in automotive and non-automotive applications all over the world. **Figure 4** illustrates that development of engine technology to utilize DME in diesel engines will have a positive effect in all sectors and across continents. Thus, a concentrated effort by different OEMs to come out with DME conversion kits or even new DME engine

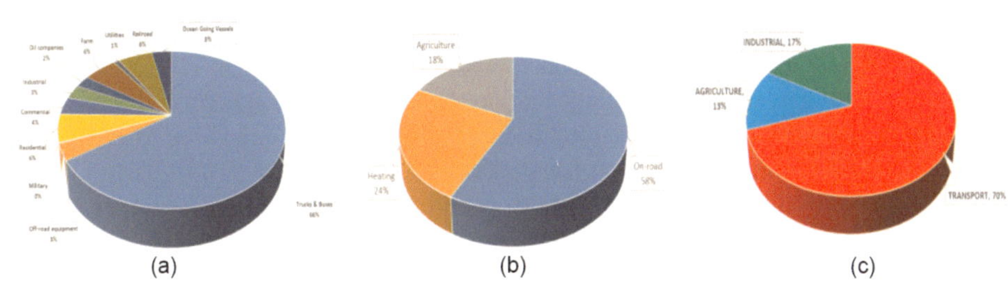


**Figure 3.**
*End-use of diesel fuel in USA, Japan and India [4–6]. (a) USA – 241 billion liters per year. (b) Japan – 60 billion liters per year. (c) India – 100 billion liters per year.*

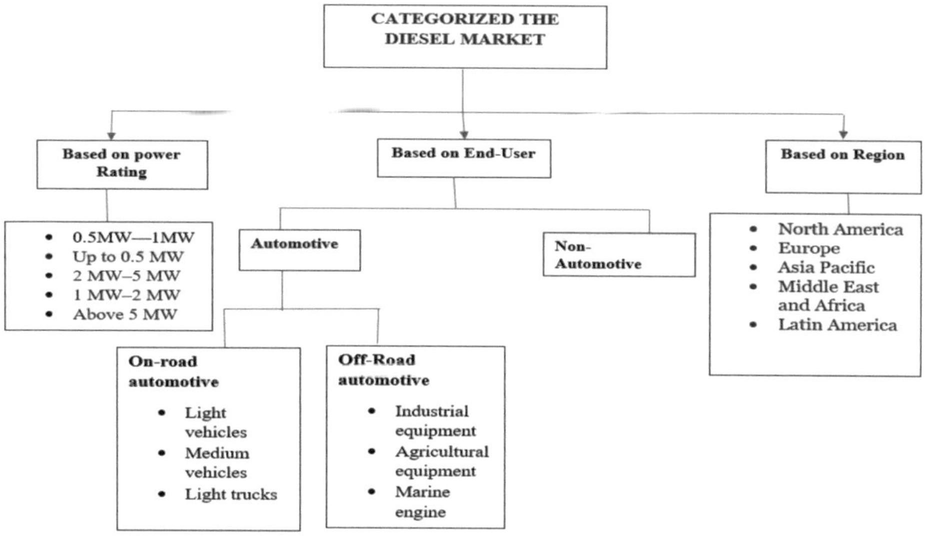


**Figure 4.**
*Categorization of the diesel engine market based on power rating and end-user [7].*

designs will not only reduce fossil fuel consumption and environmental deterioration but also has a firm economic viability. All sectors are expected to exhibit increase in use of diesel engines, however highest growth is expected from the power generation sector, trucks and busses and industrial sector. The global diesel engine market reached a value of US$ 212.4 billion in 2021 [7]. It has been predicted that the demand for diesel engines including all sectors and sizes should increase to US$ 240 billion in the year 2025 registering a compound annual growth rate of over 3% from 2020 to 2025 [8].

## 3. DME fuel production

Top Key producers of Dimethyl Ether (DME) include: a) Akzo Nobel, b) Shell, c) The Chemours Company, d) China Energy, e) Mitsubishi Corporation, f) Ferrostal GmbH, g) Grillo Werke, h) Jiutai Energy Group, i) Oberon fuels and j) Zagros. In Japan several large scale DME plants have been set-up [9]. China is the bulk producer of DME from Chinese coal and production plants have also been set-up in Trinidad and Tabago, North America, Indonesia and Uzbekistan. First bio DME production plant was constructed in Sweden. Global production of DME at present is roughly 9 million tons per year [10]. Different feedstocks can be used in the production of DME. These are natural gas, coal, waste from pulp and paper mills, forest products, agricultural by-products, municipal waste and dedicated fuel crops e.g., switch grass. Methanol dehydration is the main process for production of DME currently. Synthetic gas can however be produced by gasification of coal, biomass, natural gas reforming [11].

There are three pathways to produce DME: a. Two-step process, b. One-step process, c. Liquid-one-step process called as bio reforming. Typically, DME is produced through a two-step process with syngas the feedstock (**Table 1**). Methanol is first produced from Syngas, followed by dehydration of methanol into DME (Eqs. (1)–(4)) (**Table 1, Figure 5**) [12].

| Equation | Process Desc | Eq. |
|---|---|---|
| $CO + 2H_2 \rightarrow CH_3OH$ | Methanol formation | (1) |
| $2CH_3OH \rightarrow CH_3OCH_3 + H_2O$ | Methanol dehydration with solid-acid catalyst | (2) |
| $H_2O + CO \rightarrow CO_2 + H_2$ | Water gas shift reaction | (3) |
| $3H_2 + 3CO \rightarrow CH_3OCH_3 + CO_2$ | Net reaction | (4) |

**Table 1.**
*Two-step synthesis process of DME [12].*

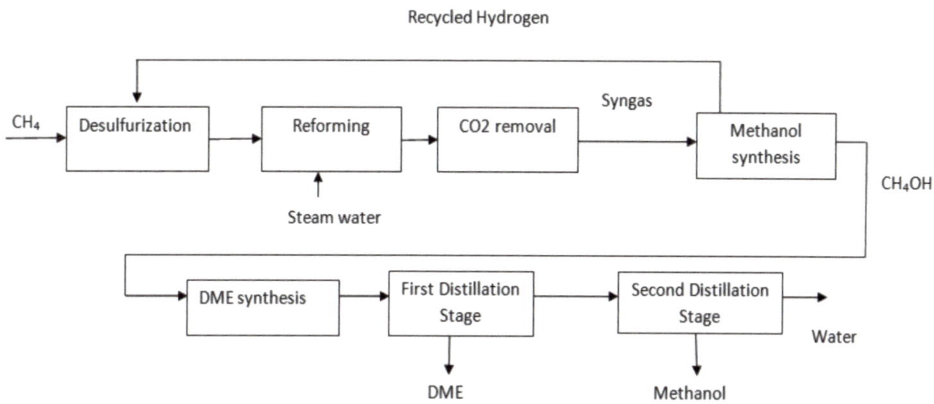


**Figure 5.**
*Two-step process for synthesis of DME [13].*

| Eq. (2) | Process Desc | Eq. |
|---|---|---|
| $3CH_4 + 3CO_2 \rightarrow 6CO + 6H_2$ | Methane-dry-reforming | (5) |
| $6CO + 6H_2 \rightarrow 2CH_3OCH_3 + 2CO_2$ | DME synthesis | (6) |
| $3CH_4 + CO_2 \rightarrow 2CH_3OCH_3$ | Net | (7) |

**Table 2.**
*Single-step synthesis of DME [12].*

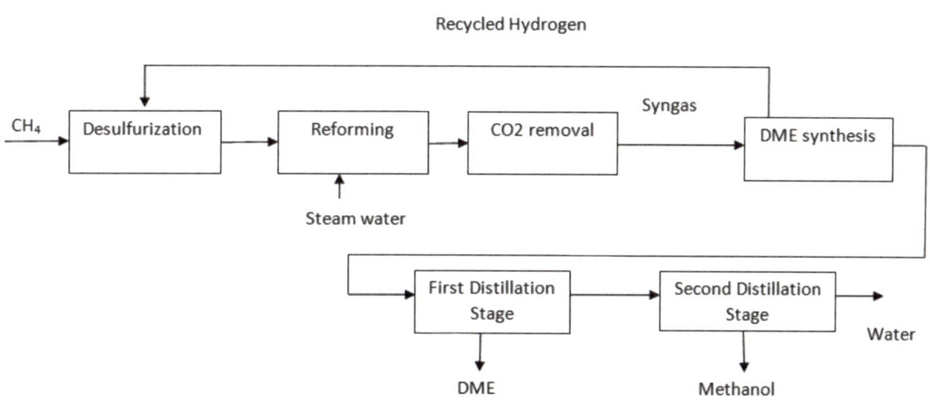


**Figure 6.**
*Single-step process for DME synthesis [13].*

In Japan, Korea and China single-step process is used most commonly (Eqs. (5)–(6)) (**Table 2** and **Figure 6**). In this process, the methanol formation, methanol dehydration and water-gas shift reaction are merged. Different feedstocks can be used in this process such as methane, scrubbed bio-gas, syngas, etc. In case of methane feedstock methane-dry-reforming is done before the single-step process. Japan Steel Company has reached production capacity of 100 t DME/day by this process (Olah, et al. 2009). Mass balance shows that for every gm DME produced 1.43 gm of CO2 is used in the process. In overall reaction, CO2 sequestration is 0.48 gm per gram of DME.

Third process for DME production is bio-reforming (**Table 3**, Eqs. (8)–(12)). For optimized process to produce methanol Metgas is used, which is 2:1 H2:CO ratio syngas. Two processes to produce Metgas are methane-steam-reforming and methane-dry-reforming step. Next steps are methanol formation and DME synthesis by dehydration of Methanol. Water produced during the dehydration of methanol is used in the methane-steam-reforming process. CO2 consumed during per gm formation of DME is 0.48 gm. There are no emissions of CO2 during bio-reforming process, however there will be CO2 emissions regarding process's energy requirements.

Both one-step and two-step DME production process are mature technologies. There are many companies which have developed single-step process for producing DME. Notable among these are Haldor-Topsoe A/S Denmark, JFE Holdings, Japan, Korea Gas Company, S. Korea, Air Products USA, NKK Japan, Oberon Fuels USA etc. Two-step DME production process has been developed by companies like Toyo Japan, Mitsubishi Gas Company Japan, Lurgi Germany, Udhe Germany. Several companies have developed novel processes and technologies for production of DME.

World's first bio-DME demonstration plant in Sweden (started in 2010) uses black liquor (waste from paper and pulp industry) to produce high-quality syngas which is used for synthesis of DME (**Figure 7**).

### 3.1 Production process developed by Oberon fuels

Oberon Fuels has developed proprietary skid-mounted, small-scale production units that convert methane and carbon dioxide to DME from various feedstocks, such as biogas from dairy manure and food waste. These small-scale plants are affordable as compared to a large plant, do not require large infrastructure and permits etc. for operation. These small-scale production units can produce 10,000 gallons (37854.12 liters) of DME per day to cater to the regional fuel markets [14].

Schematic of the plant is shown in **Figure 8**, consists of SMR, make-up syngas compressors, methanol synthesis reactors, pre-cut column and DME column and DME storage tanks.

The Oberon Fuels methane-gas-to-DME process has the following three major steps:

| Equation | Process Desc | Eq. |
|---|---|---|
| $2CH_4 + 2H_2O \rightarrow 2CO + 6H_2$ | Methane-steam reforming | (8) |
| $CH_4 + CO_2 \rightarrow 2CO + 2H_2$ | Methane-dry-reforming step | (9) |
| $4CO + 8H_2 \rightarrow 4CH_3OH$ | Methanol formation | (10) |
| $4CH_3OH \rightarrow 2CH_3OCH_3 + 2H_2$ | Methanol dehydration | (11) |
| $3CH_4 + CO_2 \rightarrow 2CH_3OCH_3$ | Net | (12) |

**Table 3.**
*Bio-reforming process for production of DME [12].*

**Figure 7.**
*Bio-DME production plant in Sweden [13].*

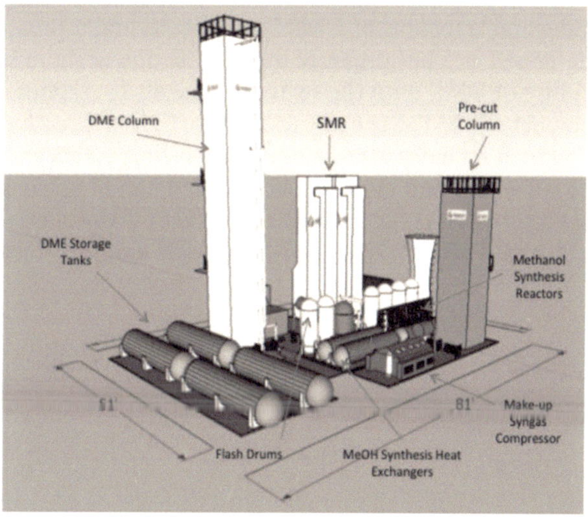


**Figure 8.**
*Schematic of a small size DME production plant developed by Oberon fuels [14].*

1. Syngas production

2. Methanol synthesis

3. Simultaneous DME synthesis and separation via catalytic distillation.

First two steps are common in large scale industrial application. Catalytic synthesis of DME with purification has been investigated in detail and has been demonstrated industrially by Oberon Fuels at Brawley California. In this plant, all production, storage, piping, tanks and valves are overground. DME produced from un-scrubbed (60% methane) HSAD (High Solid Anaerobic Digestion) of food waste, yard waste, bio-waste is called as Bio-DME. Chemically Bio-DME and DME are chemically same,

Bio-DME uses biogas (typically produced from anaerobic digester), whereas DME is produced from pipeline natural gas.

## 4. DME handling, storage and distribution

**Table 4** shows the boiling points of propane, butane and DME.

**Table 4** illustrates that the boiling point of DME is in between that of propane and butane and all three are gas at room temperature of 15°C. The storage and handling facilities for LPG can therefore be repurposed for DME at a low cost. LPG storage and handling infrastructure is widespread in many countries including India. In addition to the use of existing LPG facilities new installations need to be created for DME. In the existing LPG installations changes will be needed for seals, valves, pressure regulators, gaskets, pumps to handle DME.

### 4.1 Suitability of materials and storage and handling equipment

Storage, handling and transport of DME is as a liquid under pressure similar to LPG. DME is compressed to 5 bar pressure for liquefaction at room temperature. Material compatibility of DME with the seals and gaskets in stationary and moving parts has to be established. ASTM specification D7901 gives directions on safety and handling of DME, which includes elastomer selection for gaskets and seals (to avoid their failure). Also, all equipment for DME storage and handling have necessarily to be overground. **Table 5** provides compatibility of different elastomers with DME as per ASTM D7901. Rubber and plastics swell and deteriorate easily on contact with DME as

| | Propane (C3H8) | DME(CH3OCH3) | Butane(C2H6) |
|---|---|---|---|
| Boiling point (°C) | −42°C | −24°C | −1°C |

**Table 4.**
*Boiling points of propane, DME and butane [15].*

| Elastomer | Compatibility rating for use of DME |
|---|---|
| Natural Rubber (NR), Isoprene, Butadiene Styrene (SBR, Buna S), Butadiene (BR), Butyl (HR), Ethylene Propylene (EPR, EPDM, EP), Polyacrylate (ACM), Fluor elastomer – Di polymer, Fluor elastomer – Terpolymer | 4 |
| Polychloroprene (Neop rene, CR), Chlorosulfonated Polyethylene (CSM) | 3 |
| Polysulfide (T) | 2 |
| Silicone (VMQ), Fluor silicone (FVMQ), Perfluoro elastomer (FFKM), Polytetrafluoroethylene (PTFE), Nitrile (NBR, BUNA-N), Hydrogenated Nitrile Butadiene Rubber (HNBR) | 1 |
| Rating Legend (at room temperature)<br>1 = Little or minor effect, 0 to 5% volume swell, 2 = Minor to moderate effect, 5 to 10% volume swell,<br>3 = Moderate to severe effect, 10 to 20% volume swell, 4 = Not recommended for DME use | |

**Table 5.**
*Compatibility of DME with elastomers as per ASTM D7901 [12].*

compared to LPG. Nitrile rubber seals (NBR) and fluoro rubber seals (FKM) are used with LPG but swell in contact with DME and therefore cannot be used with DME.

Cylinder tanks for DME fuel will be similar to LPG, will be fabricated structure consisting of cylindrical shell and panel with filling valve, outlet valve, return valve, safety valve, overfill prevention device, quick coupling and fluid level gauge.

## 5. Use of DME in diesel engines

### 5.1 Fuel properties and special fuel injection system

A pressurized fuel injection system for DME is an essential requirement. Thus, the tank, fuel pumps, fuel piping and the fuel injector have to be kept under suitable pressure. Fuel pumps have to pressurize the fuel circuit to a pressure which is higher than the saturation vapor pressure of DME at the operating temperature. This will prevent the DME fuel to vaporize and cause cavitation in the fuel circuit before the fuel injection into the cylinder. Temperature of the fuel inside the fuel injector reaches to 80°C and the pressure of the fuel circuit after the fuel pumps has to be increased accordingly. A pressure higher than 30 bar is considered adequate for keeping the DME in liquified form even at higher temperatures encountered during operation of the engine. Feed-pumps will be able to pressurize the fuel to the required pressure.

ASTM standard range for viscosity of liquid fuels has a range of 1.39 to 4.2 cSt at 40°C whereas viscosity of DME is within 0.185 cSt and 0.23 cSt. Low viscosity of DME will result in leakages past clearances used for sealing like plungers and barrels, seals and gaskets and pump gears etc. Low lubricity will result in high wear and seizure of the moving parts in fuel injection system. Viscosity and lubricity enhancing additives are added to the DME to overcome these problems.

Bulk modulus of DME is less than diesel by an order of magnitude. This implies that DME is much more compressible than diesel. High compressibility of DME will result in delay in the injection timings. Injection lag will be higher as compared to diesel and therefore the ECU has to be programmed accordingly. The compression work of DME in the fuel pumps and injectors is much higher than diesel fuel and the parasitic power for pumping of fuel is higher. Large compressibility of DME also results in injection instability and this problem can be overcome by modifying the nozzle design and control of fuel temperature.

**Table 6** compares the chemical and physical properties of DME, diesel fuel and LPG (propane, butane). Diesel properties are compared to DME to understand the similarity between the two in compression-based ignition, whereas comparison of DME to LPG is required to know about the similarity in fuel handling of the two fuels.

**Table 6** shows that auto-ignition temperature of DME is lower than diesel at pressure higher than atmospheric. Also, the cetane number of DME is higher than that of diesel. Thus, DME fuel is suitable in CI engines and has high potential to replace diesel fuel. At the same time, boiling point of DME is close to propane below Zero °C and the handling, storage and distribution of DME is similar to LPG. World over, LPG is used as cooking fuel and also for transport, therefore there is adequate experience in handling and storage of LPG. Use of LPG storage and handling facilities for DME with modifications to the seals, gaskets and certain metallic parts can be done with lower efforts and cost.

DME is an oxygenated fuel with an oxygen percentage of nearly 35%. Higher Cetane number results in lower ignition delays and smaller pre-mixed combustion

| Property | DME | Diesel | Propane | Property | DME | Diesel | Propane |
|---|---|---|---|---|---|---|---|
| *Chemical structure* | CH3-O-CH3 | — | C3H8 | *Explosion limit ((% by volume of air))* | 3.4–18.6 | 0.6–7.5 | 2.1–10.1 |
| *Molar mass (g/mol)* | 46 | 170 | 44.097 | *% wt. Carbon* | 52.2 | 86.0 | 82 |
| *Lower heating value (MJ/Kg)* | 27.6 | 42.5 | 46.3 | *% wt. Hydrogen* | 13.0 | 14.0 | 18 |
| *Liquid density (Kg/m3)* | 667 | 831 | 500 | *% wt. Oxygen* | 34.8 | 0 | 0 |
| *Cetane number* | > 55 | 40–55 | — | *Critical temperature (K)* | 400 | 708 | 369.9 |
| *Stoichiometric Air-Fuel ratio* | 9.0 | 14.6 | 15.8 | *Critical pressure (MPa)* | 5.37 | 3.00 | 4.301 |
| *Auto -Ignition temperature (°C)* | 235 | 250 | 470 | *Critical density (Kg/m3)* | 259 | — | 220 |
| *Boiling point (°C)* | −20 | 180/370 | −42 | *Kinematic Viscosity of liquid (cSt)* | <1 | 3 | 4.29 |
| *Latent heat of vaporization (kJ/Kg)* | 460 (−20°C) | 250 | 372 | *Surface tension (at 298 K) N/m* | 0.012 | 0.027 | 0.007 |
| *Modulus of elasticity* | 100–1000 MPa Depend on Temp. & Press. | 1400 Mpa | 220 MPa | *Vapor pressure (at 298 K) kPa* | 530 | ≪10 | 580 |

**Table 6.**
*Comparison of properties of DME and diesel fuel [12, 15].*

phase, lower peak cylinder pressures and lower NOx formation. Absence of C-C bonds leads to sootless combustion. In the DME molecule each carbon atom is bound to three hydrogen atoms on one side and oxygen atom on the other. Bond energy of C-H is 414 kJ/mol and that of C-O bond is 359.0 kJ/mol. Higher C-H bond energy is responsible for shorter ignition delays and higher cetane number of DME.

LHV of DME is almost half that of diesel, therefore to obtain the same horsepower, flow rate of DME is about 1.7 times that of diesel. This means larger storage tanks for DME, higher diameter pipes and tubes for fuel flow, higher flow capacity of the DME pumps and the fuel injectors. Duration of Injection (DOI) of DME will be longer than diesel and the Start of Injection (SOI) has to be advanced accordingly. Lower boiling point of DME translates into faster vaporization of injected fuel in the combustion chamber. This along with lower critical temperature of DME results in superheated vapor in the combustion chamber, adequate air-fuel mixing is ensured. Large heat of vaporization also lowers the in-cylinder temperatures and lower NOx emissions.

Chain combustion reaction is possibly through one of the following competing pathways [12]:

a. C–O bond fission (pyrolysis mechanism):

$$CH_3OCH_3 = CH_3O + CH_3. \tag{13}$$

b. Hydrogen abstraction (oxidation mechanism):

$$4CH_3OCH_3 + O_2 = 4CH_3OCH_2 + 2H_2O. \quad (14)$$

$$CH_3OCH_2 = CH_2O + CH_3 \quad (15)$$

As the C-O bond energy is smaller than C-H bond, distortion of the C-O bonds in the DME molecule weakens the bonding strength and the breakage of the C-O bonds earlier. Pyrolysis is more likely to start the chain reaction at relatively low temperatures, showing as lower auto-ignition temperature.

Change of piston and cylinder heads in the existing diesel engines may not be required other than design of the fuel injection nozzle to suit the volumetric flow rate and existing piston profile. However, in order to have an optimum design of combustion chamber to suit the spray characteristics of DME changes to the piston bowl and re-location of piston rings may be needed. For best performance of the engine valve timings may also need to be modified resulting in design and development of new camshafts. For modifying existing diesel engines, it becomes necessary to retrofit a new fuel injection system right from the fuel tank, feed pumps, pressure pumps, common rails and fuel injectors.

DME is similar to LPG in terms of safety. Vapor of DME is heavier than air and settles to the ground similar to LPG. Sufficient ventilation is necessary in locations where DME is being used whether for stationary installations or transport. Ignition limit of DME is 3.4% - 18.6% by volume and therefore necessary precautions have to be implemented which will be similar to LPG. Global Warming Potential (GWP) of a molecule is its adverse effect on climate change. GWP includes both the molecules lifetime and ability to absorb radiation. DME is atmospherically not dangerous and does not contribute to global warming.

Common rail fuel injection systems have been developed for DME fueled CI engines and these engines have demonstrated good engine performances and efficiency along with significant reduction in harmful exhaust emissions. This has been made possible by having a good control of the fuel injection characteristics and temperature. The common rail concept for DME fuel have also proven effective in simple and safe fuel handling. **Figure 9(a)** illustrate a comparison of the concepts use for DME fueled diesel engines, figure is self-explanatory. **Figure 9(b)** shows a schematic of the DME fuel storage and distribution system on engine.

**Figure 9(c)** presents a comparison of the simulated injection characteristics of diesel and DME digital hydraulic operating system (DHOS) injectors. Piston lifts in both the cases are similar, whereas the fuel injection pressure for diesel injector is about 1200 bar and for the DME injector it is around 800 bar. Duration of injection (DOI) for the DME injector is higher as compared to the diesel injector and similar trend is reflected in the period of needle lift of the injector. The injection rate of fuel for the DME case is higher than diesel all along the injection and the DOI is higher as can be seen in the bottom-most figure. Higher volume of fuel flow for DME vis-a-vis diesel can be seen in the figure due to lower density and LHV of DME compared to diesel.

### 5.2 Emission characteristics

Stoichiometric combustion of DME in air yields 1.91 m of $CO_2$. This is equivalent to 66 gm of $CO_2$ per MJ (LHV) of combusted DME.

| | | |
|---|---|---|
| $CH_3OCH_3 + 3(O_2 + 3.78N_2) \rightarrow 2CO_2 + 3H_2O + 11.34N_2$ | DME combustion in air | (16) |

Stoichiometric combustion of one gram of diesel $C_{12}H_{23}$ yields 3.16 grams of $CO_2$.

| | | |
|---|---|---|
| $C_{12}H_{23} + 17.75(O_2 + 3.78N_2) \rightarrow 12CO_2 + 11.5H_2O + 67.095N_2$ | Diesel combustion in air | (17) |

Combustion of one gm of DME in air emits less $CO_2$ than combustion of one gm of Diesel, difference being 1.25 gm less $CO_2$. Experimental investigations have brought out that there is substantial reduction in particulate matter (PM), NOx and combustion noise when DME is used as a fuel in CI engines. Combustion efficiency (BSFC) of DME fuel in a CI engine is similar to diesel (**Figure 10**) and so fuel consumption can be similar on an energy basis. Also shown in **Figure 10** are comparison of the road load emissions of NOx, CO, HC and Smoke (PM) on a DME and diesel fueled engine. Smoke is undetectable and NOx, CO, HC are lower in DME engine fitted with an oxidation catalyst.

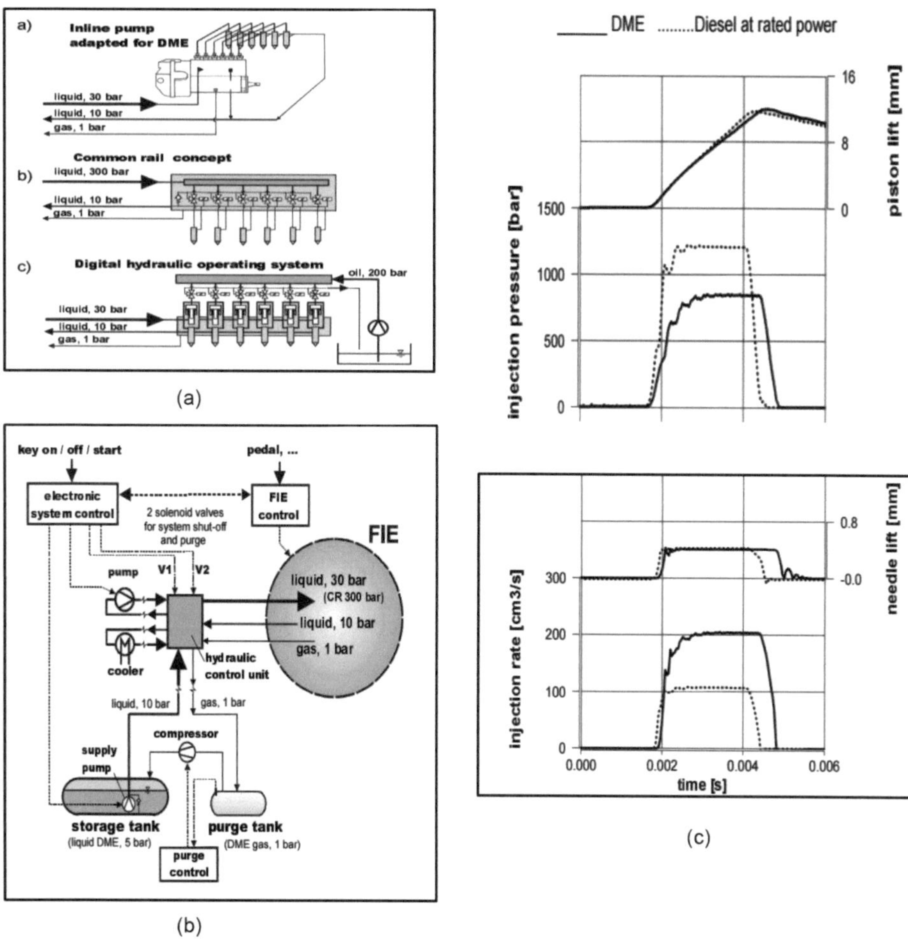


**Figure 9.**
*Fuel injection system and injection characteristics of a DME fueled engine [16]. (a) Comparison of the fuel injection concepts for DME fueled engines. (b) DME storage, handling and injection system for a DME fueled diesel engine. (c) Comparison of simulated injection characteristics of Diesel and DME DHOS injector.*

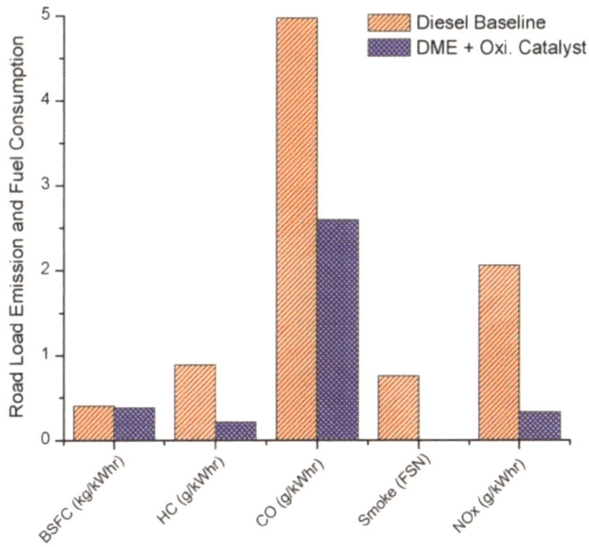


**Figure 10.**
*Comparative analysis of emission data from neat DME & mineral fueled CI engine [17].*

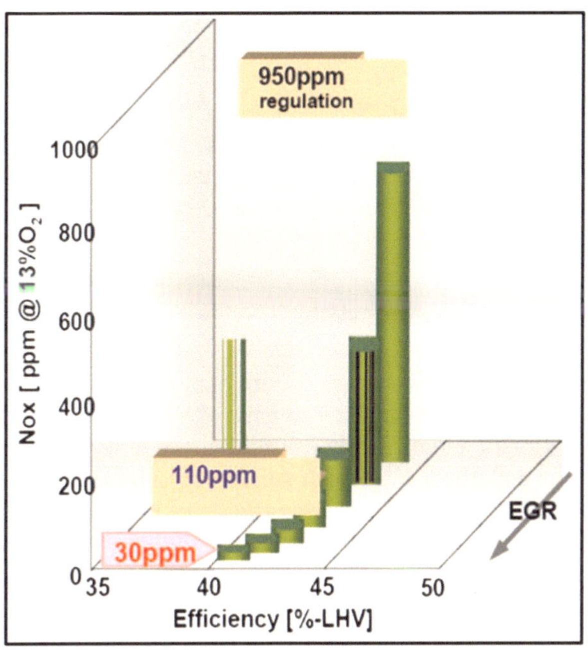


**Figure 11.**
*NOx emission data from a DME-fueled 1.15 MW diesel power generation unit [18].*

**Figure 11** gives the NOx emission data vs. engine efficiency for a 1.15 MW diesel power generation unit engine fueled with DME and with different levels of EGR. As the DME fueled engine does not produce any soot and ultra-low PM emissions, higher EGR levels can be used on the engine to reduce the NOx emissions. Against a 950 ppm NOx regulation, as the EGR is increased, very low NOx levels of 30 ppm can be achieved albeit with an engine efficiency penalty of 3%.

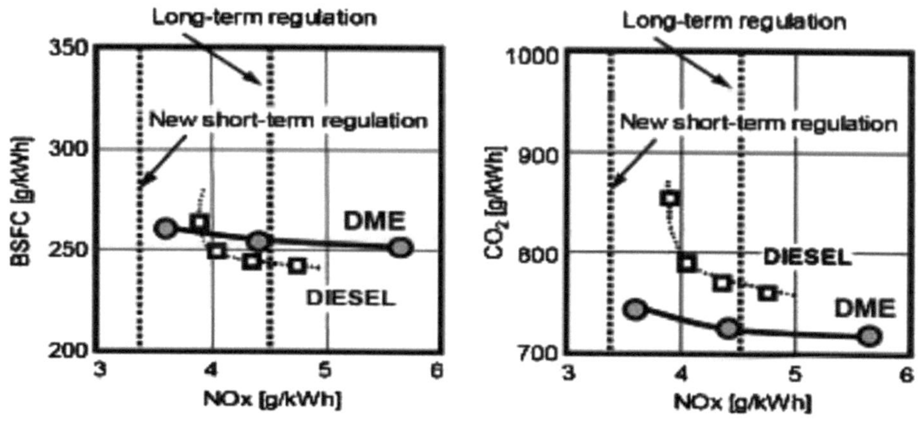


**Figure 12.**
*Comparison of fuel consumption (BSFC), NOx and CO2 emissions on a 6-cylinder 7 liter turbocharge/intercooled heavy-duty diesel engine operating in Japanese D13 mode [19].*

**Figure 12** shows emission results on 6-cylinder 7 liter turbocharged/intercooled heavy-duty diesel engine operating in Japanese D13 mode driving cycle. NOx and CO2 emissions can be reduced with DME fueled engine vis-a-vis diesel engine at comparable fuel economy. In addition, combustion noise of DME fueled engines are lesser than their diesel counterparts.

In many countries research and development of DME fueled CI engines has been carried out and commercial trails done successfully. Japan, Europe, North America, China and South Korea are the leaders in development of DME fueled engines. A brief summary of the development of DME engines is presented in **Table 7**.

## 6. Life cycle analysis of energy, emissions and GHG for DME fuel

Life cycle studies are used to compare the effect of different fuels/energy sources/transport technologies. Life-cycle analysis consists of Well-to-Pump (WTP) and Pump-to-Wheel (PTW) estimation of energy consumption and emission of pollutants and GHG impact of these pollutions. WTP path consists of a) recovery and transport of raw material, b) production of the fuel, c) transportation of fuel, and d) distribution. PTW is the vehicle operation part of the pathway. **Figure 13** shows a schematic to illustrate the Well-to-Wheel cycle for different fuel/transport technology combinations.

Many studies on life cycle energy consumption and emissions for DME fuel have been carried out globally. Lee et al. have investigated Well-to-Wheels emissions of greenhouse gases and air pollutants of di-methyl ether from natural gas and renewable feedstocks in comparison with petroleum gasoline and diesel in the United States and Europe. For this purpose they have used Greenhouse gases, Regulated Emissions and Energy use in Transportation (GREET) model developed by Argonne National Laboratory (ANL). They have used five pathways to calculate the WTW by use of DME as fuel, these are 1) fossil NG with large-scale DME plants, 2) methanol from fossil NG with large-scale plants for both methanol and DME (separately), 3) land-fill-gas (LFG) with small-scale DME plants, 4) manure-based biogas with small-scale DME plants, 5) methanol from black liquor gasification with small-scale DME plants. They

| Region/ Country | Details |
|---|---|
| Europe | • Denmark, Haldor Topsoe developed the first DME fueled vehicles in 1996, Euro 4 compliant in 1998. In Sweden Volvo developed the first DME-fueled bus in 1999, 2nd gen bus in 2005, 3rd gen DME truck in 2015 |
| North America | • Consortium was formed between Pennsylvania State University, Air Products and Chemicals Inc., the Department of Energy (DOE), Navistar International and Caterpillar between 1999 and 2001 to work on a project to convert diesel busses to DME-fueled busses |
| Japan | • National Traffic Safety & Environmental Laboratory (NTSEL) of Japan formed a consortium with Nissan diesel motors and Bosch Japan (1998–2001) to develop a heavy-duty DME fueled bus engine with mechanical fuel injection.<br>• DME-fueled vehicles have been taken up by the National Institute of Advanced Industrial Science and Technology (AIST) in collaboration with motor and oil-supply companies since 2003.<br>• In 2005, a medium-duty DME fueled truck with 7.1 liter engine was developed by AIST, Japan Oil and Gas & Metals Corporation (JOGMEC).<br>• An in-line 6-cylinder truck equipped with a diesel engine (turbocharged, intercooled, EGR, fitted with oxidation catalyst and NOx storage catalyst) has been converted to DME by Nissan diesel motors along with NTSEL.<br>• Isuzu Motors Japan has developed light and medium-duty common rail FIE DME engines. |
| China | • National Clean Vehicle Action Program was started in 2005 in which ten DME-fueled busses fitted with CI engines and mechanical fuel injection system were developed by a consortium consisting of Shanghai Motor Company, Shanghai Jiao Tong University (SJTU) and Shanghai Coking & Chemical Corporation.<br>• 2nd & 3rd generation DME vehicles fitted with common-rail fuel injection system and after-treatment devices have been developed to comply with Euro-5 emission norms.<br>• Chinese government has funded a "863 project" in which SJTU is researching new technologies for DME fueled vehicles.<br>• DME stations have been developed by ENN.<br>• In particular Shaghai city has been considering acceptance of DME as a standard fuel for trucks, taxis and busses to reduce PM2.5 pollution. |
| S. Korea | • In 2000 Korea Institute of Energy Research (KIER) took up DME engine research project.<br>• Took up a project to convert a diesel bus fitted with a 8.071 liter displacement engine to DME in 2005.<br>• A DME engine with 1.582 liter displacement and common rail fuel injection system was developed for passenger car by Hanyang University (HYU).<br>• Korea Automotive Technology Institute (KATEC) took up a project to modify a sports utility vehicle (SUV) powered by a 1.991 liter displacement engine to use DME fuel in 2009. |

**Table 7.**
*Global development of DME engines [12, 18].*

have studied DME production and use in the US and Europe in two class of vehicles (light-duty (LDV) and heavy-duty vehicles (HDV). Their studies show that WTW consumption of fossil energy and emission of GHG emissions in production and use of DME fuel is very low as compared to diesel and gasoline vehicles. Five pathways used in the production of DME are shown in **Figure 14**.

In this study a small DME production plant is assumed to have a capacity of 25 MTPD (metric ton per day) and a large-scale plant is one with a capacity of 3600 MTPD. Case NG uses the fossil NG to produce DME directly in a large scale DME production plant. NG is supplied to the DME production plant through a pipeline. In case MeOH fossil NG is first converted to methanol and thence to DME. In this case

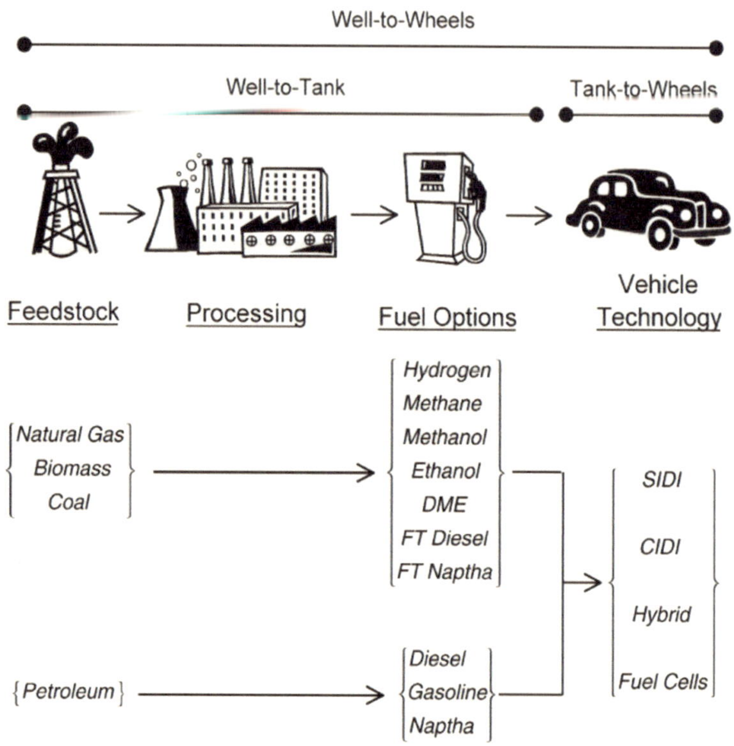


**Figure 13.**
*Well-to-wheel cycle for transportation fuels [17].*

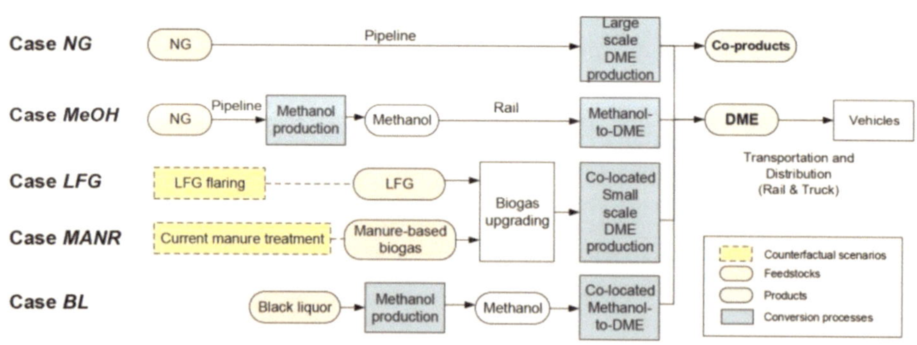


**Figure 14.**
*Five pathways for production of DME [20].*

the methanol production plant is close to the source of NG and methanol is transported to the DME production units by rail. Biogas from two sources has been considered, i.e., a) landfill gas (LFG) and b) during production and treatment of manure (MANR). Both are taken as renewable alternatives. Biogas is made up of $CH4$ and $CO2$ and is generated by anaerobic digestion (AD) of organic wastes. In both cases DME plant are small scale plants co-located with the source of biogas. Before the biogas can be fed into the DME plant it has to be cleaned in separate reactors where impurities like Sulfur compounds etc. are removed and the biogas is upgraded to the required composition.

*DOI: http://dx.doi.org/10.5772/intechopen.104969*

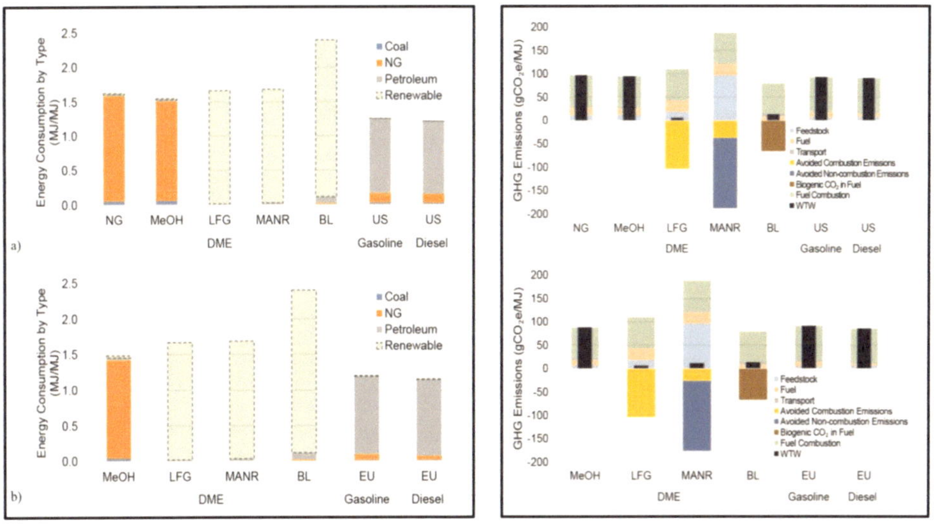


(a) (b)

**Figure 15.**
*WTW energy consumption and emissions for DME produced through different pathways. (a) Comparison of WTW energy consumption for DME production vis-a-vis gasoline and diesel production in US and EU [20]. (b) WTW GHG emissions from DME production as compared to diesel and gasoline for US and EU [20].*

**Figure 15(a)** depicts the WTW energy consumption for DME production through different pathways. DME production seems to consume more energy per MJ of fuel produced as compared to gasoline and diesel in both US and EU. This is because the conversion efficiency of raw material to DME is significantly lesser than gasoline and diesel. Although, conversion of fossil NG to DME directly or through MeOH is lesser than gasoline and diesel, however, this may also be due to scale of operation and size of plants which are much bigger and established for gasoline and diesel. Also, production of DME from renewable sources will result in zero or negligible consumption of fossil fuel.

**Figure 15(b)** shows the GHG emissions from DME production gasoline and diesel consumption and MeOH production. The emissions consist of the following components, a) For preparation and transportation of feedstock, b) Production of fuel and its transport, c) Avoided combustion and non-combustion emissions, d) biogenic CO2 in fuel and e) Fuel combustion. In the case of DME produced from bio-gas, avoided combustion and non-combustion are a major portion of the emissions inventory and in reducing the WTW emissions to very low/negative values. Thus, WTG GHG for LFG and manure based bio-gas are 6 and − 1 gCO2 e/MJ respectively and are 93% and 101% lower than US diesel. In the EU LFG and manure based bio-gas to DME process shows 6 and 12 gCO2 e/MJ of GHG emissions respectively which are 92% and 87% lower than EU diesel. If, however, regional electricity is used for production of DME then the WTW GHG emissions in the US will increase to 25 and 1 gCO2 e/MJ for LFG and manure bio-gas respectively. In the EU, the corresponding figures for DME production are 19 and 13 gCO2e/MJ from LFG and manure biogas respectively. Thus, it can be seen that the energy mix of the process has a strong impact on the WTW emissions as well.

In **Figure 16**, WTW GHG emissions vs. WTW vehicle energy consumption are plotted for some alternative fuels and petroleum-based gasoline and diesel. **Figure 17** plots the results for synthetic diesel from farmed wood, synthetic diesel from waste

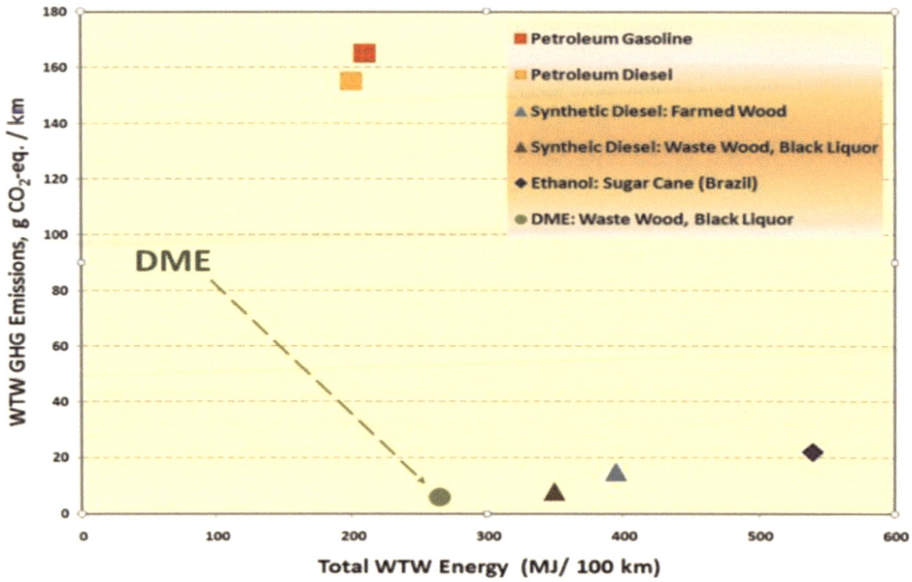


**Figure 16.**
*Well-to-wheel GHG emissions & vehicle energy consumption for some alternate fuels [18].*

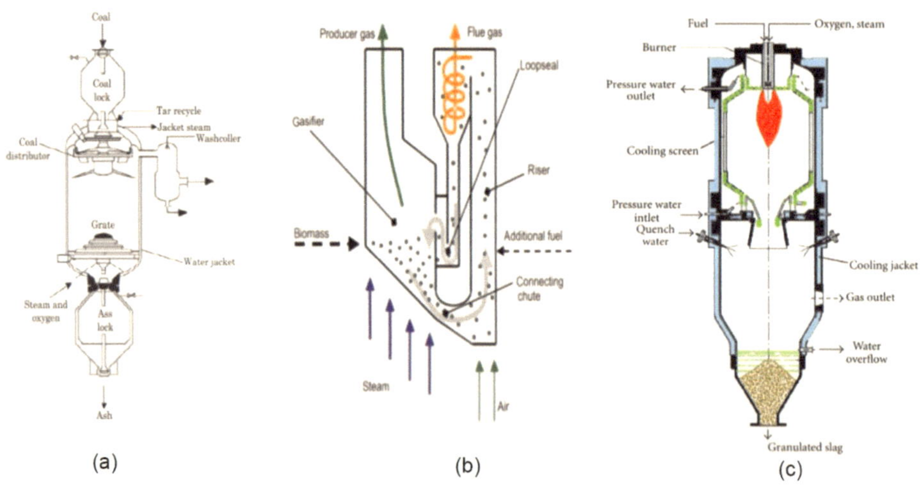


**Figure 17.**
*Different type of coal gasifiers [21]. (a) Moving Bed Gasifier concept. (b) Fluidized Bed Gasifier concept. (c) Entrained Flow Gasifier concept.*

wood and black liquor, ethanol from sugarcane (Brazil) and DME from waste wood and black liquor. WTW GHG emissions of DME produced from waste wood and black liquor are the lowest of all the fuels studied. WTW energy consumption (MJ/100 Km) for DME is second lowest after petroleum fuels. This may be due to the established production process of petroleum fuels and well-optimized engine and vehicle technologies for petroleum-based fuels. It is possible the life cycle energy consumption of DME fuel will reduce as the production technology for DME is matured and CIDI engines are designed specifically for DME fuel.

## 7. Strategy for India

### 7.1 Indigenous production

*High Ash Coal to DME* - Coal reserves in India (350 billion tons) [22] predominantly consist of high ash content (20–30% and sometimes >40%) and therefore not considered techno-economically suitable for power generation. This high ash content coal can however be converted to syngas through suitable process and catalyst and the syngas can be converted to DME through a two-step or single-step synthesis. NITI Ayog of Government of India has taken up a Methanol Mission for conversion if high-ash coal of India into methanol/DME. Under this mission Bharat Heavy Electricals Limited (BHEL), a public sector unit of government has developed an indigenous process to convert Indian high-ash coal to methanol and has set-up a pilot coal-to-methanol plant, 0.25 TPD methanol with feedstock of 1.2 TPD coal using fluidized bed gasifier. Methanol with a purity of 98–99.5% has been produced in the pilot plant. As the process matures and costs come down higher commercial scale plant of coal-to-methanol/DME can be set-up with clean coal technology to produce DME for transportation and for blending with LPG. Clean-coal technologies like carbon capture and storage and use can be used during gasification of coal to syngas. Moving bed gasifiers, fluidized bed gasifiers and entrained flow gasifiers are some of the reactors to produce syngas from coal (**Figure 18**).

*Biomass and municipal solid waste to DME* - India generates 62 million tons of municipal solid waste per year [24]. India also generates about 230 million metric tons of surplus biomass (28 GW) and about 115 million metric tons bagasse (14 GW) annually which includes agricultural residues. Organic content of the MSW (50%) [25] and the biomass/bagasse is suitable for gasification to syngas. National Chemical Laboratory has developed indigenous catalysts for conversion of methanol to DME and are in the process of developing catalysts for converting syngas to methanol/DME. This route will have the advantage of very low WTW life cycle energy and emissions footprint.

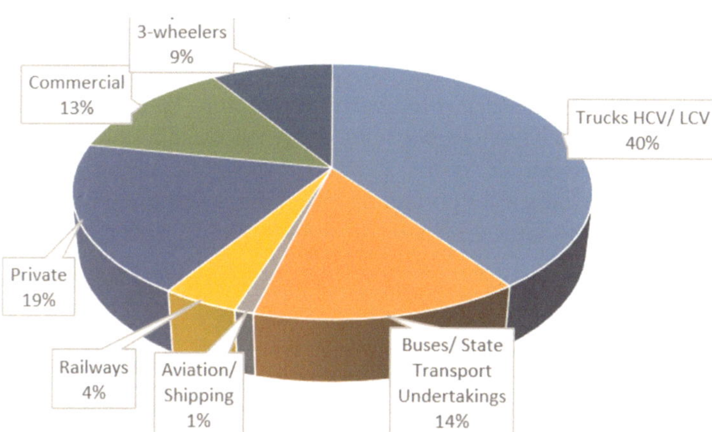


**Figure 18.**
*Transport sector, use of diesel sector (India) [23].*

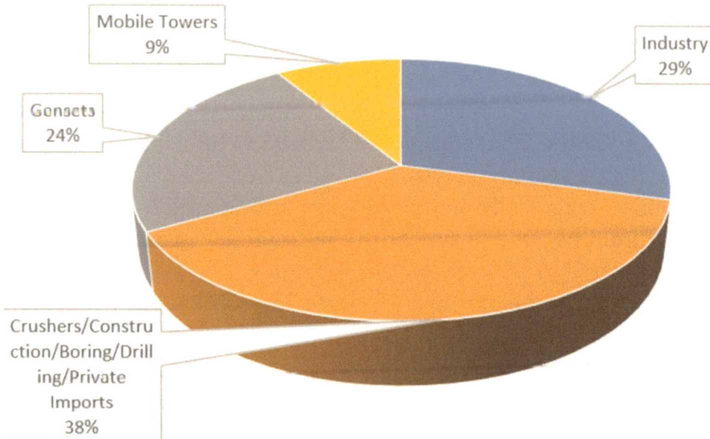


**Figure 19.**
*Use of diesel fuel by industrial sector in India [23].*

## 7.2 Conversion of diesel engines to DME

Annual consumption of diesel fuel in India is more than 100 billion liters [23]. Relative consumption of diesel fuel by different type of vehicles in the transport sector in India is shown in **Figure 17**. In the transport sector, Trucks (HCV/LCV) are the major consumers of diesel fuel (40%) followed by private cars/ SUVs (19%) and Busses/State Transport Undertakings (14%).

Thus, a concentrated effort to convert diesel trucks and busses to use DME as fuel will have a major effect on reduction of GHG in India. Conversion of private diesel vehicles will be difficult as these vehicles are powered by different types and make of diesel engines. **Figure 19** displays the details of the Industrial sector regarding consumption of diesel fuel. In this sector, equipment such as construction, boring, drilling, earth-moving etc. are the largest users of diesel fuel (38%) followed by Industry (29%) and Gensets (24%). It may not be possible to take up conversion of the construction and earth moving equipment diesel engines to operate on DME due to the nature of the work involved or diesel use in furnaces etc., however conversion of gensets to DME can be taken up and will have a positive effect on the GHG emissions and the reduction of use of fossil diesel.

Conversion of diesel engines in the agricultural sector (13% consumption of diesel fuel) to DME will also be beneficial for India in terms of reduction of GHG and diesel fuel. Agriculture equipment like tractors, diesel pumps, agricultural implements, tillers, harvesters, thrashers etc. operate in limited areas and the provision of DME dispensation facilities to cater to the demands of a particular area may be economically feasible.

# 8. Conclusion

Decarbonizing of the transport and other engineering sectors has become an urgent necessity to stop further global warming. Electrification of power trains by using batteries, fuel cells, electric transmission etc. promise to drastically reduce and even reverse the global warming trend. Battery and fuel cell technologies are however still in the evolution phase and for them to match the power density, energy density, low-cost and the reliability of liquid fuels-based IC engines is expected to take at least another two to three decades. Meanwhile the IC engines technology has reached a high level of maturity, sophistication, knowledge and economy of scale. IC engines-based power trains will continue to dominate the transportation, industrial, power generation and other sectors for next three decades. Therefore, it is imperative that solutions are found to make IC engines-based power trains clean and energy efficient. Compression Ignition (CI) engines with diesel fuel are used in many heavy-duty applications like long-distance trucks, busses, agricultural equipment, earth moving and construction equipment. Conversion of these diesel engines to use an alternative clean fuel like DME will aid in protecting the environment and conservation of energy. Primarily this will involve development of a pressurized fuel injection system and matching the fuel injection parameters to the physical and chemical properties of DME. Sufficient developments have been done in the past in development of fuel injection systems for DME fueled CI engines, therefore with the collaboration of the engine OEMs and the fuel equipment manufacturers diesel engines conversion to DME can be taken up in an economical way and in a time-bound manner. Impact of such conversions will be seen in the form of reduction of WTW energy consumption and harmful emissions.

## Author details

Anirudh Gautam[1*] and Ankita Singh[2]

1 Urban Transport and High-Speed Directorate, Research Designs and Standards Organization, Ministry of Railways, Government of India, Lucknow, India

2 Institute of Technology and Management, GIDA, Gorakhpur, India

*Address all correspondence to: ag.srestha@gmail.com

IntechOpen

© 2022 The Author(s). Licensee IntechOpen. This chapter is distributed under the terms of the Creative Commons Attribution License (http://creativecommons.org/licenses/by/3.0), which permits unrestricted use, distribution, and reproduction in any medium, provided the original work is properly cited.

## References

[1] https://mordorintelligence.com/industry-reports/diesel-power-engine-market

[2] https://prod.iea.org/reports/key-world-energy-statistics-2021/final-consumption

[3] https://www.eia.gov/outlooks/ieo/pdf/IEO2021_ChartLibrary_full.pdf

[4] https://www.dieselforum.org/policy/primer-on-fuels-diesel-and-advanced-renewable-biofuels

[5] https://lb-aps-frontend.statista.com/statistics/1253301/japan-diesel-fuel-consumption-by-use/

[6] https://pib.gov.in/newsite/printrelease.aspx?relid=102799

[7] https://www.imarcgroup.com/diesel-engine-market

[8] https://www.arizton.com/market-reports/diesel-engine-market-size-analysis

[9] https://www.fortunebusinessinsights.com/dimethyl-ether-market-104309

[10] https://www.aboutdme.org/FAQ

[11] Mii T, Uchida M. Fuel DME Plant in East Asia, Business Planning & Development Department, Toyo Engineering Corporation (TEC), Japan. In: Proceedings of 15th Japan Joint Symposium. Dhahran, Saudi Arabia; 2005. Available from: https://www.scribd.com/document/57924037/Fuel-DME-Plant-in-East-Asia#:~:text=In%20China%20DME%20plants%20for%20fuel%20use%20already,under%20detailed%20feasibility%20study%20to%20introduce%20around%202008

[12] California Dimethyl Ether Multimedia Evaluation, Tier I. The University of California, Davis, The University of California, Berkley, California Environmental Protectoin Agency, Multi-media Working Group; 2015. Available from: https://www.arb.ca.gov/fuels/multimedia/meetings/dmetierireport_feb2015.pdf

[13] Marcello De Falco, Associate Professor, University UCBM, Rome, Italy. Available from: http://www.oil-gasportal.com/dimethyl-ether-dme-production-2/?print=pdf

[14] https://oberonfuels.com/technology/oberon-process/

[15] https://www.duncanseddon.com/docs/pdf/dme-dimethyl-ether.pdf

[16] Gill et al. Production Feasible DME Technology for Direct Injection CI Engines. SAE Technical Paper 2001-01-2015; Warrendale, PA, U.S.A: SAE International; 2001

[17] Semelsberger TA, Borup RL, Greene HL. Dimethyl ether (DME) as an alternative fuel. Journal of Power Sources. 2006;**156**:497-511

[18] Fleisch TH, Basu A, Sills RA. Introduction and advancement of a new clean global fuel: The status of DME developments in China and beyond. Journal of Natural Gas Science and Engineering. 2012;**9**:94e107. DOI: 10.1016/j.jngse.2012.05.012

[19] Arcoumanis et al. The potential of di-methyl ether (DME) as an alternative fuel for compression-ignition engines: A review. Fuel. 2008;**87**:1014-1030

[20] Lee U, Han J, Wang M, Ward J, et al. Well-to-wheels emissions of greenhouse

gases and air pollutants of dimethyl ether from natural gas and renewable feedstocks in comparison with petroleum gasoline and diesel in the United States and Europe. SAE International Journal of Fuels and Lubricants. 2016;**9**(3). DOI: 10.4271/2016-01-2209

[21] Coal 2021 Analysis and forecast to 2024, International Energy Agency (IEA). 2021

[22] https://pib.gov.in/PressReleasePage.aspx?PRID=1813245

[23] https://pib.gov.in/newsite/printrelease.aspx?relid=102799

[24] Ministry of New and Renewable Energy. Power Generation From Municipal Solid Waste Twentieth Report. New Delhi: Lok Sabha Secretariat; 2016

[25] www.cpcb.nic.in

Chapter 6

# Molecular Contribution of Fatty Acid Esters in Biodiesel Fueled CI Engines

*Gokul Raghavendra Srinivasan, Safak Yildizhan, Shalini Palani, Lakshmanan Thangavelu and Ranjitha Jambulingam*


## Abstract

This present chapter set one's sight on understanding the contribution of fatty acid ester (FAE) molecules in deciding the performance, emission, and combustion characteristics of their biodiesel in CI engine. For this purpose, both produced waste animal fat-oil (WaFO) biodiesel and their characterized FAEs, blended in calculated proportions with neat diesel were tested individually under same testing conditions. Preliminary findings confirmed the significant contribution of FAEs in deciding the overall engine characteristics of WaFO biodiesel; and were influenced by their fundamental molecular properties like chain length, and degree of unsaturation. Superior combustion characteristics were accounted by early initiation of combustion by saturated FAEs; followed by prolonged combustion of unsaturated FAEs using fuel bound oxygen content. Meanwhile, mixed performance characteristics were explained by its long chained saturated and unsaturated FAEs, which imparted their higher density and viscosity, and reduced calorific value than neat diesel. Emission characteristics reported reduced CO and HC emission, and increased $CO_2$ and $NO_X$ emissions citing the equally balanced concentration of both long chained saturated and unsaturated FAEs, which favored complete combustion using its oxygen molecules. Besides assessing engine characteristics, WaFO biodiesel was evaluated for its fuel properties as per ASTM standards, along with neat diesel.


**Keywords:** biodiesel combustion, fatty acid esters, cetane number, degree of unsaturation, carbon chain length

## 1. Introduction

Use of biodiesel in CI engines for commercial and industrial purposes has been increasing steadily, very soon after various government policies stressed on the shift to renewable energy resources [1]. This paved progressive pathway for many



IntechOpen

researchers to focus on improvising the performance of these biodiesel in CI engines; and also, simultaneously monitoring for controlled levels of exhaust emissions. Hence, numerous suggestions have been proposed for enhancing these engine characteristics which includes introducing blends [2, 3], adding fuel or chemical additives [4, 5], dual fuel mode [6, 7], fuel preheating [8]; and even modifying engine parameters like varying injection pressure, injection timing and introducing exhaust gas recirculation [9, 10]. However, operating engine on biodiesel blended with neat diesel is regarded as most economic and efficient technique; and in general, report increased rate of fuel consumption, carbon dioxide ($CO_2$) and nitrogen oxide (NOx)emissions, and reduced thermal efficiencies and carbon monoxide (CO) emissions [11–13]. Though, many biodiesel report similar trends; variation in their results arises with differences in the fuel properties, besides their testing conditions. Hence, one requires fundamental understanding of various factors influencing the performance of biodiesel in CI engine; especially the role of fuel and its properties in deciding these parameters. In fact, these fuel properties are macroscopic entities; and in turn are governed by the chemical compounds available in it, along with their molecular properties. In other words, these parameters are primarily influenced by the fatty acid esters available in the biodiesel [14]; and supporting this, Srinivasan et al. [15] reported that engine characteristics of any biodiesel is an outcome of coordinated behavior of its FAE molecules in CI engine, especially contributed by its dominant FAEs [15].

However, one requires better understanding of these FAEs in order to study their influence and contribution in CI engine. To begin with, FAEs are the fundamental units of biodiesel, and are made up of commonly known fatty acids, in form of carboxylates at one end and an alkyl chain at its alcohol moiety bridged through an Ester functional group [16]. In general, palmitic acid, oleic acid, stearic acid, linoleic acid, linolenic acid and myristic acid are the most frequently reported fatty acids [17]; whilst, alcohol includes methanol, ethanol, isopropanol and butanol, etc. [18, 19]. Furthermore, palmitic acid and oleic acid are the most commonly occurring saturated and unsaturated FAs, respectively, followed by linoleic acid (unsaturated) and stearic acid (saturated) [20].

Moreover, these fatty acid esters collectively constitute to molecular structure of the biodiesel and contribute to the overall fuel properties of biodiesel based on their molecular properties [21, 22]. Moving further, long chain saturated FAEs, predominantly produced from saturated triglycerides report, increased kinematic viscosity, cetane number, calorific value in addition to reduced density than compared to their counterpart unsaturated FAEs. As a result, these FAEs has tendency to produce higher thermal efficiencies and reduced concentration of incomplete combustion products as a result of its complete oxidation [11, 21, 23–25]. On the other hand, unsaturated FAEs, with single or multiple unsaturated bonds in their FA moieties report lower cetane number and calorific value, thereby resulting in poor thermal efficiencies, in-cylinder pressure and heat release rates. Adding to this, high exhaust gas temperatures (EGTs) followed by increased concentration of $NO_X$ emission [14, 26, 27], are also contributed by these unsaturated FAEs; especially by the alkyl esters of oleic acid and linoleic acid. In specific, $NO_X$ emission of any biodiesel increases with addition of unsaturated bonds in their FA moiety; and is accounted by the increased adiabatic flame temperature upon combustion inside the cylinder [28]. Besides, adding an aliphatic ($-CH_2$) group to the alcohol moiety simply enhanced the cetane number of the biodiesel; however, the concentration of particulate matter increased by two fold [29].

From these studies it is clearly evident that, FAEs have a significant role in deciding the overall engine characteristics of its biodiesel; yet, it lacks sufficient results necessary for explaining the contribution of FAEs, in case of a multiple feedstock based biodiesel. With these understandings of FAEs and necessity for this underdone work, this present chapter focus on studying the influence of dominant fatty acids esters in deciding the overall engine characteristics of a biodiesel produced from the homogeneous mixture of different waste animal fats and fish oil, blended in equivalent proportion.

## 2. Waste animal fat-oil (WaFO) biodiesel production

Waste animal fats and oil used in this study includes beef tallow, chicken fat, mutton suet and pork lard, and fish fat oil; and were rendered from wide variety of animal wastes. For instance, beef tallow was rendered from tannery fleshing and meat processing wastes; while, chicken fat, mutton suet and pork lard were rendered from their respective wastes collected from different slaughter house units. Besides, waste fish fat oil was directly procured from the leather tanneries associated with oil tanning process; in its existing form. Here, waste fats (tallow, suet, lard and chicken fat) were rendered from their respective wastes using dry rendering technique; which involved with autoclaving each waste individually, at 120°C and 2 bar pressure [30]. Post rendering, each fat was filtered to remove any solid residues, washed with distilled water to remove suspended residues; and was dehydrated to remove residual moisture content by heating at 110°C. Likewise, similar pre-treatment process was followed for waste fish fat oil. Following that, the pre-treated waste fats and oil were refined to remove phospholipids by means of degumming using orthophosphoric acid as explained by Srinivasan et al., [22, 30].

Post refining, tallow, suet, lard, chicken fat and fish fat oil were mixed in equivalent proportion; and was blended into a homogeneous feedstock (WaFO). The blended waste Fat-oil (WaFO) was esterified by refluxing it with ethanol and 1 wt.% of concentrated sulfuric acid (conc. $H_2SO_4$), in order to reduce its overall free fatty acid (FFA) content so as to avoid formation of soap during transesterification. For the production of WaFO biodiesel, the WaFO sample was transesterified by following the under mentioned reaction parameters: (i) oil to ethanol molar ratio: 1:8; (ii) catalyst concentration: 0.75 wt.% of potassium hydroxide (KOH); (iii) reaction temperature: 72°C; (iv) reaction time: 150 mins. Here, the volume of ethanol was calculated from the optimized molar ratio using a simple equation Eq. (1), which correlates the molecular weight and density of triglycerides and ethanol [30].

$$V_{alchol} = \frac{V_{sample} * m * \rho_{TG} * M_{alchol}}{\left[92.17 - 3 + \left[3\left(\sum_{i=1}^{n} M_{FA} * x_i\right) - 17\right]\right] * \rho_{alchol}} \qquad (1)$$

Completing the reaction, resultant mixture was decanted in a separating funnel for 24 h; when the residual glycerol got separated from the WaFO biodiesel and settled down at the bottom. Lastly, the separated biodiesel was washed with hot distilled water successively to remove residual ethanol and glycerol, soaps, and salts; and was dried at 110°C to remove moisture content from it.

## 3. Evaluation of fuel properties for WaFO and WaFO biodiesel

For the purpose of Characterization of fatty acids, the WaFO was processed into test sample as per the standard preparation technique [31], while, the WaFO biodiesel was tested directly, in a Gas chromatography-Mass Spectrometer (GC–MS); and was studied from their spectral data. Accordingly, WaFO reported oleic acid, palmitic acid, stearic acid as its dominant FAs; with their concentration as 35.41%, 24.24% and 16.15%, respectively. In the same way, WaFO biodiesel reported the ethyl esters of characterized dominant FAs, with their concentration as 35.63%, 27.73% and 18.34%, respectively. Summing up, WaFO Biodiesel was made up of 51.4% of saturated FAEs and 49.18% of unsaturated FAEs; and suggested that the resultant biodiesel was evenly balanced with both saturated and unsaturated FAEs, which reflected in its molecular formula ($C_{19}H_{37}O_2$).

Following that, the WaFO biodiesel was assessed for its fuel compatibility with neat diesel, and suitability in CI engines by evaluating its fuel properties in accordance with ASTM D6751 standards. To begin with, Density of WaFO biodiesel was measured using a simple hydrometer, as specified in ASTM D1298 method; and was reported to be 4.14% higher than neat diesel. Next up, ASTM D445 method was followed to measure the kinematic viscosity of WaFO biodiesel using a calibrated glass-viscosity tube, and was found to be 27.96% higher than neat diesel. Again, flash and fire point

| Properties | Diesel | WaFO biodiesel | ASTM standards | Permissible range |
|---|---|---|---|---|
| Density, kg/m$^3$ | 837 ± 7.6 | 871.68 ± 5.12 | D1298 | — |
| Specific gravity | 0.84 ± 0.008 | 0.872 ± 0.006 | D1298 | 0.86–0.90 |
| Kinematic viscosity, mm$^2$/s | 3.72 ± 0.24 | 4.76 ± 0.21 | D445 | 1.90–6.0 |
| Flash point, °C | 64 ± 2.5 | 144 ± 2.6 | D93-16 | 130 min |
| Fire point, °C | 72 ± 2.2 | 153 ± 2.45 | D93-16 | — |
| Cloud point, °C | 0 ± 1 | 1.5 ± 1 | D2500 | −3 to 12 |
| Pour point, °C | −15 ± 1.5 | −2.7 ± 1.5 | D7346-15 | −15 to 10 |
| Cetane Number | 50 ± 1.4 | 63.87 ± 1.6 | D613 | 47 (min) |
| Calorific value, MJ/kg | 42.6 ± 0.1 | 37.87 ± 0.1 | D240 | 35 to 43 |
| Saponification value, mg KOH | — | 191.38 ± 1.1 | D5558 | — |
| Acid Value, % | — | 0.11 ± 0.02 | D664 | 0.80 max |
| Iodine value, g $I_2$ | — | 53.26 ± 0.92 | D5554 | 120 max |
| Carbon, wt.% | 85.16 ± 1.14 | 76.71 ± 1.08 | D5291 | — |
| Hydrogen, wt.% | 14.26 ± 0.75 | 12.54 ± 0.73 | D5291 | — |
| Oxygen, wt.% | 0 | 10.75 ± 0.52 | D5291 | — |
| Sulfur, wt.% | 9.87 ± 0.48 | 2.67 ± 0 .22 | D5453 | — |
| Phosphorus, wt.% | 0.12 ± 0.02 | 0.001 | D4951 | — |
| Molecular formula | $C_{16}H_{28}$ | $C_{19}H_{37}O_2$ | — | — |
| Molecular weight, g/mol | 220.39 | 297.5 | — | — |

**Table 1.**
*Fuel properties of WaFO biodiesel evaluated as per ASTM standards along with neat diesel and their permissible range.*

of WaFO biodiesel were reported to be 80 and 81OC higher than neat diesel, respectively; and were tested in Pensky Martens closed-cup apparatus as described in ASTM D93–16 method. And, cetane number of WaFO biodiesel, evaluated according to ASTM D613 method, was found to be 27.74% greater than neat diesel. Here, higher density, kinematic viscosity, flash point and cetane number for WaFO biodiesel than neat diesel were contributed by the long carbon chained FAEs like ethyl oleate, ethyl palmitate and ethyl stearate; yet, remained significantly lower due to the presence of unsaturated ethyl oleate in it [21, 25, 30, 32].

In contrast, calorific value of WaFO biodiesel was reported 11.1% lesser than neat diesel, upon tested inside a bomb calorimeter as per ASTM D240 method; and this reduction was clarified by its fuel bound oxygen molecules and absence of sulfur content, which fails to contribute a significant share towards its calorific value [33]. Looking into its chemical properties, saponification value and iodine vale of WaFO biodiesel was found to be 191.38 mg KOH/gm and 53.26 g $I_2$/100 gm, on account of its increased concentration of unsaturated FAEs. Meanwhile, the acid value of WaFO biodiesel was estimated as 0.11% by using ASTM D664 method, which acknowledged the effective conversion of FFAs and monoglycerides into fatty acid esters. Lastly, analytical data related to chemical composition of WaFO biodiesel stated its average molecular weight to be 35% higher than neat diesel; whose carbon and hydrogen content was estimated to be 9.92 and 12.06% lesser than the latter fuel. Moreover, WaFO biodiesel exhibited 10.75% of oxygen content available in it; and is regarded as an oxygenated biofuel in view of this fuel bound oxygen content. **Table 1** summarizes the fuel properties of WaFO biodiesel and neat diesel evaluated as per ASTM standards, along with their permissible range and testing methods.

## 4. Engine characteristics of WaFO and its dominant FAEs

The evaluation of performance, emission and combustion characteristics of WaFO biodiesel was carried out in a Kirloskar TV1 single cylinder CI engine equipped with in-built water cooling system, with **Table 2** consolidates the product specifications of the test engine and flue gas analyzer used in this present study [15, 32]. Here, the parameters tested for this present study includes performance characteristics (specific fuel consumption and brake thermal efficiency), emission characteristics (mon- and di- oxides of carbon and nitrogen, unburnt Hydrocarbon emission, and exhaust gas temperature), and combustion characteristics (maximum in-cylinder pressure, ignition delay, heat release rate). For purpose of testing, two different types of samples

| **Kirloskar engine TV 1 specifications** | | **AVL DI GAS 444 N (five gas analyzer)** | |
|---|---|---|---|
| **Type: four stroke, single cylinder water cooled** | | **Measurement** | **Resolution** |
| Rated power | 5.2 kW | CO [0–15% Vol] | 0.0001% Vol |
| Rated speed | 1500 rpm | HC [0–20000 ppm Vol] | 1 ppm/10 ppm |
| Bore diameter (D) | 87.5 mm | $CO_2$ [0–20% Vol] | 0.1% Vol |
| Stroke (L) | 110 mm | $O_2$ [0–25% Vol] | 0.01% Vol |
| Compression ratio | 17.5:1 | $NO_X$ [0–6000 ppm Vol] | 1 ppm Vol |

**Table 2.**
*Product specification of test engine and flue gas analyzer [15, 32].*

have been used in this study and are named as follows: blend samples and ester samples. In specific, blend samples consist of B10, B20 and B30 samples, with 10%, 20% and 30% of biodiesel blended in neat diesel, respectively; and will be used for assessing the trends of biodiesel's performance in engine. On the other hand, ester samples consist of characterized dominant FAEs, ethyl oleate, ethyl palmitate and ethyl stearate, blended in the concentration with respect to B20 blend; and are named as oleate blend, palmitate blend, stearate blend. For better understanding, the blending of blend and ester samples are represented in form of mathematical correlations Eqs. (2) and (3) [15]; and are used for calculating the volume of diesel and biodiesel/ester required for making the necessary blends.

$$\text{Amount of diesel (ml)} = V_{overall} * \{1 - [\varphi_E * \psi_B]\} \quad (2)$$

$$\text{Amount of biodiesel (ml)} = V_{overall} * \{\varphi_E * \psi_B\} \quad (3)$$

Here, B20 blend was identified as ideal proportion for understanding the influence of FAEs in deciding the engine characteristics of WaFO biodiesel; and was acknowledged due to the increased performance of any biodiesel at their 20% blend [34]. In addition, blending ester samples reduced the technical challenges associated with low temperature crystallization and increased viscosity, besides their cost. **Table 3** reports the overall engine characteristics of WaFO biodiesel blends, along with neat diesel averaged over their engine loads. For ensuring accuracy in results, all the experimental runs were performed in triplicates and are reported in form of mean ± standard error, wherever applicable.

### 4.1 Combustion characteristics

#### *4.1.1 Maximum/peak In-cylinder pressure (Pmax)*

In general, in-cylinder pressure inside the cylinder signifies the degree of homogenous mixing of injected fuel with air, and helps in enhancing the rate of combustion. From **Table 3** and **Figure 1**, both blend and ester samples reported higher in-cylinder pressure against neat diesel sample owing to their higher cetane number, which shortened their ignition delay (ID), thereby allowing them to get combusted using

| Parameters | Unit | Diesel sample | B10 blend | B20 blend | B30 blend |
|---|---|---|---|---|---|
| Pmax | Bar | 51.6 ± 1.24 | 56.8 ± 1.19 | 58 ± 1.22 | 59.4 ± 1.32 |
| iHRR | $kJ/m^3.deg$ | 56.3 ± 1.58 | 62.28 ± 1.62 | 63.1 ± 1.72 | 64.4 ± 1.64 |
| ID | °CA | 19.2 ± 0.52 | 16.8 ± 0.57 | 16.2 ± 0.56 | 15.6 ± 0.6 |
| SFC | kg/kW-hr | 0.31 ± 0.02 | 0.36 ± 0.03 | 0.38 ± 0.02 | 0.4 ± 0.02 |
| BTE | % | 32.3 ± 0.47 | 30 ± 0.51 | 28.1 ± 0.49 | 27.3 ± 0.48 |
| CO Emission | % | 0.25 ± 0.03 | 0.17 ± 0.02 | 0.15 ± 0.04 | 0.13 ± 0.03 |
| $CO_2$ Emission | % | 5.8 ± 0.57 | 6.8 ± 0.59 | 7.3 ± 0.62 | 7.7 ± 0.61 |
| $NO_X$ Emission | PPM | 581.3 ± 10.3 | 703.8 ± 11.4 | 734.9 ± 11.2 | 771.1 ± 10.6 |
| HC Emission | PPM | 57.8 ± 2.5 | 49.6 ± 3.11 | 52.4 ± 2.97 | 55 ± 3.2 |
| EGT | °C | 208.8 ± 5.16 | 252.2 ± 6.07 | 263.8 ± 4.27 | 273.6 ± 7.12 |

**Table 3.**
*Engine characteristics of blend samples, averaged over the engine load.*

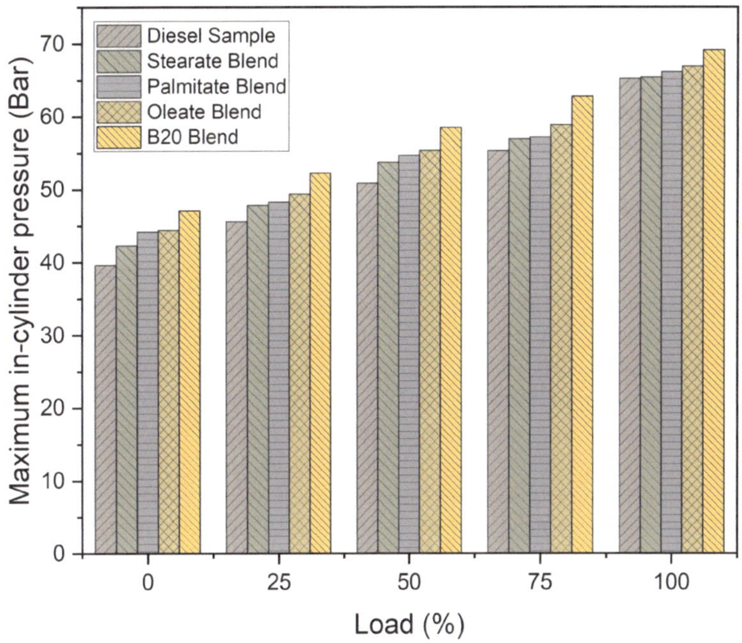


**Figure 1.**
*Maximum in-cylinder pressure of WaFO B20 blend and ester samples.*

their fuel bound oxygen content [32, 35, 36]. Accordingly, B10 blend reported 10.33%, B20 blend reported 12.64% and B30 blend reported 15.46%, higher peak in-cylinder pressure than compared to neat diesel. Likewise, stearate blend reported 3.27%, palmitate blend reported 5.1% and oleate blend reported 6.75%, higher peak in-cylinder pressure than compared to neat diesel.

Upon comparing ester samples with Biodiesel (B20) blend, oleate blend reported minimal variation in peak in-cylinder pressure by 5.52%, followed by palmitate blend and stearate blend reporting 6.72% and 8.3%, respectively. Here, the reduced peak pressure for palmitate and stearate blend signifies their early start of combustion (SOC) citing their shortened ID, besides their reduced concentration. In contrast, oleate blend reported marginal reduction in peak pressure, citing its unsaturation, which reduced its cetane number and prolonged its ID. This prolonged time delay accumulated a significant amount of fuel during premixed burn phase, and got combusted using the available fuel bound oxygen during the diffusion combustion phase [37]. Correlating this, presence of saturated FAEs (ethyl palmitate and ethyl stearate) in WaFO biodiesel initiated the early SOC during the premixed combustion phase, because of their higher cetane number; and provided sufficient activation energy for initiating the combustion of unsaturated FAEs (ethyl oleate, etc.) during the controlled combustion phase. Moreover, in-cylinder pressure increased with engine load for both blend and ester samples, considering the increasing amount of fuel combusted, intending to meet the energy demand of the engine.

### *4.1.2 Instantaneous heat release rate (iHRR)*

More often, heat release rate curve briefs out about the time line of the combustion stroke, indicating the Start Of Injection (SOI), Ignition Delay (ID), Start of

Combustion (SOC); and ultimately, the amount of heat released during the combustion of fuel [38]. From **Table 3** and **Figure 2**, both blend and ester samples happened to report higher iHRR than neat diesel citing their early initiation of combustion and its prolonged duration, which provided adequate time for the accumulated low volatile fuel to undergo combustion during both premixed phase and diffusion combustion phase [39]. In addition, fuel bound oxygen played a crucial role in ensuring the complete oxidation of these FAEs in blend and ester samples. Comparatively, B10 blend reported 11.36%, B20 blend reported 12.97% and B30 blend reported 15.47%, higher heat release rate than compared to neat diesel. In like manner, stearate blend reported 3.75%, palmitate blend reported 5.73%, and oleate blend reported 6.82%, higher heat release rate than compared to neat diesel.

Relative to Biodiesel (B20) blend, oleate blend reported minimal variation in iHRR (by 5.39%), followed by palmitate blend (6.41%) and stearate blend (8.12%), respectively. From above comparison, it was evident that HRR of oleate blend remained higher owing to its unsaturation content, resulting in prolonged ID and reduced premixed combustion phase; helping the accumulated low volatile fuel to oxidize completely using its fuel bound oxygen during the diffusion combustion phase. In contrast, palmitate blend exhibited higher iHRR because of its saturation content, which required less activation energy, and minimal ID; thereby initiating early combustion and providing enough energy for the progressing combustion. Similar trend was reported for stearate blend; however, it remained lower than all other ester samples due to the reduced concentration of ethyl stearate in the diesel blend. Collectively, it can be inferred that saturated FAEs (ethyl palmitate and ethyl stearate) were responsible for the activities during the premixed combustion phase, especially the early ignition of WaFO biodiesel. Following this, unsaturated FAEs (ethyl oleate) were found to be playing crucial role in enhancing the overall HRR through their

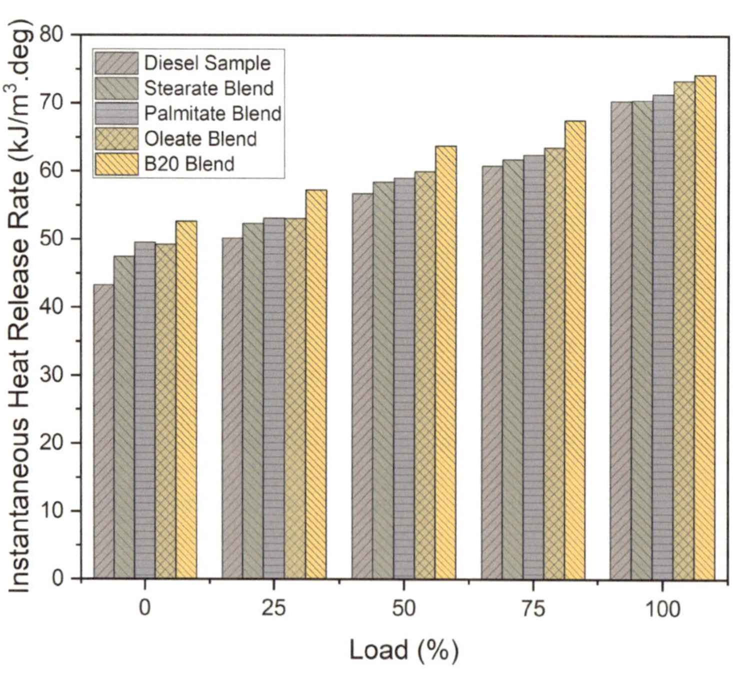


**Figure 2.**
*Instantaneous heat release rate of WaFO B20 blend and ester samples.*

delayed combustion during diffusion combustion phase, thereby liberating high amount of heat energy. Like Pmax, HRR also increased with engine load for both blend and ester samples, considering the increasing amount of fuel combusted, in order to meet the energy demand of the engine.

### *4.1.3 Ignition delay (ID)*

Ignition delay of the fuel signifies the delay period noted between the SOI and SOC; and is always represented in terms of crank shaft angle. From **Table 3** and **Figure 3**, both blend and ester samples reported reduced ID due to their high cetane number; and played a significant role in initiating the combustion well before the neat diesel. As a matter of fact, this ID is widely influenced by both physical and chemical delay; but is predominantly influenced by chemical delay [40]. Accordingly, variation in ID between neat diesel and B10 blend, B20 blend and B30 blend were found to be 2.4°, 3° and 3.6° CA BTDC, lower than the former. In the same manner, variation in ID between neat diesel and oleate, stearate and palmitate blend were reported to be 0.4°, 1° and 1.2° CA BTDC, lower than the diesel sample.

Amongst ester samples compared with B20 biodiesel blend, oleate blend reported 2.6° CA BTDC, stearate blend reported 2° CA BTDC, and palmitate blend reported 1.8° CA BTDC, higher ID. It follows that, both palmitate and stearate blends exhibited shortened ID owing to their higher cetane number because of their higher saturation. Yet, higher delay period than B20 (biodiesel) blend was explained by the reduced availability of ethyl palmitate and ethyl stearate in their blend samples. On contrary, oleate blend reported longer ID than other ester samples due to their low cetane number, accounting its unsaturation and increased availability; besides its high viscosity. Eventually, WaFO biodiesel reported shortened ID because of its saturated

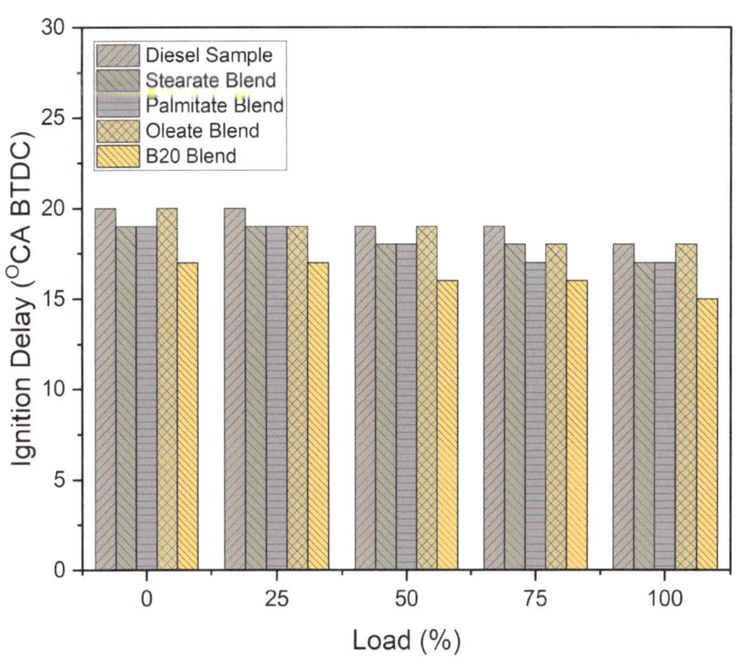


**Figure 3.**
*Ignition delay of WaFO B20 blend and ester samples.*

FAEs (ethyl palmitate and ethyl stearate) which exhibited early SOC, and initiated the combustion of their unsaturated counterparts. Adding to this, the unsaturated FAEs (ethyl oleate) themselves had higher CN than diesel, which allowed it to initiate early SOC. Here, ID of test samples reduced with increasing engine load, citing the increased availability of fuel. Especially, both blend and ester samples reported lower ID in view of more amount of fuel injected, which indirectly signified increased cetane number.

## 4.2 Performance characteristics

### *4.2.1 Specific fuel consumption (SFC)*

In general, Specific fuel consumption reports about the fuel requirement of the engine, for producing 1 unit of power [41, 42]. From **Table 3** and **Figure 4**, it can be noted that diesel sample reported lowest SFC amongst all test samples owing to its superior calorific value, and low density. As well, absence of long to very long carbon chained molecules in the diesel simply reduced its viscosity, which enhanced its rate of atomization and vaporization. Supporting this, B10 blend reported 18.77%, B20 blend reported 23.20% and B30 blend reported 29.87%, higher SFC than neat diesel; whereas, ester samples reported higher SFC by 5.8%, 9.66%, 13.67% for stearate blend, palmitate blend and oleate blend, respectively.

In comparison with B20 (biodiesel) blend, oleate blend reported 7.71%, palmitate blend reported 10.97%, and stearate blend reported 14.10%, lower SFC. Here, oleate blend reported highest SFC amongst other ester samples owing to its unsaturation, resulting in reduced calorific value, which demanded more fuel to meet the energy equivalence demand. Besides, increased density and kinematic viscosity favored poor

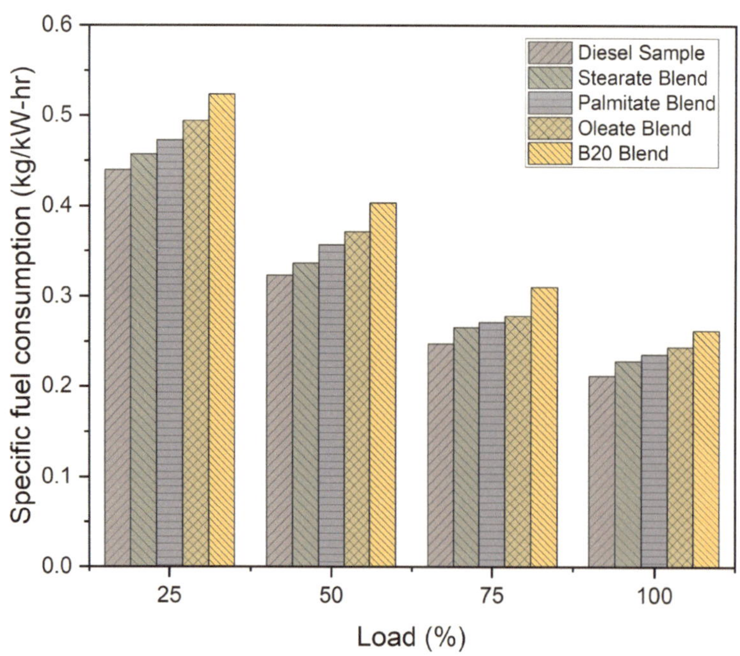


**Figure 4.**
*Specific fuel consumption rate of WaFO B20 blend and ester samples.*

atomization and vaporization, thereby leading to poor combustion; and again demanded surplus fuel to satisfy the energy demand. Meanwhile, both stearate blend and palmitate blend reported low SFC due to their slightly higher calorific value, which helped in deriving maximum heat energy output. Inspite of long carbon chains contributing to their increased density and viscosity, these samples reported low rate of fuel consumption citing their reduced availability in the blend sample and superior calorific value of diesel, itself. Summing up, both saturated (ethyl palmitate and ethyl stearate), and unsaturated FAEs (ethyl oleate) are responsible for the increased SFC of WaFO biodiesel, accounting their long carbon chains. Also, unsaturation in the WaFO biodiesel had negative impact on its overall calorific value, thus consuming more fuel to produce the equivalent work. Oddly, trend of SFC curve reduced with increasing engine load for all test samples, suggesting that the brake power increased along with engine load [43].

#### *4.2.2 Brake thermal efficiency (BTE)*

In common practice, the capability of the engine to produce actual mechanical work output by converting the stored chemical energy in the fuel is signified by its brake thermal efficiency; and correlates brake power with the fuel power [32]. From **Table 3** and **Figure 5**, compared with neat diesel, lower BTE was reported for B10 blend by 10.29%, B20 blend by 13.05% and B30 blend by 15.5%; and, for stearate blend by 2.47%, palmitate blend by 4.68%, and oleate blend by 7.62%. Here, high BTE for diesel, inspite of low cetane number, was explained by its superior calorific value and low volatility, which allowed it to undergo complete combustion especially during its diffusion combustion phase; inspite of its lack of fuel bound oxygen content [44].

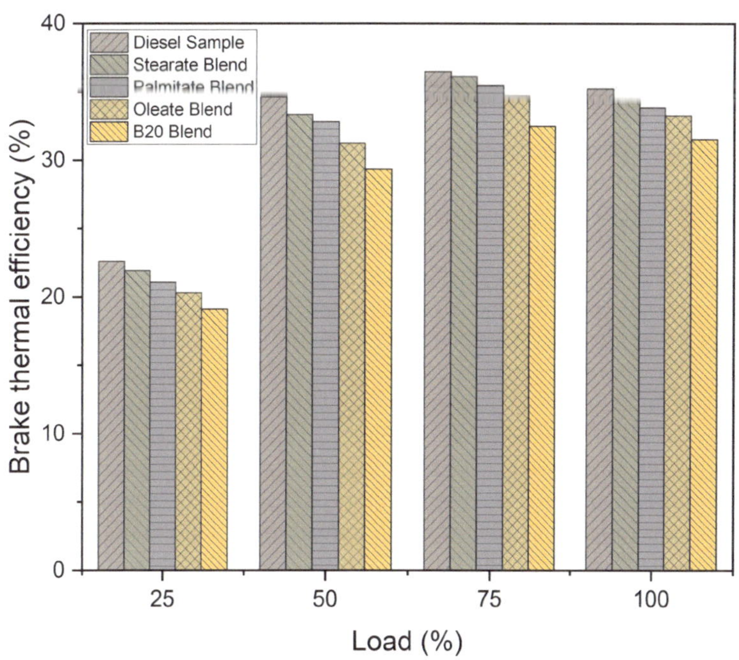


**Figure 5.**
*Brake thermal efficiencies of WaFO B20 blend and ester samples.*

Compared to B20 (biodiesel) blend, oleate blend reported 6.24%, palmitate blend reported 9.65%, and stearate blend reported 12.22%, higher BTE. Here, stearate blend exhibited highest BTE amidst other ester samples because of its increased calorific value; and reduced availability of ethyl stearate in the blend sample, which had a significant effect on its resultant viscosity. Moreover, shortened ID of ethyl stearate provided it sufficient time to get combusted during premixed combustion phase, and supply sufficient energy for the accumulated diesel to get combusted rapidly during the diffusion combustion phase; thereby resulting adequate amount of heat energy. Likewise, palmitate blend also reported similar phenomenon; however, reduced BTE was explained by its increased concentration than stearate blend. Unlike this, oleate blend reported lowest BTE amongst ester samples citing its unsaturation, inferior calorific value, and increased rate of viscosity; hence, requiring more amount of fuel for energy equivalence. However, higher BTE than B20 blend was explained by the reduced availability of ethyl oleate in the blend sample, and its efficacy to undergo complete oxidation using its fuel bound oxygen. Comparing these results, it can be inferred that saturated FAEs (ethyl palmitate and ethyl stearate) initiated combustion during the premixed phase, and provided sufficient activation energy for initiating the combustion of unsaturated FAEs (ethyl oleate) during the diffusion combustion phase. Besides, in view of early SOC due to shorted ID, FAEs in WaFO reported early ignition and underwent complete oxidation using its fuel bound oxygen content; thus, reporting similar BTE like neat diesel. Again, BTE of all test samples increased with engine load, considering the increasing amount of fuel combusted, in order to meet the energy demand of the engine [45].

### 4.3 Emission characteristics

#### *4.3.1 Carbon monoxide (CO) emission*

In general, CO emission is considered as secondary by-product during combustion; and its presence in exhaust gas signifies incomplete combustion of fuel inside engine cylinder. Infact, CO emission arises in case of poor atomization, improper air-fuel mixing, deprived oxygen content, insufficient time for completion of combustion, and even engine's operating conditions; in addition to fuel's molecular properties like unsaturation, C/H ratio, and even aromaticity [46]. From **Table 3** and **Figure 6**, both blend and ester samples reported lower CO emission against neat diesel because of their fuel bound oxygen content, which was responsible for the completion of their oxidation; and, leaving behind only a small portion of partially combusted CO emissions. Relatively, CO emission remained reduced for B10 blend by 35.66%, B20 blend by 45.76% and B30 blend by 52.04%; and, for palmitate blend by 10.52%, stearate blend by 22.31%, than compared to neat diesel. In contrast, oleate blend reported higher CO emission (by 15%), than compared to neat diesel sample.

As compared with B20 (biodiesel) blend, stearate blend reported 45.86%, palmitate blend reported 70.69%, and oleate blend reported 122.45%, higher CO emission. Here, both palmitate and stearate blends reported higher CO emissions; and was explained by their reduced availability and long carbon chained molecules, inspite of their shortened ID and fuel bound oxygen content. Furthermore, oleate blend reported highest CO emission amongst other test samples on account of its unsaturated double bond in its FA moieties [40, 47]. Besides, delayed combustion encouraged the rapid combustion of accumulated fuel during diffusion combustion phase, thereby increasing the CO concentration. Summarizing this, WaFO biodiesel reported reduced CO emission in view of its fuel bound oxygen molecules in their FAEs; yet, it

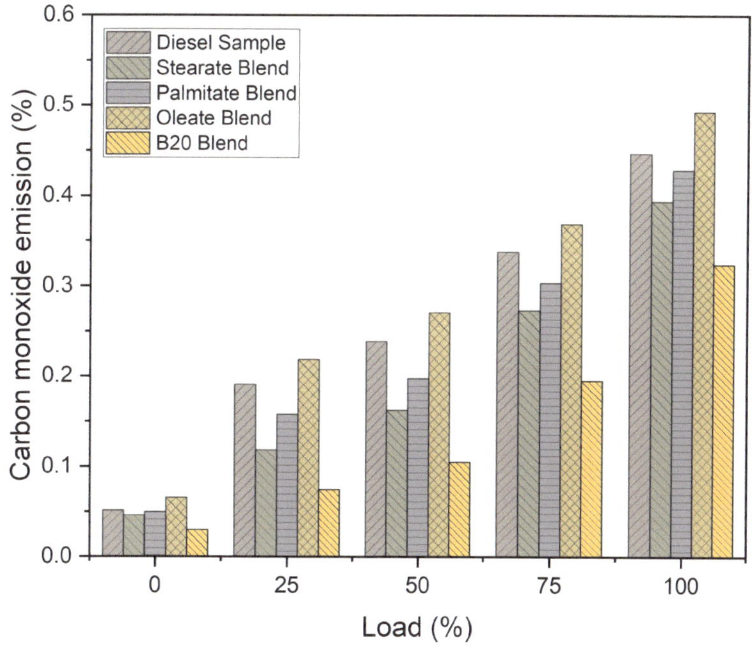


**Figure 6.**
*Carbon monoxide emission of WaFO B20 blend and ester samples.*

reported significant traces of CO due to its unsaturated FAEs (ethyl oleate). To be noted, saturated FAEs (ethyl palmitate and ethyl stearate) ensured complete oxidation of WaFO biodiesel by providing sufficient activation energy for its unsaturated counterparts. Again, CO emissions of both blend and ester samples increased along with engine load, and were explained by the increasing amount of fuel injected into the cylinder to meet the energy demand of the engine.

### *4.3.2 Carbon dioxide ($CO_2$) emission*

Unlike CO emission, $CO_2$ emission is considered as the primary product during combustion; and its presence in exhaust gas signifies the completion of fuel's combustion inside the engine cylinder. Again, concentration of $CO_2$ emission is influenced by the fuel's molecular properties like unsaturation, C/H ratio, and even aromaticity; besides the operating condition of the engine [46]. From **Table 3** and **Figure 7**, both blend and ester samples reported higher $CO_2$ emission than diesel sample citing the presence of their fuel bound oxygen molecules and their higher cetane number; which prolonged its combustion duration for their complete oxidation. Equally important, higher concentration of $CO_2$ emission was also contributed by the long carbon chains in their FAE molecules. Accordingly, B10 blend reported 23.59%, B20 blend reported 35.76% and B30 blend reported 45.58%, higher $CO_2$ emission than compared to neat diesel. Likewise, stearate and palmitate blends reported increased $CO_2$ emission by 6.23% and 15.41% higher $CO_2$ emission, respectively; meanwhile, oleate blend reported lower $CO_2$ emission by 6.97%.

Upon comparing ester samples with Biodiesel (B20) blend, palmitate blend reported 12.33%, stearate blend reported 18.72%, and oleate blend reported 28.32%, lower $CO_2$ emission. As a matter of fact, both palmitate and stearate blends reported

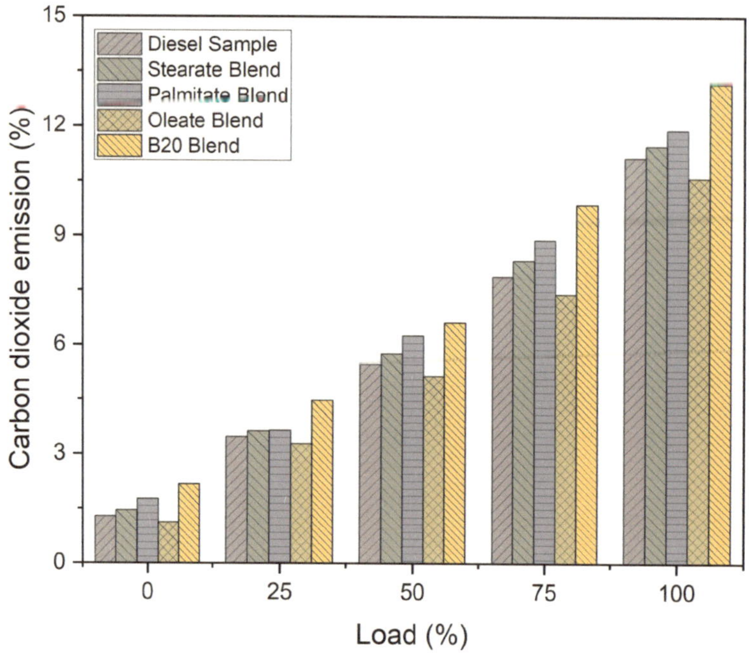


**Figure 7.**
*Carbon dioxide emission of WaFO B20 blend and ester samples.*

higher $CO_2$ concentration than oleate blend on account of their reduced availability and saturation, which improvised their overall effectiveness of combustion. Especially, palmitate blend reported its $CO_2$ emission closer to B20 blend, stating its higher concentration than stearate blend; and its ability to initiate early combustion, thereby providing sufficient time for the accumulated diesel to combust completely. Meanwhile, oleate blend reported lowest $CO_2$ emission amongst all test samples on account of its unsaturation and increased availability, which reduced the effectivity of atomization thereby combusting poorly [48]. In addition, delayed SOC allowed it to combust rapidly which hindered its complete oxidation, thereby forming incomplete combustion products. Summing up, higher concentration of $CO_2$ emission for WaFO biodiesel, inspite of its unsaturation was explained by the presence of its saturated FAEs (ethyl palmitate and ethyl stearate), which initiated early SOC and ensured the progress of combustion of the unsaturated FAEs (ethyl oleate). Again, $CO_2$ emissions increased along with engine load, and were also explained by the increasing amount of fuel injected into the cylinder to meet the energy demand of the engine.

### *4.3.3 Nitrogen oxide ($NO_X$) emission*

Often, $NO_X$ emission in exhaust gas is also regarded as secondary by-product during combustion; however, it arises when engine reports high operating temperatures, especially high exhaust gas temperatures. In relevance to that, $NO_X$ emissions due to high EGTs are explained by higher cetane number and fuel bound oxygen content inducing prolonged combustion; besides the viscosity of fuel [30, 32]. From **Table 3** and **Figure 8**, both blend and ester samples reported higher NOx emission due to their shortened ID, and increased viscosity; which increased the overall duration of combustion, and liberate sufficient heat energy fairly enough for producing $NO_X$ emission. In

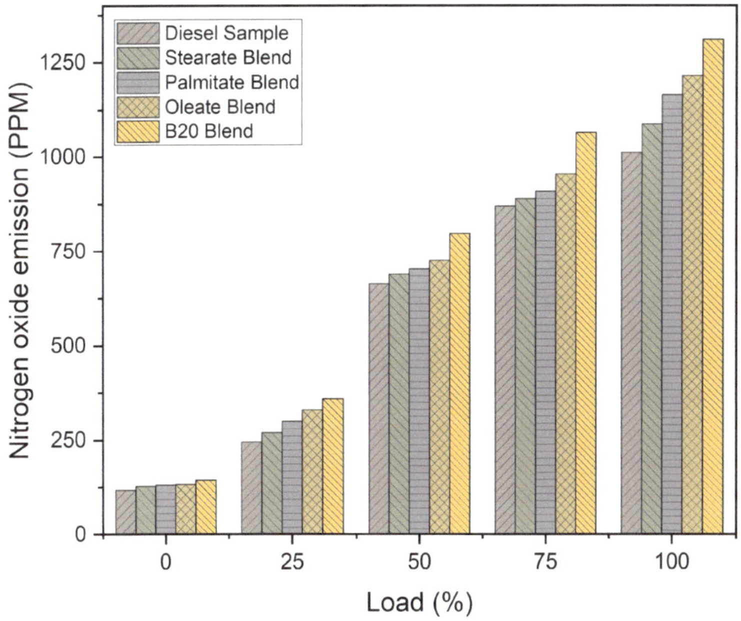


**Figure 8.**
*Nitrogen oxide emission of WaFO B20 blend and ester samples.*

addition, calorific value of these test samples also contributed to this harmful emission. Supporting this, $NO_X$ emission was increased by 22.17% for B10 blend, 28.2% for B20 blend and 36.06% for B30 blend; and 6.55% for stearate blend, 11.8% for palmitate blend, and 17.3% for oleate blend, than compared to neat diesel.

In comparison with B20 (biodiesel) blend, oleate blend reported 8.53%, palmitate reported 12.66% and stearate blend reported 16.57%, lower $NO_X$ emission. In specific, palmitic and stearate blend reported lower $NO_X$ emission than B20 blend signifying their early SOC due to shortened ID; and provided sufficient activation energy for initiating the combustion of diesel during diffusion combustion phase. Yet, these samples reported reduced $NO_X$ emission because of their volatility. On the other hand, oleate blend exhibited higher $NO_X$ emission owing to its increased availability, high viscosity, and reduced cetane number which led to its accumulation in event of its delayed SOC. Besides, rapid combustion of this accumulated fuel liberated high temperature inside the cylinder, and produced high $NO_X$ emission. Outlining these results, higher $NO_X$ emission of WaFO biodiesel was influenced by its unsaturated FAEs (ethyl oleate), which liberated very high temperatures inside the cylinder, thereby forming high $NO_X$ emissions. Interestingly, saturated FAEs (ethyl palmitate and ethyl stearate) also liberated very high temperatures during premixed phase, besides contributing activation energies to unsaturated FAEs, thus favoring $NO_X$ formation. Like other emissions, $NO_X$ emissions of both blend and ester samples increased with engine load on account of more fuel being combusted inside the engine to meet the energy demand, thereby delivering their equivalent work and heat.

#### *4.3.4 Exhaust gas temperatures (EGT)*

EGT from the engine defines the progress of combustion inside the cylinder; and is dependent on the engine's operating conditions and properties of fuel used.

Conventionally, fuel reporting delayed SOC, with prolonged duration exhibits higher EGTs; and these high temperatures contribute to NOx emissions [49, 50]. From **Table 3** and **Figure 9**, both blend and ester samples exhibited higher EGTs accounting their higher cetane number and fuel bound oxygen content; which favored higher rate of combustion and liberated large amount of heat. Moreover, viscosity and calorific value of these samples also contributed to their high EGTs. Supporting this, B10 blend reported 20.97%, B20 blend reported 26.49% and B30 blend reported 31.52%; and stearate blend reported 6.1%, palmitate blend reported 10.62%, and oleate blend reported 15.46%, higher EGTs than compared to neat diesel.

Amongst ester samples compared with B20 biodiesel blend, oleate blend reported 8.48%, palmitate blend reported 12.23% and stearate blend reported 15.76%, lower exhaust gas temperature. Especially, palmitate and stearate blends combusted earlier due to shortened ID, which forced the highly volatile, accumulated diesel to combust rapidly, and limiting the heat generation and EGT. In case of oleate blend, low cetane number allowed it to undergo prolonged combustion, and assisted the diesel for combustion during diffused combustion and after burning phase; thus liberating large amount of heat and increase its EGT. On the whole, WaFO biodiesel with significant amount of unsaturated FAEs (ethyl oleate) exhibited prolonged combustion accompanied with high rate of combustion using their fuel bound oxygen molecules; and liberated high EGTs [29]. Meanwhile, saturated FAEs (ethyl palmitate and ethyl stearate) contributed to a minimal amount to EGT, accounting their early ignition and supplying of activation energy to the unsaturated FAEs; thus contributing minimal to EGTs. Like NOx emissions, EGT of both blend and ester samples increased with engine load on account of more fuel being combusted inside the engine to meet the energy demand, thereby delivering their equivalent work and heat.

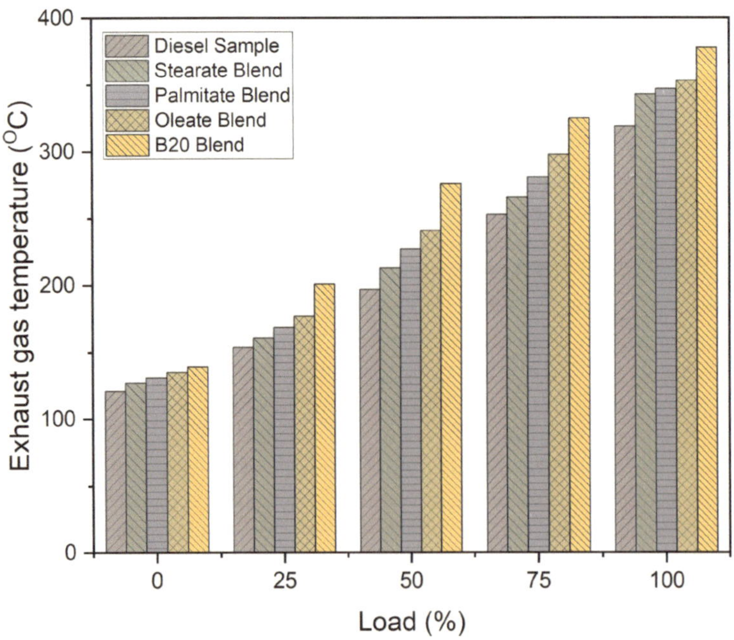


**Figure 9.**
*Exhaust gas temperature of WaFO B20 blend and ester samples.*

#### *4.3.5 Unburnt hydrocarbon (HC) emission*

Unburnt hydrocarbons in exhaust gas signifies the inability of the fuel to get completely combusted near the cylinder wall, owing to reduced flame temperatures near the fuel-rich zones and poor combustion kinetics and quenched flame [51]. From **Table 3** and **Figure 10**, both blend and ester samples displayed reduced HC emission in event of complete oxidation using their fuel bound oxygen content. Adding to this, these oxygen molecules helped in liberating high flame temperatures, and propagated throughout the cylinder and combusted unburnt hydrocarbons. Unfortunately, neat diesel reported traces of HC emission as a consequence of its rapid combustion, owing to its high volatility which reduced the adiabatic flame temperature near the cylinder walls. In comparison with diesel blend, B10 blend reported 17.75%, B20 blend reported 10.91% and B30 blend reported 5.89%; and oleate blend reported 23.71%, palmitate blend reported 27.15%, and stearate blend reported 34.72%, lower HC emissions.

Relatively, stearate blend reported lowest HC emission (by 27.22%), followed by palmitate blend (18.54%) and oleate blend (14.66%), than compared to B20 (biodiesel) blend. Supporting this, palmitate and stearate blends exhibited lower HC emission, and was clearly evident that presence of oxygen in these samples reduced their HC emission. Explaining this, these saturated FAEs initiated early combustion and provided sufficient temperature inside the cylinder for ensuring complete oxidation of diesel [27]. Whilst, oleate blend reported higher HC emission than other ester samples because of its increased availability and unsaturation; which resulted in poor atomization and vaporization, and reduced the effectiveness of combustion (i.e. in complete combustion) for the liquid droplets present at the localized zone with reduced flame temperatures. Consolidating these results, it can be concluded that WaFO biodiesel

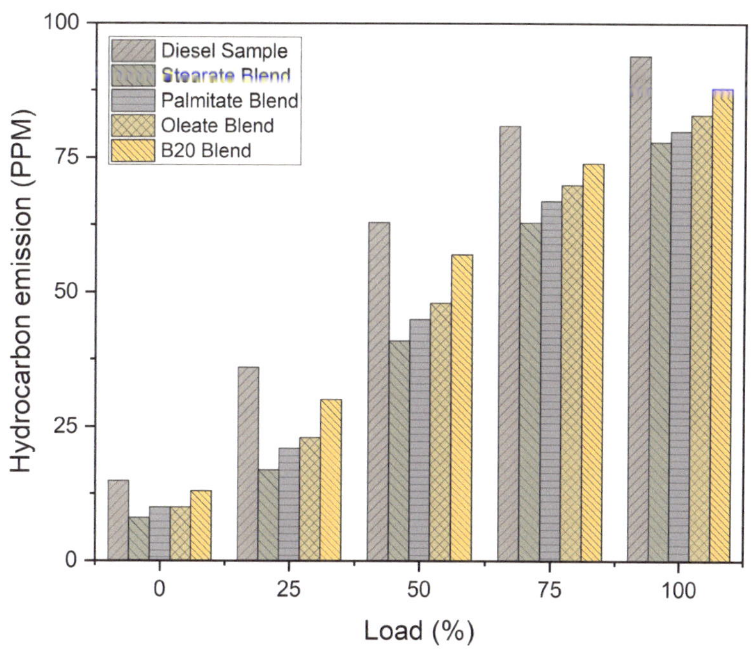


**Figure 10.**
*Hydrocarbon emission of WaFO B20 blend and ester samples.*

combusted completely using its fuel bound oxygen content. Interestingly, unsaturated FAEs (ethyl oleate) in WaFO biodiesel reduced its rate of atomization, thus forming micro fuel droplets; however, they were combusted by the heat energy supplied by the saturated FAEs (ethyl palmitate and ethyl stearate). Here, HC emission of test samples increased with engine load; yet, HC emission of blend and ester samples remained lower than diesel sample even at higher loads due to high engine temperatures, besides their high cylinder pressures [52].

## 5. Conclusions

Thus, this present chapter strongly concludes that the overall engine characteristics of a biodiesel is contributed by its FAEs; and are influenced by their molecular properties including their chain length and unsaturation. Accordingly, engine characteristics, which includes their performance, combustion and emission characteristics of WaFO biodiesel were influenced by its dominant FAEs, and following were the key conclusions deduced from the above study:

i. Ethyl palmitate and ethyl stearate were identified as dominant saturated FAEs, which were responsible for initiating early combustion due to their higher cetane number, and contributing to higher efficiencies owing to their high calorific values. On the other hand, ethyl oleate was characterized as the dominant unsaturated FAEs, and was acknowledged for prolonging the combustion duration due to its unsaturation and need for high activation energy.

ii. High cylinder pressure and heat release rate were explained by early SOC during premixed phase by saturated FAEs, which initiating the combustion of unsaturated FAEs using fuel bound oxygen, and liberated large amount of heat and temperature.

iii. Inspite of complete oxidation of both saturated and unsaturated FAEs, biodiesel reported reduced thermal efficiencies and increased fuel consumption rate in view of their inferior calorific value than neat diesel. In addition, slightly higher density and viscosity also setback biodiesel's overall performance in engine.

iv. Increased concentration of completely combusted products and reduced concentration of incompletely combusted products were acknowledged by the fuel bound oxygen content of biodiesel. Overall emission characteristics of biodiesel were improved by prolonged combustion of unsaturated FAEs, which was further improvised by saturated FAEs.

Based on these conclusions, it is again evident that FAEs decide the overall engine characteristics of their biodiesel; and this work can be used as a preliminary guideline for deciding the idle feedstock for producing biodiesel, which meets the requirement of the engine's output and application.

## Author details

Gokul Raghavendra Srinivasan[1,2]*, Safak Yildizhan[3], Shalini Palani[1], Lakshmanan Thangavelu[4] and Ranjitha Jambulingam[1]*

1 $CO_2$ Research and Green Technologies Centre, Vellore Institute of Technology, Vellore, India

2 R&D Department, Steamax Envirocare Private Limited, Delhi, India

3 Automotive Engineering Department, Çukurova University, Adana, Turkey

4 Department of Mechanical Engineering, SRM Institute of Science and Technology, Chennai, India

*Address all correspondence to: gokusrinivasan@gmail.com and ranjitha.j@vit.ac.in

IntechOpen

© 2022 The Author(s). Licensee IntechOpen. This chapter is distributed under the terms of the Creative Commons Attribution License (http://creativecommons.org/licenses/by/3.0), which permits unrestricted use, distribution, and reproduction in any medium, provided the original work is properly cited.

## References

[1] Lerman LV, Gerstlberger W, Lima MF, Frank AG. How governments, universities, and companies contribute to renewable energy development? A municipal innovation policy perspective of the triple helix. Energy Research & Social Science. 2021;**71**: 101854. DOI: 10.1016/j.erss.2020.101854

[2] Kamaraj R, Rao YK, Balakrishna B. Biodiesel blends: A comprehensive systematic review on various constraints. Environmental Science and Pollution Research. 2021:1-6. DOI: 10.1007/s11356-021-13316-8

[3] Durbin TD, Collins JR, Norbeck JM, Smith MR. Effects of biodiesel, biodiesel blends, and a synthetic diesel on emissions from light heavy-duty diesel vehicles. Environmental Science & Technology. 2000;**34**(3):349-355. DOI: 10.1021/es990543c

[4] Mujtaba MA, Kalam MA, Masjuki HH, Gul M, Soudagar ME, Ong HC, et al. Comparative study of nanoparticles and alcoholic fuel additives-biodiesel-diesel blend for performance and emission improvements. Fuel. 2020;**279**:118434. DOI: 10.1016/j.fuel.2020.118434

[5] Khan H, Soudagar ME, Kumar RH, Safaei MR, Farooq M, Khidmatgar A, et al. Effect of nano-graphene oxide and n-butanol fuel additives blended with diesel—*Nigella sativa* biodiesel fuel emulsion on diesel engine characteristics. Symmetry. 2020;**12**(6): 961. DOI: 10.3390/sym12060961

[6] Akar MA, Kekilli E, Bas O, Yildizhan S, Serin H, Ozcanli M. Hydrogen enriched waste oil biodiesel usage in compression ignition engine. International Journal of Hydrogen Energy. 2018;**43**(38):18046-18052. DOI: 10.1016/j.ijhydene.2018.02.045

[7] Lakshmanan T, Nagarajan G. Experimental investigation of port injection of acetylene in DI diesel engine in dual fuel mode. Fuel. 2011;**90**(8): 2571-2577. DOI: 10.1016/j.fuel.2011.03.039

[8] Kodate SV, Yadav AK, Kumar GN. Combustion, performance and emission analysis of preheated KOME biodiesel as an alternate fuel for a diesel engine. Journal of Thermal Analysis and Calorimetry. 2020;**141**(6):2335-2345. DOI: 10.1007/s10973-020-09814-5

[9] Kannan GR, Anand R. Effect of injection pressure and injection timing on DI diesel engine fuelled with biodiesel from waste cooking oil. Biomass and Bioenergy. 2012;**46**:343-352. DOI: 10.1016/j.biombioe.2012.08.006

[10] Balasubramanian D, Hoang AT, Venugopal IP, Shanmugam A, Gao J, Wongwuttanasatian T. Numerical and experimental evaluation on the pooled effect of waste cooking oil biodiesel/diesel blends and exhaust gas recirculation in a twin-cylinder diesel engine. Fuel. 2021;**287**:119815. DOI: 10.1016/j.fuel.2020.119815

[11] Selvam DJ, Vadivel K. Performance and emission analysis of DI diesel engine fuelled with methyl esters of beef tallow and diesel blends. Procedia Engineering. 2012;**38**:342-358. DOI: 10.1016/j.proeng.2012.06.043

[12] Gad MS, El-Shafay AS, Hashish HA. Assessment of diesel engine performance, emissions and combustion characteristics burning biodiesel blends from jatropha seeds. Process Safety and Environmental Protection. 2021;**147**:518-526. DOI: 10.1016/j.psep.2020.11.034

[13] Palani Y, Devarajan C, Manickam D, Thanikodi S. Performance and emission

characteristics of biodiesel-blend in diesel engine: A review. Environmental Engineering Research. 2021. DOI: 10.4491/eer.2020.338

[14] Szybist JP, Song J, Alam M, Boehman AL. Biodiesel combustion, emissions and emission control. Fuel Processing Technology. 2007;**88**(7): 679-691. DOI: 10.1016/j.fuproc.2006.12.008

[15] Srinivasan GR, Shankar V, Jambulingam R. Experimental study on influence of dominant fatty acid esters in engine characteristics of waste beef tallow biodiesel. Energy Exploration & Exploitation. 2019;**37**(3):1098-1124. DOI: 10.1177/0144598718821791

[16] Srinivasan GR, Jambulingam R. Comprehensive study on biodiesel produced from waste animal fats-a review. Journal of Environmental Science and Technology. 2018;**11**(3): 157-166. DOI: 10.3923/jest.2018.157.166

[17] Canakci M, Sanli H. Biodiesel production from various feedstocks and their effects on the fuel properties. Journal of Industrial Microbiology and Biotechnology. 2008;**35**(5):431-441. DOI: 10.1007/s10295-008-0337-6

[18] Verma P, Sharma MP, Dwivedi G. Impact of alcohol on biodiesel production and properties. Renewable and Sustainable Energy Reviews. 2016;**56**:319-333. DOI: 10.1016/j.rser.2015.11.048

[19] Sanli H, Canakci M. Effects of different alcohol and catalyst usage on biodiesel production from different vegetable oils. Energy & Fuels. 2008; **22**(4):2713-2719. DOI: 10.1021/ef700720w

[20] Timms RE. Physical chemistry of fats. In: Elsevier, editor. Fats in Food Products. Boston, MA: Springer; 1994. pp. 1-27. DOI: 10.1007/978-1-4615-2121-1

[21] Knothe G. Dependence of biodiesel fuel properties on the structure of fatty acid alkyl esters. Fuel Processing Technology. 2005;**86**(10):1059-1070. DOI: 10.1016/j.fuproc.2004.11.002

[22] Srinivasan GR, Jambulingam R. Theoretical prediction of thermo-physical properties of beef tallow based biodiesel. FME Transactions. 2020;**48**(3):600-610. DOI: 10.5937/fme2003600R

[23] Blangino E, Riveros AF, Romano SD. Numerical expressions for viscosity, surface tension and density of biodiesel: Analysis and experimental validation. Physics and Chemistry of Liquids. 2008; **46**(5):527-547. DOI: 10.1080/00319100801930458

[24] Shu Q, Yang B, Yang J, Qing S. Predicting the viscosity of biodiesel fuels based on the mixture topological index method. Fuel. 2007;**86**(12–13):1849-1854. DOI: 10.1016/j.fuel.2006.12.021

[25] Knothe G. A comprehensive evaluation of the cetane numbers of fatty acid methyl esters. Fuel. 2014;**119**:6-13. DOI: 10.1016/j.fuel.2013.11.020

[26] Gopinath A, Puhan S, Nagarajan G. Effect of unsaturated fatty acid esters of biodiesel fuels on combustion, performance and emission characteristics of a DI diesel engine. International Journal of Energy & Environment. 2010;**1**(3)

[27] Benjumea P, Agudelo JR, Agudelo AF. Effect of the degree of unsaturation of biodiesel fuels on engine performance, combustion characteristics, and emissions. Energy & Fuels. 2011;**25**(1):77-85. DOI: 10.1021/ef101096x

[28] Schönborn A, Ladommatos N, Allan R, Williams J, Rogerson J. Effect of the molecular structure of individual fatty acid alcohol esters (biodiesel) on

the formation of NOx and particulate matter in the diesel combustion process. SAE International Journal of Fuels and Lubricants. 2009;**1**(1):849-872. DOI: 10.4271/2008-01-1578

[29] Hellier P, Ladommatos N, Allan R, Rogerson J. The influence of fatty acid ester alcohol moiety molecular structure on diesel combustion and emissions. Energy & Fuels. 2012;**26**(3):1912-1927. DOI: 10.1021/ef2017545

[30] Srinivasan GR, Shankar V, Chandra Sekharan S, Munir M, Balakrishnan D, Mohanam A, et al. Influence of fatty acid composition on process optimization and characteristics assessment of biodiesel produced from waste animal fat. Energy Sources, Part A: Recovery, Utilization, and Environmental Effects. 2020;**1**:1-9. DOI: 10.1080/15567036.2020.1771477

[31] Christie WW. Preparation of ester derivatives of fatty acids for chromatographic analysis. Advances in Lipid Methodology. 1993;**2**(69):e111

[32] Srinivasan GR, Palani S, Munir M, Saeed M, Thangavelu L, Mohanam A, et al. Engine characteristics study on beef tallow biodiesel produced by ethanol based co-solvent transesterification. Energy Sources, Part A: Recovery, Utilization, and Environmental Effects. 2020 Oct;**8**:1-21. DOI: 10.1080/15567036.2020.1826014

[33] Demirbas A. Effects of moisture and hydrogen content on the heating value of fuels. Energy Sources, Part A: Recovery, Utilization, and Environmental Effects. 2007;**29**(7):649-655. DOI: 10.1080/009083190957801

[34] Öner C, Altun Ş. Biodiesel production from inedible animal tallow and an experimental investigation of its use as alternative fuel in a direct injection diesel engine. Applied Energy. 2009;**86**(10):2114-2120. DOI: 10.1016/j.apenergy.2009.01.005

[35] Agarwal AK, Dhar A, Gupta JG, Kim WI, Choi K, Lee CS, et al. Effect of fuel injection pressure and injection timing of Karanja biodiesel blends on fuel spray, engine performance, emissions and combustion characteristics. Energy Conversion and Management. 2015;**91**:302-314. DOI: 10.1016/j.enconman.2014.12.004

[36] Raman LA, Deepanraj B, Rajakumar S, Sivasubramanian V. Experimental investigation on performance, combustion and emission analysis of a direct injection diesel engine fuelled with rapeseed oil biodiesel. Fuel. 2019;**246**:69-74. DOI: 10.1016/j.fuel.2019.02.106

[37] Puhan S, Saravanan N, Nagarajan G, Vedaraman N. Effect of biodiesel unsaturated fatty acid on combustion characteristics of a DI compression ignition engine. Biomass and Bioenergy. 2010;**34**(8):1079-1088. DOI: 10.1016/j.biombioe.2010.02.017

[38] Xue J. Combustion characteristics, engine performances and emissions of waste edible oil biodiesel in diesel engine. Renewable and Sustainable Energy Reviews. 2013;**23**:350-365. DOI: 10.1016/j.rser.2013.02.039

[39] Zhu L, Cheung CS, Huang Z. Impact of chemical structure of individual fatty acid esters on combustion and emission characteristics of diesel engine. Energy. 2016;**107**:305-320. DOI: 10.1016/j.energy.2016.04.030

[40] Jambulingam R, Shankar V, Palani S, Srinivasan GR. Effect of dominant fatty acid esters on emission characteristics of waste animal fat biodiesel in CI engine. Frontiers in Energy Research. 2019;**7**:63. DOI: 10.3389/fenrg.2019.00063

[41] Ray SK, Prakash O. Biodiesel extracted from waste vegetable oil as an alternative fuel for diesel engine: Performance evaluation of kirlosker 5 kW engine. In: Chattopadhyay J, Singh R, Prakash O, editors. Renewable Energy and Its Innovative Technologies. Singapore: Springer; 2019. pp. 219-229. DOI: 10.1007/978-981-13-2116-0_18

[42] Patil VV, Patil RS. Investigations on partial addition of n-butanol in sunflower oil methyl ester powered diesel engine. Journal of Energy Resources Technology. 2018;**140**:1. DOI: 10.1115/1.4037372

[43] Kumar TS, Kumar PS, Annamalai K. Experimental study on the performance and emission measures of direct injection diesel engine with Kapok methyl ester and its blends. Renewable Energy. 2015;**74**:903-909. DOI: 10.1016/j.renene.2014.09.022

[44] Sanli H, Canakci M, Alptekin E, Turkcan A, Ozsezen AN. Effects of waste frying oil based methyl and ethyl ester biodiesel fuels on the performance, combustion and emission characteristics of a DI diesel engine. Fuel. 2015;**159**:179-187. DOI: 10.1016/j.fuel.2015.06.081

[45] Sekhar SC, Karuppasamy K, Vedaraman N, Kabeel AE, Sathyamurthy R, Elkelawy M, et al. Biodiesel production process optimization from *Pithecellobium dulce* seed oil: Performance, combustion, and emission analysis on compression ignition engine fuelled with diesel/biodiesel blends. Energy Conversion and Management. 2018;**161**: 141-154. DOI: 10.1016/j.enconman.2018.01.074

[46] Abdel-Rahman AA. On the emissions from internal-combustion engines: A review. International Journal of Energy Research. 1998;**22**(6):483-513. DOI: 10.1002/(SICI)1099-114X(199805)22:6<483::AID-ER377>3.0.CO;2-Z

[47] Chukwuezie OC, Nwakuba NR, Asoegwu SN, Nwaigwe KN. Cetane number effect on engine performance and gas emission: A review. American Journal of Engineering Research. 2017;**6**:56-67

[48] Abdul Malik MS, Shaiful AI, Mohd Jaafar MN, Mohamad SA. Combustion and emission characteristics of coconut-based biodiesel in a liquid fuel burner. Energies. 2017;**10**(4):458. DOI: 10.3390/en10040458

[49] Abu-Hamdeh NH, Alnefaie KA. A comparative study of almond biodiesel-diesel blends for diesel engine in terms of performance and emissions. BioMed Research International. 2015. DOI: 10.1155/2015/529808

[50] Coniglio L, Bennadji H, Glaude PA, Herbinet O, Billaud F. Combustion chemical kinetics of biodiesel and related compounds (methyl and ethyl esters): Experiments and modeling–advances and future refinements. Progress in Energy and Combustion Science. 2013;**39**(4):340-382. DOI: 10.1016/j.pecs.2013.03.002

[51] Correa SM, Arbilla G. Carbonyl emissions in diesel and biodiesel exhaust. Atmospheric Environment. 2008;**42**(4): 769-775. DOI: 10.1016/j.atmosenv.2007.09.073

[52] Azad AK, Rasul MG, Giannangelo B, Ahmed SF. Diesel engine performance and emission study using soybean biodiesel blends with fossil diesel. In: Elsevier, editor. Exergy for a Better Environment and Improved Sustainability. Cham: Springer; 2018. pp. 137-155. DOI: 10.1007/978-3-319-62575-1_10

Chapter 7

# Feasibility of Biodiesel Production in Pakistan

*Juma Sahar, Muhammad Farooq, Anita Ramli and Abdul Naeem*

## Abstract

Pakistan's energy is mainly dependent on the imported fossil fuels as the explored fossil fuels of the country are insufficient to meet the country's current energy needs. Meanwhile, these fossil fuels have negative environmental consequences and are too expensive to electrify remote areas. To address the country's serious energy shortages, Pakistan's Alternative Energy Development Board (AEDB) has suggested to introduce energy mix to meet the increasing energy demand and fuel the economy. Renewable energy endorsing unique environmentally friendly nature, constant supply, wider availability and ease of integration into existing infrastructure. Biodiesel is considered the best and most easily accessible source of energy among all renewable energy resources. However, there is still substantial room for development of renewable energies in Pakistan. This literature review examines the availability of biomass resources in Pakistan and their potential for meeting the country's rapidly growing energy demand, boosting Country economy and creates new employments in the near future.

**Keywords:** biodiesel, renewable energy, fossil fuels, feedstocks, bifunctional heterogeneous catalysts

## 1. Introduction

Energy has been a fundamental need of a human society. On the other hand, energy consumption has increased exponentially due to rapid growth in population and modernization [1]. The population of world has grown after Second World War, from two billion to seven billion in the 21st century [2, 3]. Currently fossil fuels are the major source for the primary energy of the world (**Figure 1**) [4, 5].

According to the International Energy Outlook 2013 set by the U.S Energy Information Administration [6, 7], the total energy consumed in 2010 was $5.5282 \times 10^{20}$ J, which is predicted to rise further to $8.6510 \times 10^{20}$ J by 2040. Accordingly, the total world energy consumption will grow by 56% between 2010 and 2040; as given in **Figure 2**. The mismatch between the energy supply and energy demand has increased dramatically all over the world.

The limited fossil fuelsand the associated problems such as energy security environmental issueshave emphasized the need for sustainable, reliable renewable energy sources.



IntechOpen

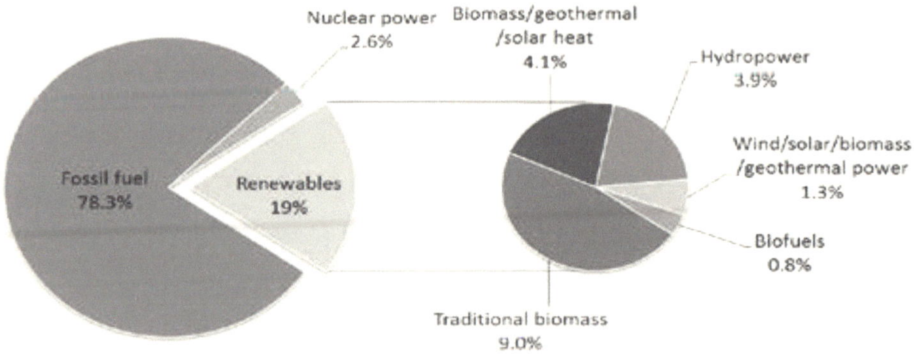


**Figure 1.**
*Global energy consumption in 2013 [3].*

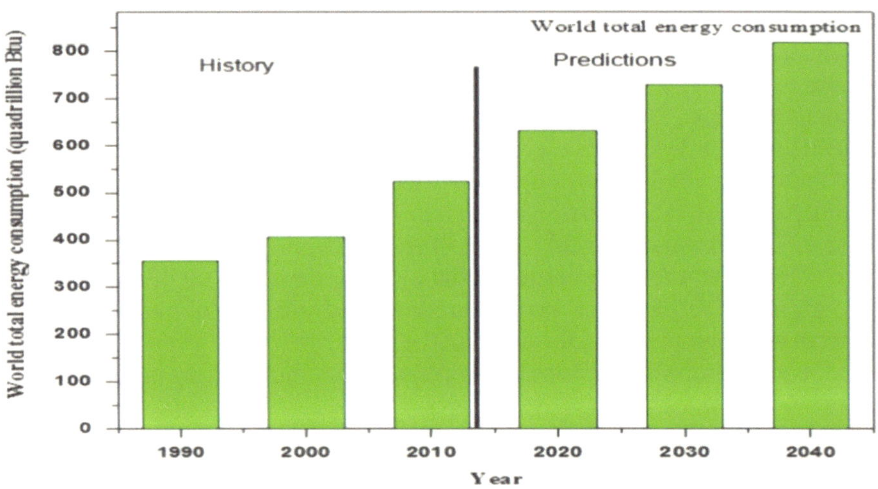


**Figure 2.**
*Total world energy consumption, history and projection [7].*

In the view of the current energy scenario, renewable energy sources could be fantastic choice for the world to meet the increasing energy demand and socio economic development. Renewable energy sources are gaining much attention due to their non-toxicity, biodegradability and low emissions profile as compared to petro diesel [8, 9]. According to US energy information administration, there are seven countries Paraguay (100), Iceland (100%), Costa Rica (99%) Norway (98.5%) Austria (80%), Brazil (75%) and Denmark (69.4%) in the World to have or very near to 100 percent renewable energy sources. Resources of renewable energy are available on large scale such as hydropower, solar, biomass, wind and geothermal energy **Figure 1**. The fossil fuels substitution with renewable energy sources will have very positive effect on greenhouse gases emissions. It has been reported that 2% replacement of fossil fuels with renewable energy sources will result in 1.8% reduction of emissions of $CO_2$ while replacement of 100% will lead to 90% reduction [10]. In the

*DOI: http://dx.doi.org/10.5772/intechopen.101967*

current energy scenario, renewable energy sources could be a fantastic choice for the World to meet the increasing energy demand. Among them, biodiesel is considered to be the most reliable and consistent source of renewable energy supply.

## 2. Biodiesel as an emerging energy resource

Biodiesel may be defined as an oxygenated, non-toxic, biodegradable, eco-friendly and sulfur-free alternative diesel oil. Chemically biodiesel may be defined as a fuel that is composed of mono-alkyl esters of long chain fatty acids obtained from renewable sources such as animal fats, vegetable oilsthat comply the ASTM and European quality standards. Different natural oils are used for the production of biodiesel such as coconut, rapeseeds, soybeans and waste cooking oil (**Figure 3**).

## 3. Biodiesel production technologies

Several efforts have been made to produce derivatives of vegetable oil that can approximate the performance and properties of hydrocarbon-based diesel fuels. The problems associated with the vegetable oil to be used as diesel fuel are high viscosity, low stability against oxidation and the subsequent reactions of polymerization,

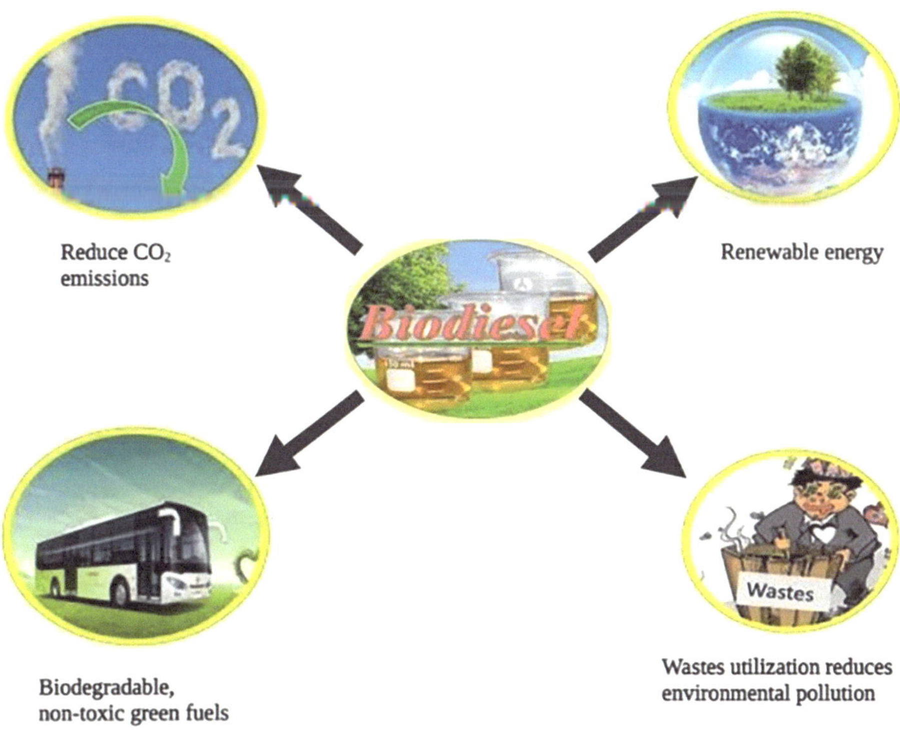


**Figure 3.**
*Advantages of biodiesel [11].*

low volatility due to which incomplete combustion occurs, resulting in the formation of high amount of ash [12]. Different process can be used in order to change these properties such as direct use or blending, micro emulsion, pyrolysis (thermal cracking) and the most conventional process is the transesterification.

### 3.1 Direct use or blending

In beginning of 1980, there was a considerable discussion about the use of vegetable oil as a fuel. The concept of using food as a fuel was explained in 1981 by Bartholomew, demonstrating that petroleum should be the alternative fuel for combustion rather than the vegetable oil. Direct use of vegetable oils has been considered impractical and not satisfactory for both direct and indirect diesel engines. The high viscosity, free fatty acid content, acid composition and the formation of gum due to polymerization and oxidation during storage and combustion are the obvious problems.

Ma et al. [13] highlighted two severe problems such as incomplete combustion and oil deterioration associated with the direct use of vegetable oil as a fuels. Therefore, it will be significant to dilute the vegetable oils with some materials such as diesel fuels, ethanol or solvents to reduce the density and viscosity of vegetable oils.

Bilgin et al. [14] reported that 4% ethanol addition to diesel fuel increased the brake torque, brake thermal efficiency and brake power while decreasing the consumption of brake specific fuel. As the ethanol boiling point is less than the diesel fuel, ethanol could assist the process of combustion through an unburned blend spray.

### 3.2 Pyrolysis or thermal cracking

Generally, pyrolysis may be defined as the thermochemical decomposition of feedstock at medium (300–800°C) to high temperatures (800–1300°C) in an inert atmosphere. Pyrolysis means a chemical change that is caused by the application of thermal energy in the absence of oxygen or air or by the application of heat in the presence of catalyst that results in the bonds cleavage and formation of various small molecules [15]. Being a type of destructive distillation, it is performed in an inert atmosphere in the temperature range of 300–1300°C. Based on the operating conditions, pyrolysis may be classified into three subclasses such as conventional pyrolysis that occur in the temperature range of 550 K–900 K, (400–500°C) fast pyrolysis occurring in 850 K–1250 K (400–650°C) and the flash pyrolysis occurs in the 1050 K–1300 K (700–1000°C) range of temperature. Pyrolysis is the process used for the synthesis of fuel from triglycerides, vegetable oil, animal fats or natural fatty acids. Fast pyrolysis is used for the bio-oil production. Vegetable oils can be cracked to improve cetane number and reduce the viscosity. The products obtained as a result of cracking include carboxylic acids, alkanes, alkadienes, alkenes and aromatics in various proportions. Rape seed oil, cotton seed oil, soybean oil and other oils with the use of appropriate catalyst were successfully cracked to get biofuel.

### 3.3 Micro-emulsions

Micro-emulsions are isotropic, translucent or clear, thermodynamically stable dispersion of water, oil, surfactants, co-surfactants (amphiphilic molecule) for stabilization. In micro-emulsions, the droplet diameters range from 100 to 1000 Å (10 nm–100 nm). A micro-emulsion can be made of vegetable oils with an ester and dispersant (co-solvent) or vegetable oil with alcohol and surfactant with or without diesel fuels [16].

Alcohols such as ethanol or methanol are frequently used as a viscosity lowering additives. Whereas higher alcohols are used as surfactants. The alkyl nitrates are also used as cetane improvers. It has been reported that micro-emulsion can results in the reduction of viscosity, increase in cetane number and good spray characters in the biodiesel. However, continuous use of micro-emulsified diesel causes problems in engine such as formation of carbon deposits, injector needle sticking and incomplete combustion.

### 3.4 Transesterification (alcoholysis)

Transesterification is a process that involves the reaction of triglycerides such as vegetable oil, with alcohol in the presence of a catalyst to produce 3 moles of fatty acid esters and one mole of glycerol [17]. Catalyst is used to increase the rate and yield of the reaction. The reaction is reversible. Excess alcohol is used to shift the equilibrium to the product side. Suitable alcohols such as methanol, ethanol, propanol, butanol and amyl alcohol are used for the transesterification reaction. Among these methanol and ethanol are most frequently used because of their low cost and physical and chemical advantages (polar and shortest chain alcohol). The fatty acid methyl ester (FAME) obtained by this process can be used as an alternative fuel for diesel engines [18]. The catalyst used for transesterification may be acid or base (homogeneous or heterogeneous) and lipase enzymes. Transesterification reaction depends on various factors such as catalyst concentration, nature of the feedstock, molar ratio of alcohol-oil, agitation

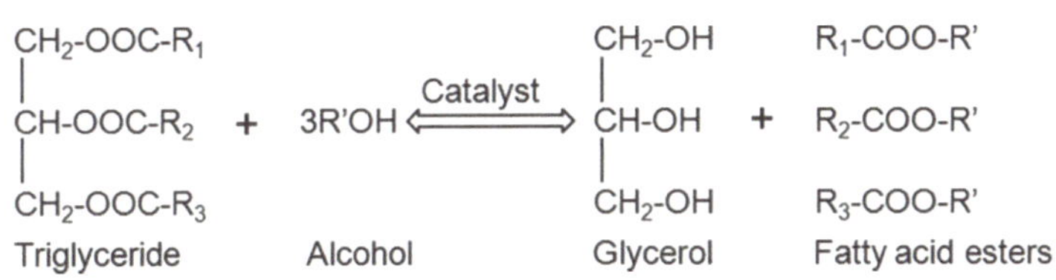


**Figure 4.**
*Transesterification reaction [20].*

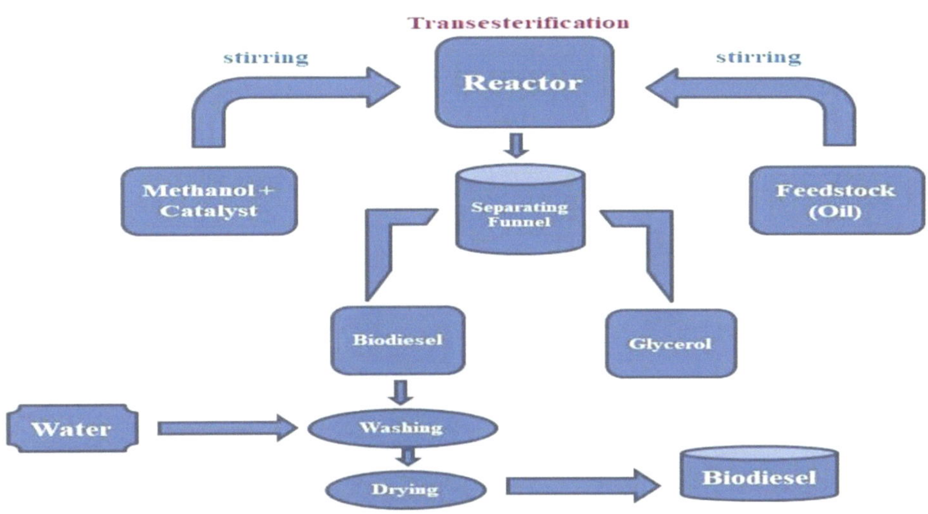


**Figure 5.**
*Schematic representation of transesterification.*

rate, temperature, reaction time, amount of free fatty acids and moisture content [19]. Transesterification is a reversible reaction and proceeds by mixing the reactants under heat. In this process, 1 mole of triglyceride react with 3 moles of alcohol gives 3 mole of fatty acid alkyl ester and 1 mole of glycerol in a sequence of three reversible reactions where the triglyceride are converted to diglycerides and then to monoglycerides as shown in **Figure 4**. From each step, one molecule of alkyl ester is produced (**Figure 5**).

## 4. Catalysts used for biodiesel production

The catalysts used in the transesterification reaction, are extremely important to the group. The presence of a catalyst speeds up the reaction, increasing the yield of the final product. These catalysts are classified into two major categories: homogeneous catalysts and heterogeneous catalysts, each of which can further be divided into subgroups. The classification is shown in **Figure 6**.

### 4.1 Homogeneous transesterification

#### *4.1.1 Homogeneous base-catalyzed transesterification*

The base catalysts used for the process of Transesterification include KOH, NaOH, carbonates and corresponding potassium and sodium alkoxides such as sodium ethoxide, sodium methoxide, sodium butoxide and sodium propoxide. The alkaline catalyzed Transesterification reactions are 4000 times faster than acid catalyzed Transesterification reactions. As compared to acidic catalyst, the base catalyst are less corrosive to industrial equipments, hence alkaline catalysts are mostly employed in commercial. However, the base catalysedTransesterification reaction is affected significantly by the presence of free fatty acid (FFA) and moisture content in the feed-stock. Therefore, the glycerides and alcohol used for Transesterification must be substantially anhydrous. It has been recommended that the FFA contents should be less than 2%, whereas the moisture content below 0.5 wt%. As the value of FFA is inversely proportional to the conversion efficacy, therefore small amount of water and high FFA contents present in animal fats and vegetable oils results in the deactivation of the catalyst and cause saponification (soap formation), which consequently decrease

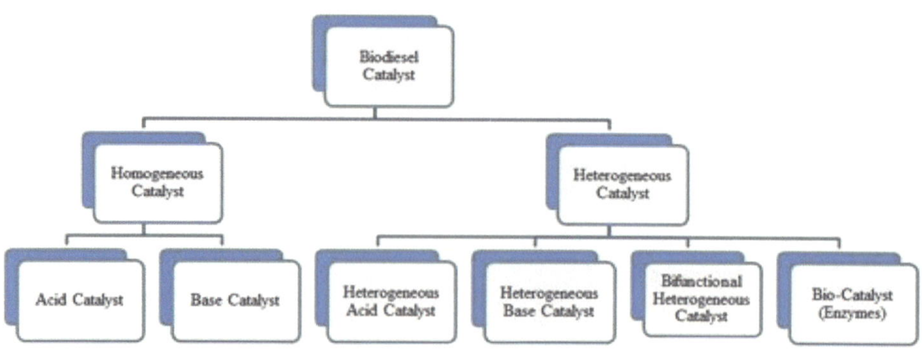


**Figure 6.**
*Catalysts used for biodiesel production.*

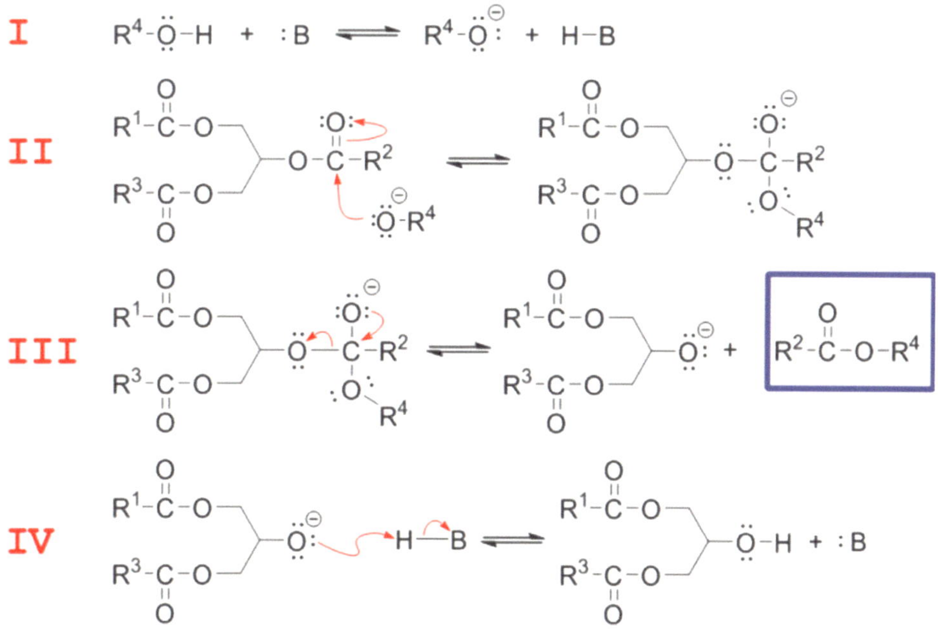

**Figure 7.**
*Mechanism for base-catalyzed transesterification [22].*

the biodiesel yield and renders the separation of glycerol and ester [21]. So, low free fatty acid content in triglycerides is required for base catalyzed Transesterification. Homogeneous acid catalyst is then referred for Transesterification.

#### *4.1.2 Mechanism*

Generally, the mechanism of base-catalyzed Transesterification of animal fats or vegetable oils involves four steps [13, 21]. In the first step, the base react with the alcohol gives an alkoxide and protonated catalyst. In the second step, nucleophilic attack of the alkoxide at the carbonyl group of the triglycerides and generates a tetrahedral intermediate. In the third step, alkyl ester and corresponding anion of diglyceride is produced. The final step involves the deprotonation of the catalyst to regenerate the active species that is able to start another catalytic cycle by reacting with the second molecule of the alcohol. Same mechanism is followed by the diglycerides and monoglycerides to convert to a mixture of alkyl esters and glycerol. The mechanism is summarized in the **Figure** 7.

### 4.2 Homogeneous acid-catalyzed transesterification

Mineral acids such as $H_2SO_4$, HCl and $H_3PO_4$are widely used for the acid catalyzed transesterification reaction. Acid catalysts are recommended for the oils that have higher free fatty acid contents such as waste oil or palm oil [23]. Such types of oils are first treated with acid catalyst (esterification) before the basic transesterification in order to convert the free fatty acids to esters. In this case, the FFA is esterified until the free fatty acid content becomes lower than 0.5% [24] In acid catalysis the oil is treated with acid catalyst and gives biodiesel and water but the water must be

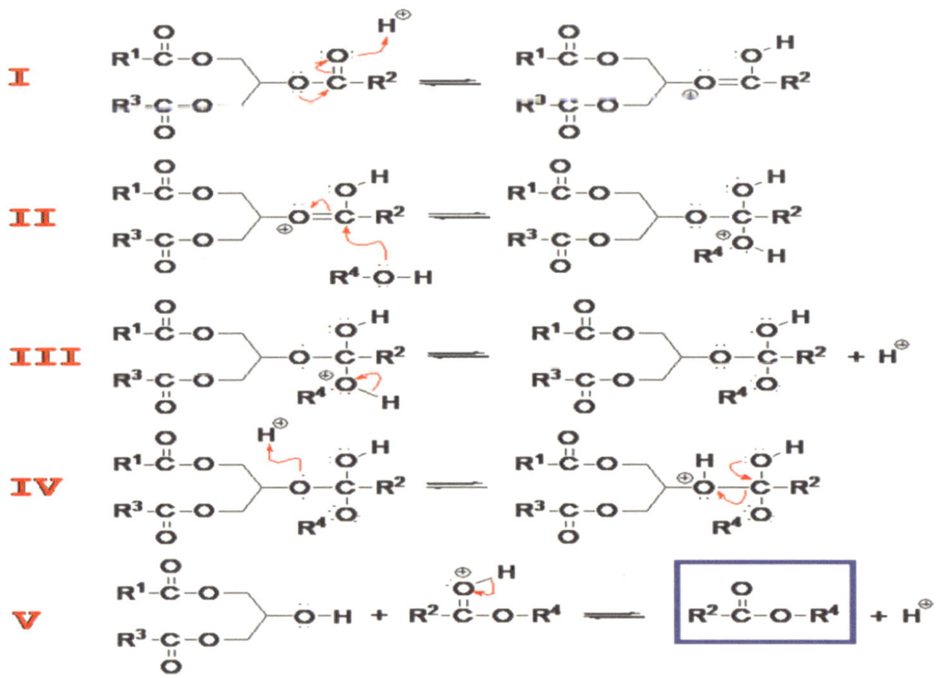


**Figure 8.**
*Mechanism for acid-catalyzed transesterification [25].*

removed immediately because it will results in the soap] formation in base catalyzed transesterification.

*4.2.1 Mechanism*

In the acid catalyzed transesterification, the protonation of carbonyl group of the ester results in the formation of carbocation, which after a nucleophilic attack of the alcohol produces a tetrahedral intermediate. This intermediate then eliminates the glycerol to form a new ester and to regenerate the catalyst. This mechanism is related to a monoglyceride. However, this reaction can be extended to di- and triglycerides (**Figure 8**).

### 4.3 Bio-catalyst (enzyme) catalyzed Transesterification

In enzyme catalyzed Transesterification, the reaction is catalyzed by various lipases such as candida rugasa, candida Antarctica, immobilized lipase (lipozyme RMIM) pseudomonas cepacia, pseudomonas spp. Or rhizomucarmiehei. The yield of biodiesel greatly depends on the type of enzyme used [23]. 60% biodiesel yield was achieved from transesterification of soyabean oil using commercially avalaible-imobalized lipase (Lipozyme RMIM) [26, 27]. More importantly sufficient time is required for the enzyme catalyzed Transesterification as compared to base catalyzed Transesterification. However, the various parameters such as pH, temperature, solvent, type of micro-organism that generate enzyme etcmust be optimized to achieve the industrial goals. This process is highly selective, more efficient, produces less side products or waste i.e., environmentally favorable and involves less consumption of energy because reaction can be carried out in mild conditions [28].

Arumugam et al. [29] used the sardine oil (byproduct of fish industry) as a low cost feedstock for the production of biodiesel. The FFA content of the oil was high (32mgKOH/G of oil) and the lipase enzyme immobilized on activated carbon was used for the Transesterification. Various reaction conditions were optimized such as methano/oil ratio 9:1, water content 10 v/v% and temperature 30°C. Reusability of the catalyst was studies for 5 cycles and 13% drop in FAME yield occurred.

### 4.4 Heterogeneous Transesterification

In heterogeneous catalysis, the phase of the catalyst is different from the phase of the reactants. Heterogeneous catalysts are very important in various fields such as industrial bulk chemical production, synthesis of selective chiral molecueles and energy [30]. Various process problems associated with homogeneous Transesterification, such as regeneration or separation of the catalyst, soap formation, disposal of byproducts, treatment of waste effluents and corrosion in case of acid catalyst have been solved by the use of heterogeneous Transesterification. Heterogeneous catalysts they are easily recovered at the end of the reaction by decantation or filteration, reusablility, show potential activity, selectivity, longer catalyst lifetimes and cost effective green process [31]. Interestingly heterogeneous catalysts could be used in certain harsh conditions such as high temperature and pressure. Heterogeneous catalysts may be solid base catalyst or solid acid catalyst.

Heterogeneous catalysts can be designed to bring out entrapment and grafting of the active molecules on the surface or inside the pores of the solid support such as alumina, silica or ceria. Mixed metal oxides [32], transition metal oxides [33], ion exchange resin [34], Alkali earth metal oxides [35] and alkali metal compounds supported on zeolite or alumina [36] have been used in different chemical reactions such as aldol condensation, isomerization, oxidation, Michael condensation, Knoevenagel condensation, and transesterification [37].

#### *4.4.1 Heterogeneous base catalyst*

Heterogeneous base catalysts are used to overcome the constraints such as saponification that hinders the glycerol separation from the layer of methyl ester associated with the homogeneous base catalysts. These catalysts show superior catalytic activities under mild conditions and are non-corrosive, environmentally friendly, have less disposal problems and easily separated from the reaction mixture [38, 39]. Moreover, the properties of these catalysts can be tuned accordingly to enhance activity, selectivity and longer catalyst lifetime. Various metal-based oxides such as alkali metal, alkaline earth metals and transition metal oxides can be used as a base catalyst for the biodiesel production from oils by trans-esterification process. The structure of metal oxides consists of cations (positive metal ions) that possess Lewis acid characteristics and anions (negative oxygen ions) that possess Brønstedbase characteristics. The combination of Lewis acid and Bronsted base characteristics make them potential catalyst for transesterification reaction.

##### *4.4.1.1 Alkaline earth and alkali metal-based catalyst*

Alkaline earth metal oxides such as CaO, MgO, BaO, BeO and SrOhave successfully been used as a catalysts for biodiesel production by many researchers.

Calcium oxide is favored ecofriendly material that haslonger life time because it is cheap catalyst, moderate reaction conditions and high activity. Generally, calcium hydroxide and calcium nitrate are used as precursors for the CaO production. Recently, several calcium-rich waste materials such as mollusk shell and bones, chicken eggshells have been used for CaO synthesis to minimize the biodiesel production cost, problem of waste disposal.

Demirbas [40] described the supercritical conditions effect on the sunflower oil catalytic Transesterification in the presence of 3 wt% of CaO with 60–120 mesh size, 40: 1 of alcohol/oil molar ratio, at pressure of 24 MPa and 252°C The author reported 98.9% yield of methyl ester in reaction time of 26 min.

#### *4.4.1.2 Mixed metal-based catalyst*

Mixed metal oxides consist of two or more type of metal cations. Oxides may be binary, ternary and quaternary and so on with respect to the presence of the number of different metal cations [41]. Mixed metal-based oxides are mainly used as basic catalyst depending on the mixture of the catalyst. More importantly, the basicity of these catalysts can be tuned by changing their chemical composition and procedure for synthesis. Similarly, activation energy, type of synthesis method and structure of the catalyst have a strong impact on the final basicity of the mixed metal oxides.

It has been reported that, calcining MgO with $ZrO_2$ gives a bimetallic oxide MgO-$ZrO_2$having high basicity character and is almost unaffected by dissolution. Similarly, MnO, CuO and CuO supported on $Al_2O_3$ have been investigated in transesterification reaction at room temperature, yielded upto 97%. $Al_2O_3$-ZnO mixed oxide and rare earth oxides were studied but require high temperature for biodiesel production from vegetable oils. Calcium bimetallic oxides such as $CaCeO_3$, $CaZrO_3$, $CaMnO_3$, $CaTiO_3$ and $Ca_2Fe_2O_5$ have also been investigated for the transesterification at 60°C, which displayed good activity and reusability [42, 43].

Xie et al. [44] used the Zinc aluminate catalyst ($ZnAl_2O_4$) in a batch processing for the biodiesel production from waste cooking oil. More than 95% ester yield was obtained at temperature greater than 150 C, alcohol to oil molar ratio 40:1, stirrer speed of 700 rpm, reaction time of 2 h and varying the catalyst amount in the range of 1–10 wt%. The catalyst was reused for 3 cycles and the yield reduced after the 3 run. The authors reported that the decrease may be due to the carbon deposition on the surface catalyst or loss of tiny particles of the catalyst during the process of recovery.

Basic catalyst may have several problems during the process of transesterification because they are sensitive to free fatty acid content. If the free fatty acid content is higher than 2 wt %, soap formation occurs resulting in decrease in the yield of biodiesel. The downstream purification process raises problems such as producing a large amount of wastewater [45].

### *4.4.2 Heterogeneous acid catalysts*

#### *4.4.2.1 Metal oxides/mixed metal oxides*

Metal oxides such as FeTiO, ZrFeO, ZrFeTiO and Cesium-doped heteropolyacid have been used successfully as solid acid catalysts for the Transesterification of oil using ethanol and methanol as a solvent. Acid catalysts are insensitive to water content and free fatty acid (FFAs) present in the feedstock and is a are preferable method for cheaper feedstock [45].

Alhassan et al. [46] developed Ferric-manganese-based solid catalyst by impregnating the support material of sulfated zirconia with $Fe_2O_3$-MnO. The catalyst wascalcined for 3 h at 600°C. The synthesized catalyst was then used for the waste cooking oil Transesterification. The author found 96.5% yield of biodiesel under optimum reaction conditions of oil to alcohol molar ratio of 1:20, at 180°C temperature and catalyst loading of 3 wt%. The yield of the catalyst remained the same (96.5%) for 6 runs but decreased upto 87% upon the seven run. They reported that the decrease may be due to blockage of the energetic centers as a result of the accumulation of triglycerides in the pores of the catalyst.

#### *4.4.2.2 Heteropoly acid derivatives*

Heteropolyacids and their salts are also used as solid acid catalysts for the biodiesel production. HPAs withKeggin structure can be prepared very easily as compared to other HPAs. They possess high thermal stability and are preferably used for production of biodiesel from different feedstocks. Keggin-type HPA has a low specific surface area, which can be overcome using appropriate supportive material. Similarly, HPAs supported on the carriers are used in biodiesel production because of their structural mobility and superacidity.

Sakthivel et al. [47] used the tungstophosphoric acid (HPW) and MCM-48-supported HPW catalysts for the esterification of long chain fatty acids and alcohol in supercritical $CO_2$ (sc-$CO_2$) medium. High yield was obtained in the supercritical $CO_2$ medium due to the rapid diffusion of reactants and products in the MCM-48 channels and high contact of the reactants with the catalyst.

Acidic catalyst may have several problems such as very slow reaction rate, corrosive to reactors and pipelines. Normally, high reaction temperature, high oil to methanol molar ratio and long reaction time are required [45].

### *4.4.3 Heterogeneous bifunctional (acid: base) catalysts*

As the alkali catalyzed transesterification of the feedstock with higher FFA contents can produce low yield of biodiesel, because the FFA reacts with the alkali catalyst and produce the foam that results in separation and emulsification problems [48]. To solve this problem, a two steps catalytic process for the biodiesel production is recommended. In the first step, the free fatty acid contents of the feedstock are esterified using the acidic catalyst such as ferric sulfate or sulfuric acid. In the second step, biodiesel are produced by the transesterification using the basic catalyst such as CaO or ZnO. The problem of the catalyst removal in the first step can be avoided by neutralizing the acid catalyst by using the extra alkaline catalyst in the second step. But the use of extra catalyst can increase the overall cost of the biodiesel production. The residues of the acidic or alkaline catalyst in the products of biodiesel can cause the engine problems because the acidic catalyst can attack the metallic parts of the engine. On the other hand, basic catalyst can produce higher level of incombustible ash. Therefore, both the catalyst must be removed properly from the biodiesel to avoid the aforementioned problems [49, 50]. Further, it can be concluded that there is substantial room for the development of an efficient and effective catalyst for profitable biodiesel technology (**Figure 9**).

Recently, bifunctional heterogeneous catalysts has been introduced to solve the drawbacks adhere with the solid base/acid catalyst and develop more economical biodiesel technology. The bifunctional heterogeneous solid catalyst can be used as an alternative for the biodiesel production that can promote both esterification and Transesterification simultaneously [52].

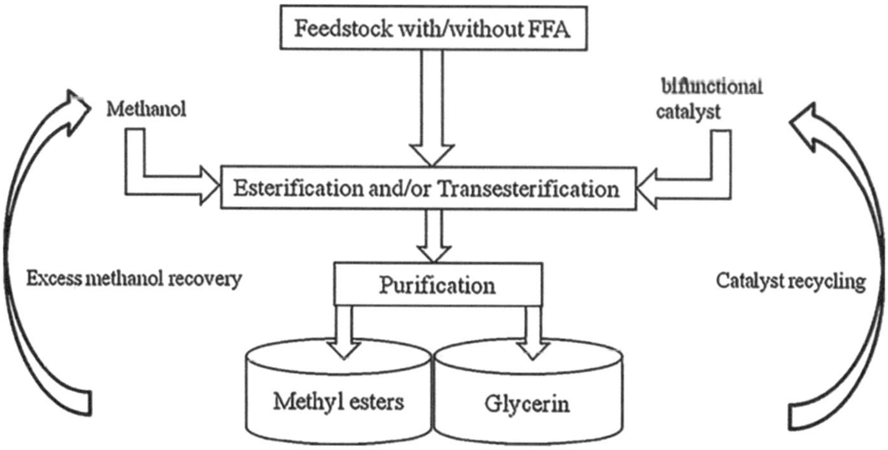


**Figure 9.**
*Schematic representation of operating principle of bifunctional catalyst [51].*

In recent years, bifunctional heterogeneous catalysts have been used widely for the production of industrial fine chemicals. The bifunctionality concept has been designed to drive complex reactions through the advance approach of combining two hostile functions, such as acid and base, with cooperative interactions between their active sites precisely positioned functional groups [53]. Therefore, bifunctional heterogeneous catalyst can perform simultaneous esterification and transeseterification of free fatty acids and triglycerides respectively without being affected by the water content present or produced during the formation of biodiesel [54].

*4.4.3.1 Mechanism*

Generally, heterogeneous reactions involve three steps such as adsorption, surface reaction and desorption [55]. In the first step, carbonyl group of free fatty acids (FFA) adsorbs on acid sites while methanol adsorb on the basic site of the catalyst to produce carbocation and oxygen anion for esterification and transesterification respectively. In the second step, at the surface of the catalyst, nucleophilic attacked carbocation and oxygen anion at each methanol hydroxyl group and triglyceride carbonyl group for esterification and transesterification reactions, respectively. The nucleophilic attack would generate tetrahedral intermediate. Finally, the product (FAME) is formed from desorption of hydroxyl group and alkyl triglycerides from catalyst surface after breaking the -OH and -C-O- bond respectively, while the deprotonated catalyst regenerated the active species for starting another catalytic cycle. Glycerol, $H_2O$, are produced as by-product during esterification and transesterification reactions (**Figure 10**).

*4.4.3.2 Transition metal-based catalysts*

Transition metals such as Ni, Fe and Co based compounds have been extensively investigated as bifunctional heterogeneous catalyst for biodiesel production. The TiO and MnO have shown good catalytic activity for biodiesel production. These catalysts have been used for the simultaneous esterification of FFAs and transesterification of triglycerides under continuous flow conditions by using low grade feedstock with high fatty acids contents ofupto 15%.

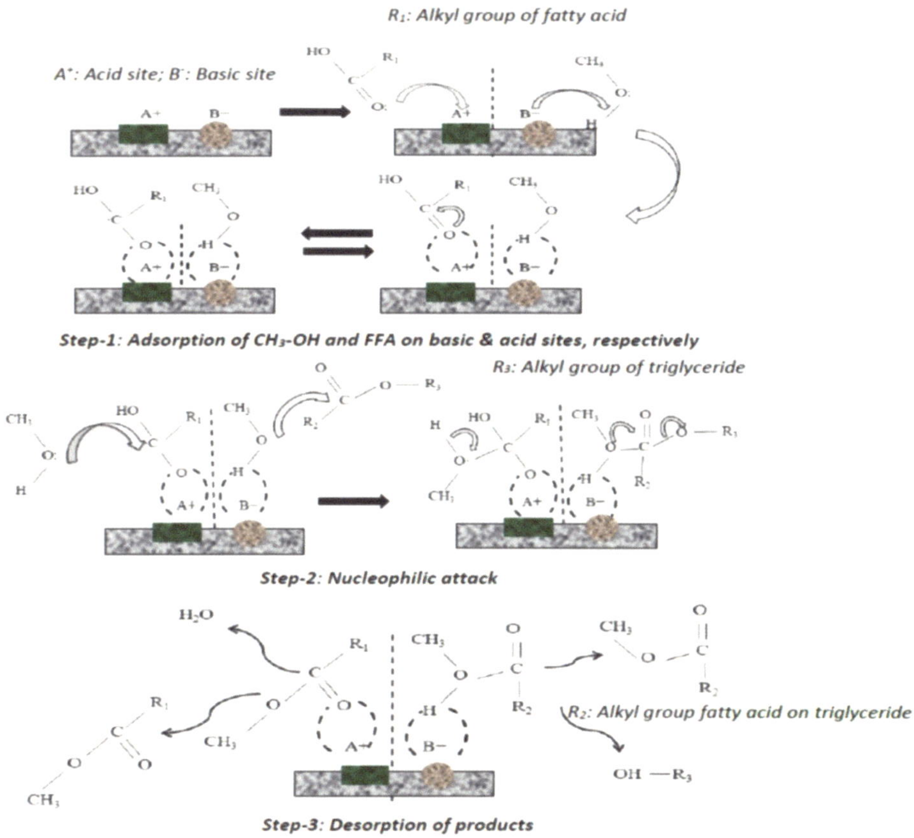


**Figure 10.**
*Mechanism for esterification and transesterification reactions on a bifunctional heterogeneous catalyst [56].*

Cannilla et al. [57] used a novel MnCeOx system for the transesterification of refined sunflower with the methanol. The performance of such catalyst was compared with that of common acid supported catalyst. The results showed that MnCeOx system have a superior activity especially by operating at low temperature i.e., ≤120°C. The catalytic performance was the result of synergic role played by the presence of both base/acid character and textural porosity.

#### *4.4.3.3 Mixed metal oxides*

Mixed metal oxides have shown potential applicationsin terms of their catalytic activity in various reactions due to their increased active acidic or basic sites and large surface area. As a result of these characteristic, the mixed metal oxides can simultaneously catalyze the esterification and transesterification and increases the yield of reaction under mild reaction condition [32].

Many researchers have investigated the catalytic activity of mixed metal oxide for biodiesel production. Furata et al. [58] prepared the $Al_2O_3/ZrO_2/WO_3$ solid catalyst by co-precipitation method for biodiesel production from soybean oil. The catalyst was compatible for both esterification and transesterification at 250°C temperature and alcohol to oil molar ratio of 40:1, provided 90% methyl ester yield.

## 5. Feedstock used for the production of biodiesel

The feedstock is one of the key factor that plays vital role in the economics of the biodiesel technology. More than 350 oil-bearing crops have been identified as a potential feedstock for the production of biodiesel. The feedstock should fulfill two main requirements (i) large production scale (ii) low production cost [59]. The feedstock availability for the production of biodiesel depends upon the geographical location, local soil conditions, regional climate and agricultural practices of any country. The suitability of feedstock depends upon various factors such as oil yield per hectare, production cost, oil content of the seeds and relevant product properties of the oil. It has been found that, the cost of the feedstock is about 75% of overall production cost of biodicsel [60]. Therefore, selection of cheapest feedstock is a major problem and high relevant to the biodiesel industry. Biodiesel feedstocks are generally categorized into four classes as shown (**Figure 11**).

### 5.1 Vegetable oils

#### *5.1.1 Edible vegetable oils*

Resources of edible oil such as peanut [62], soybeans [63], sunflower [64], rapeseed [65], safflower, coconut and palm oil are extensively utilized for biodiesel production and are classified as first generation biofuels because these were the first crops used for production of biodiesel [66]. Many countries of the World such as USA, Malaysia and Germany, have well off plantations of these vegetable oils. Currently, more than 95% of the world biodiesels are produced from the edible oils where rapeseed oil contributes 84%, sunflower 13%, 1% palm oil, 2% soybean and others. However, economic and social problems such as food versus fuel crisis and various environmental issues (such as destruction of vital soil resources), usage and deforestation of the available arable land are adhere with use of edible oils.

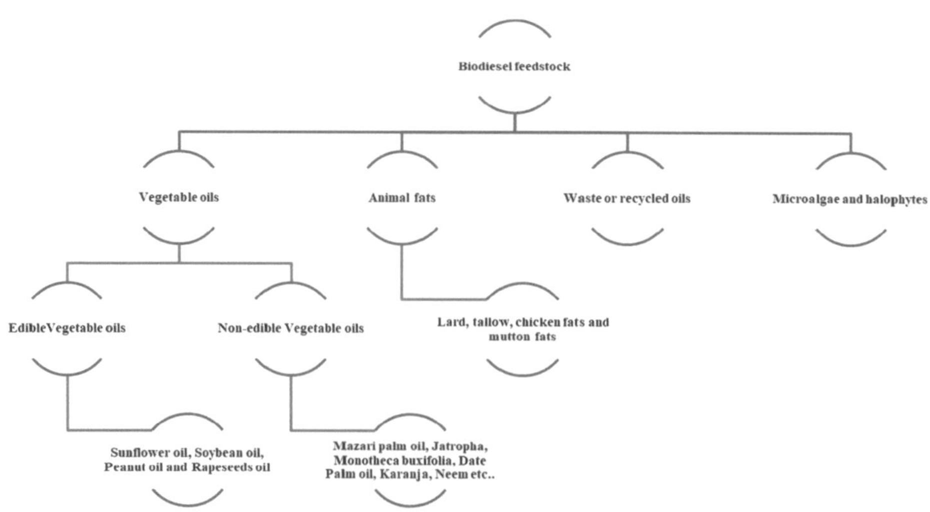


**Figure 11.**
*Feedstocks used for biodiesel production [61].*

*DOI: http://dx.doi.org/10.5772/intechopen.101967*

### *5.1.2 Non-edible vegetable oils*

Due to the presence of some toxic components in the non-edible vegetable oils, they are not suitable to be used for human food. The use of non-edible vegetable oil for the production of biodiesel would pave the ways to overcome the economic, social and environmental problems and tackle the energy crises worldwide [60]. Non-edible vegetable crops are grown on the lands that are largely unproductive, located in poverty-stricken areas and in degraded forests. These plants can also be planted on fallow lands, cultivator's field boundaries and in public land such as roads, railways and irrigation canals. Plants of non-edible feedstocks are well adapted to arid, semi-arid conditions require low moisture and fertility. Moreover, these plants can grow and propagated through cutting or seeds [67]. As these plants oilsdo not compete with food therefore the seed cake may be used as fertilizers for soil enrichment. Therefore, from economic and social prospective, edible oils must be replaced by some suitable feedstock for biodiesel production. Hence, non-edible feedstocks for biodiesel production could be considered as sustainable and alternative fuels.

#### *5.1.2.1 Mazari palm (*Nannorrhops ritchiana*)*

Mazari is the local name for dwarf palm (*Nannorrhops ritchiana*), belongs to the family of Arecaceae. It is a small gregarious, shrubby and tufted palm with blue-green to gray-green fan-like leaves having several stems growing slowly and connected to form a single base. It is one of the most versatile palms that can survive in intense winds, blazing heat and snowy cold with almost water-free environment. It is native to southwestern Asia, from Southeast of the Arabian Peninsula to east through Iran and Afghanistan to Pakistan. In Pakistan, it is mostly found on either side of Suleiman range in sandy soil depressions with the height ranging from 600 to 1100 m. In Khyber Pakhtunkhwa it is found in Totakan, Jandia (Kalpani, District Mardan), Swat, Kohat, Anbar, Bannu, Kurram Agency, North and south Wazirizstan, Orakzai, DI Khan, in Punjab, KotAddu, Qasoor, Gujrat, in Balochistan Musa Khail, Loralai, Khuzdar, Harnai, etc (**Figure 12**).

Mazarifibres are widely used for making ornamental products, ropes, mates, banns, different commodities for mosques, trays, baskets, grain bins, brooms, cupboards, hand fans and decoration pieces etc. (shown in **Figure 13**) [68, 69].

**Figure 12.**
*Mazari palm seeds.*

**Figure 13.**
*Different products of mazari palm.*

Fresh and dried leaves both are used for making products. Raw mazari production in the Pakistan is about 37,315 tons. Baluchistan is the biggest producer of the mazari with an average annual production of 27,265 tons [70]. In 1991, the total exports of the products prepared from mazari by rural people were 126 milion rupees. Main buyer of these products are the local people because most of the products are used for domestic purposes and also these fascinating products attract both domestics and international tourists. **Figure 14** shows the main buyer of the products.

The fruits of *Nannorrhops ritchiana* are edible. Young leaves of mazari palm are sweet in taste and used as laxative in livestock. It can be used for the treatment of diarrhea and dysentery as well as gastrointestinal diseases [72, 73]. Cytotoxic and

**Figure 14.**
*Main buyer of mazari palm products [71].*

antifungal activities have been evaluated from the crude extracts of these plants. Fruits of mazari are orange to brown while seeds are brown in color. Recently, the seeds has been used for the extraction of oil that can be used for the biodiesel production by the process of Transesterification. The mazari palm seeds contain average oil content of 15%. Based on the data, mazari palm oil could be one of the potential feedstock for biodiesel in Pakistan to overcome the energy crisis and minimize the energy gap.

*5.1.2.1.1 Jatropha*

Jatropha curcas is the bionomical name of Jatropha, belongs to spurge family. It is commonly known as Barbados, Purging or Physic nut. The height of Jatropha plant is about 6 m and is a flowering plant. The plant matures in 9–10 months and yield 2–3 times per year. On maturation, green rounded shaped seeds appeared on the plants and then turn into light blue or purple colored hard shells. The oil bearing mass located inside the shells known as meat or kernels. Oil content in the seeds varies from 20–60% by weight [74, 75]. J. curcas oil could be a valuable feedstock for the production of biodiesel in Pakistan (**Figure 15**).

Jatropha is a multipurpose drought resistant plant that is widely distributed in the wild or semi-cultivated areas in South East Asia, Pakistan, India and Central and South America. It is well adapted to arid and semi-arid conditions [76]. Jatropha is rich source of hydrocarbons. Therefore, it is considered as commercial source for biofuel production all over the world. Jatropha oil contains 42% oleic, 35% linoleic, 14% palmatic and 6% stearic acid by composition [77].

In Pakistan, certain institutions are promoting Jatropha cultivation at the nursery level in various locations across Baluchistan, Punjab, and Sindh. In nurseries, these cultivated plants ranged in age from a few weeks to 18 months [78]. However, after three years of private sector efforts in2008, oil bearing crop cultivation increased from 2 acres to over 400 acres. PSO (Pakistan State Oil) took a step in this direction in 2008, planting 20,000 saplings in farms. They've recently increased the number of samples taken for each transplantation, up to 20,000 or more. PSO's initiatives aimed to plant more than 6 million trees produce 24 million kg of oil bearing seeds, and produce 7.2 million L of biodiesel worth 345 million PKR at a unit price of PKR 48 $L^{-1}$ [79].

Other interested parties, such as the Karachi Forest Department and the Pakistan Army, have also successfully planted Jatropha plants in Sindh [80]. So far, the Forest Department has been successful in cultivating 3000 samples on a trial basis in

**Figure 15.**
*Jatropha curcas [68].*

Malir Cantonment in 2010 for the cultivation of Jatropha seeds supplied by PS [81]. Similarly, the Pakistan Agricultural Research Council (PARC) and KijaniEnergy, a Canadian company, are interested in establishing large scale Jatropha cultivation for the production of biodiesel on marginal lands [79]. Kijani Energy invested approximately US$ 150 million in2009, resulting in the use of 200,000 acres of land for Jatropha cultivation in Umerkot, Khairpur, Tharparker, Cholistan, and Sanghar.

#### *5.1.2.1.2* Monotheca buxifolia *(gurguray)*

*Monotheca buxifolia* (gurguray) is a member of genus Monotheca belongs to Sapotaceae and is evergreen plant found in hilly areas of Northern Pakistan particularly in the Dir District, District Kurram (Ex-FATA) and Karak. It is also distributed in mountains of Afghanistan, South-east Saudi Arabia and Northern Oman. The plant bears small fruits locally called Gurgura that can be used as fresh as well as dried [82, 83]. *Monotheca buxifolia* is mainly used for fodder, small timber, fuel, roof thatching materials and fence around cultivated fields due to its thorny nature. The fruits of *Monotheca buxifolia* is hematanic, purgative, laxative, antipyretic, vermicidal, referegerant, therefore used for the treatment of gastro-urinary disorders. The gurgura leaves contain terpenoids, anthraquinones, reducing sugars, cardiac glycosides, flavonoids, saponins, tannins and poly-phenolic compounds [84, 85]. The seeds contain about 20% of the oil. So these seeds can be used for the biodiesel production by Transesterification method. Therefore could be a potential feedstock to subside edible oils for fuel (**Figure 16**).

#### *5.1.2.1.3 Date or date palm (*Phoenix Dactylifera *L.)*

Date or date palm is a flowering plant species belongs to the palm family Arecaceae cultivated for its edible sweet fruit. It is a dioecious having separate male and female plants. It is a source of human nutrition rich with dietary fibers, carbohydrates, lipids, proteins, some vitamins and mineral matter [86]. For millennia, the date palm tree has been cultivated in the Middle East and North Africa, and it is thought to be the world's oldest domesticated fruit tree. Because of the variety of resources it provides, it has traditionally been the most valuable fruit crop in harsh arid or desert

**Figure 16.**
Monotheca buxifolia *(gurguray)*.

environments where water scarcity and extreme temperatures are common. Date palm trees are now grown in semi-arid climates and other parts of the world, including southern Europe, Australia and America. There are now over 100 million date palm trees in the world with around 2000 cultivars [87, 88]. A palm tree produces 500 kg of fresh dates per year on average, with production beginning at 5 years and lasting up to 60 years. Date production and consumption have increased rapidly, from 1.88 million t in 1965 to 3.43 million tons in 1990 and 8.46 million tons in 2016, with Middle Eastern and African countries dominating production [89]. It's a pitted fruit with a seed in the centre surrounded by a fleshy pericarp as shown in **Figure 17**.

The date seeds are very hard ranging from 5 to 15 mm in length with oblong shape with a ventral groove. The weight is about 11–18% of the total fruit mass and contain 4–13% of oil. Based on these digits, an estimated 1.3 million tons of date seeds and 127,000metric tons of date seed oil (similar amount of biodiesel) could be annually produced. In 2015, the total annual production of biodiesel was 38,700 tons in the Middle East and Africa [90, 91]. Date production in the world reached 9.07 million metric tons in 2019, up from 8.4 million metric tons in 2017. Similarly, date palm is widely distributed in different areas of Baluchistan, Sindh, KPK and Punjab. It has been

**Figure 17.**
*Date fruit and seeds [87].*

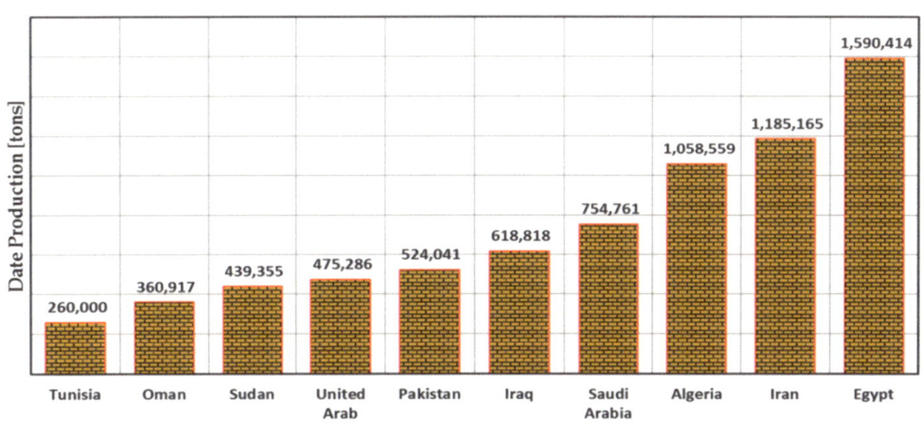


**Figure 18.**
*Top 10 global date-producing countries [87].*

reported that the annual production of date seed is around 600,000 metric ton per year in Pakistan [92]. These seeds are used as feed for animals in some areas. However, most of these seeds degrade without any proper utilization. Therefore, the use of date seeds as biodiesel feedstock could be a promising to concern energy solution (**Figure 18**).

*5.1.2.1.4 Karanja (*Pongamia Pinnata*)*

Karanja (*Pongamia Pinnata*) is a medium sized evergreen tree, naturally found in India, Pakistan, Bangladesh, Malaysia, Vietnam, Sri Lanka, Thailand, Florida, Australia, South East Asia. Karanja has been successfully introduced to humid tropical regions of the world and parts of China, Australia, USA and New Zealand. Karanja tree is similar to the neem tree and is highly tolerant to salinity which can grow in different soil textures such as sandy, stony and clay [76]. Historically, in India and neighboring countries, karanja has been used as a source of traditional medicines, green manure, animal fodder, water-paint binder, timber, fish poison, pesticides and fuel. Recently, it has been identifiedas a viable source of oil for biofuel industries. About 9 kg-90 kg seeds pods are obtained from one tree, yielding upto 40% oil, per seed and about 8 kg–24 kg kernel is obtained from one tree and yields 30–40% oil [93]. The yield of kernels per tree has been reported to be about 8 to 24 kg with composition: 27.5% oil, 19% moisture, and 17.4% protein. The oil of karanja is reddish brown and rich in oleic acid and unsaponifiable matter (**Figure 19**) [94].

Many researchers have utilized karanja oil as feedstock for biodiesel production. It has been reported that the biodiesel obtained from karanja shows excellent properties such as low acid value, lower viscosity and higher flash point. Naik et al. [95] followed two steps process for the production of biodiesel from karanja oil with 20% free fatty acid. First, acid-catalyzed esterfication was applied using 0.5% (w/w) $H_2SO_4$, 6:1 methanol to oil ratio at 65°C. The acid treated oil was later transesterifiued with KOH using 1% (w/w) potassium hydroxide, 6:1 methanol to oil ratio to lower the FFA content. The yield of biodiesel obtained by dual step process from karanja oil was 96.6–97% at 65°C.

*5.1.2.1.5 Neem (*Azadirachta indica*)*

Neem (*Azadirachta indica)* tree belongs toMeliaceae family and is a multipurpose evergreen tree which can be grown in almost all kind of soils such as saline, clay,

**Figure 19.**
*Karanja (*Pongamia Pinnata*) [93].*

**Figure 20.**
*Neem (*Azadirachta indica*).*

alkaline, shallow soils, stony, dry, and even on solid having high calcareous soil. Neem is native to Pakistan, India, Srilanka, Malaysia, Burma, Japan, Indonesia and tropical regions of Australia. It can survive in arid, semi-arid climate with maximum temperature of 49°C and rainfall as low as 250 mm. The seeds of the neem contain 20–50 wt% oil of green to brown colored. In Pakistan neem plant is widely spread in Khyber Pakhtunkhwa, Punjab and Sindh. These trees were planted as part of a project initiated by the Sindh Government in 2008, in response to the Ex-President of Pakistan Asif Ali Zardari's directives to encourage the planting of Neem trees in the province. The Sindh government has set aside Rs.7 billion to plant 10,000 Neem trees on both sides of the National Highway and the Superhighway [96].

Muthu et al. [97] produced the neem methyl ester from the neem oil in the presence of catalysts by two steps process of esterfication and Transesterification. Sulfated Zirconia was used as solid acid catalyst for esterfication, while alkali catalyst i.e., KOH was used for Transesterification. Optimum conversion of free fatty acid was achieved with 1 wt% of sulfated zirconia (acid) catalyst, at 65°C temperature, 9:1 methanol/oil ratio and 2 h reaction time. The acid value of the raw oil was reduced by 94% (21.76 mg KOH/g) which show the successful conversion. The authors noted that when the pretreated oil was transesterified in the presence of KOH, 95% conversion efficiency was achieved (**Figure 20**).

## 5.2 Microalgae for biodiesel production

Microalgae are eukaryotic or prokaryotic photosynthetic micro-organism that can grow rapidly and live in harsh conditions due to their unicellular or simple multicellular structure [98]. Examples of eukaryotic micro-organisms are green algae i.e., chlorophytaand diatoms i.e., bacillariophyta and prokaryotic micro-organisms are cyanobacteria. Microalgae are present in all existing ecosystem of the earth, not only in aquatic but also terrestrial ecosystem that lives in a wide range of environmental conditions [99]. Interestingly, it is observed in small ponds and ditches in the villages and towns become fully green within a week during the rainy season in Pakistan. Although in Pakistan, the cultivation of oleaginous microalgae is in its infancy, however several species of algae are reported in the literature that can further process or cultivated for the production of oil [100]. Microalgae can provide feedstock for several types of renewable fuels such as methane, biodiesel, ethanol and hydrogen. Biodiesel produced from algae contains no sulfur, reduce emissions of particulate matter, hydrocarbons, CO and SOx. However, NOx emissions may be higher in some types of engine.

Furthermore, a Pakistani researcher at Japan's Mie University claims that the country could benefit from using its 27–28 million acre saline lands for algal farming, which would create jobs and benefit the rural community [101]. Four algae strains suitable for cultivation in Pakistan's deserts have been identified by other researchers. Other researchers have identified four strains of algae that are suitable for cultivation in Pakistan's deserts and produce acceptable lipid yields, i.e. 40% by weight *Haematococcus pluvialis, Microcoleus vaginatus, Chlamydomona sperigranulata*, *Synechocystis* [102].

To produce biodiesel, researchers at the National University of Sciences and Technology (NUST) cultivated Chlorella vulgaris in a closed photo-bioreactor (20 L) in a controlled environment and characterized its properties. At 5000 and 9000 psi and 50 and 80°C, the highest biodiesel yield (more than 99%) was achieved. The biodiesel produced was found to be of ASTM D6751 quality [102].

## 5.3 Waste cooking oil as feedstock of biodiesel

The term waste cooking oil (WCO) refers to vegetable oil that has been used in production of food and no longer viable for its intended use. Sources of waste cooking oil are domestic, industrial and commercial products [103]. Waste cooking oils are problematic waste streams that need to manage properly because if WCO is disposed improperly, down streams of the kitchen, the oil solidifies and cause blockages of sewer pipes [98, 104]. Degraded waste cooking oil gets into sewage system and causes corrosion to metal and concrete elements [105]. Thus, the waste cooking oil can be used as an effective feedstock for the biodiesel production via Transesterification [99].

In Pakistan, waste cooking oil sources include hotel chains, confectioneries, restaurants and domestic cooking. Pakistan is basically an agricultural country and has diverse ecological conditions, so the people mainly depend upon the agricultural products. Plants and crops that yield edible oils for cooking purposes are cultivated on extensive scale in the country. These oils are used in local shops, hotels, huts and every home of Pakistan [80]. Pakistani people use meat of cows, buffaloes, camels, goats, poultry on large scale and use fats for cooking purposes. These all are the major sources for collection of waste cooking oil.

## 5.4 Waste animal fats

Animal fats and vegetable oils are of two types of biological lipid materials that are made up of mainly triacylglycerides (TAGs) and less diacyglycerides DAG and monoacylglycerides (MAGs) [106]. Fats and oil have similar physical properties and chemical structures such as hydrophobicity, water-insolubility and solubility in nonpolar organic solvents. However, the high fatty acids content in fats and their different distributions make it different from oil. Oils are generally liquid at room temperature while fats and greases are solids due to their high content of saturated fatty acids (SFA). Different waste animal fats such as tallow (mutton tallow from sheep and beef tallow from domestic cattle), pork lard (rendered pork fat), chicken fats and grease. Since, many animal meat processing facilities, rendering companies of collecting and processing of animal mortalities, large food service and processing facilities create large amount of waste animal fats (WAFs), that will be a great opportunity to produce biodiesel from these very cheap raw material [107]. The use of these waste animal fats as a feedstock for biodiesel production will eliminate the need of their disposal.

## 6. Pakistan's energy scenario

Pakistan is the world's sixth largest country in terms of population, (213 million) and an annual growth rate of 2%.A significant portion (63%) of this population lives in rural areas, while 37 percent live in urban areas [108]. The recent economic growth and an ever-increasing population, has resulted in an increase in energy consumption. The country still depends on conventional resources of fossil oil.

Various initiatives to promote renewable energy in Pakistan have been taken over the years, but their outcomes are still pending due to a lack of sound policy [109]. Recently, Alternative Energy Development Board (AEDB), was established in 2003 [78], in Pakistan to improve green technologies that can reduce greenhouse gas emissions and promote renewable technologies through a variety of projects that have been recognized on an international level by the International Solar Energy Society (ISES) and the World Wind Energy Association (WWEA) [110].

There is a significant gap between Pakistan's energy production and energy demands, which is being bridged by the import of fossil fuels and requires substantial state revenue to be spent on these imports. Pakistan imported 13.57 Milliontons of oil equivalent (MTOE) of petroleum during fiscal year 2014–2015, ultimately putting tremendous pressure on the economy by increasing the import bills [111]. Transportation and power generation are the main fossil fuel consuming sectors in Pakistan. Fuel price increases frequently, leading to increases in transportation costs and utility bills for both public and private consumers and pose socioeconomic challenges for the country. At present, Pakistan's indigenous resources account for only up to 15 percent of the country's energy requirements [112]. Pakistan spends approximately 60% of its currency exchange on importing fuels to meet energy needs, and these import bills can be significantly reduced if indigenous alternative energy resources are used appropriately [110].

Pakistan's government is searching for cost-effective, environmentally friendly alternative energy sources in order to address current energy crises and maintain economic stability [108].

## 7. Recommendations from the perspective of Pakistan

The use of agricultural residues as a renewable energy resource in Pakistan can provide a sustainable way to enhance the country's energy mix in order to meet ever-increasing energy needs. Energy production through suitable and efficient technologies can have multiple positive economic impacts on Pakistan, (1) by saving huge investments in energy imports, (2) by reducing harmful gas emissions in order to protect the environment and (3) by empowering the people of the country in terms of social aspects [111]. It can provide multiple job opportunities to people working in the agricultural, transportation and daily wagering sectors. Furthermore, public awareness campaigns emphasizing the importance of renewable energy resources, as well as basic education on how to effectively manage these resources, should be launched [113]. This can be achieved by distinct financial assistance programs should be made available to encourage business investments in the renewable energy production sector [114].

Various important steps and measures must be taken as soon as possible, such as the establishment of generous research and development programs at the Country's Universities and research institutions, with a focus on research activities involving renewable resources in the country.

## 8. Conclusion

This review presents an extensive analysis of the potential of biomass for renewable energy production in Pakistan. It also emphasizes the availability of local biomass resources as well as state-of-the-art of biomass conversion technologies. Heavy reliance on imported fossil fuels and global climate change are key factors contributing to Pakistan's economic problems. To address these issues, relying on locally available renewable energy sources is a promising and cost-effective financial solution. The transportation sector is a major importer of petroleum fuels, accounting for the majority of the total import bill. Biodiesel and bio-ethanol, can supplement HSD/petrol, transportation fuels. To overcome this issue biodiesel production with full utilization of its by-products can provide a sustainable and environmentally friendly replacement of mineral high speed diesel (HSD).

Moreover, comprehensive detail of the locally abundantly available feedstocks for biodiesel production has also been discussed in this chapter. Overall, this study further concludes that Pakistan has the immense potential to produce economical viable biodiesel from the locally available feedstocks.

## Author details

Juma Sahar[1]*, Muhammad Farooq[1], Anita Ramli[2] and Abdul Naeem[1]

1 National Centre of Excellence in Physical Chemistry, University of Peshawar, Peshawar, Khyber Pakhtunkhwa, Pakistan

2 Department of Fundamental and Applied Sciences, UniversitiTeknologi Petronas, Malaysia

*Address all correspondence to: jsahar14@yahoo.com

IntechOpen

© 2022 The Author(s). Licensee IntechOpen. This chapter is distributed under the terms of the Creative Commons Attribution License (http://creativecommons.org/licenses/by/3.0), which permits unrestricted use, distribution, and reproduction in any medium, provided the original work is properly cited.

## References

[1] Avni A, Blázquez M. Can plant biotechnology help in solving our food and energy shortage in the future? Current Opinion in Biotechnology. 2011;**22**:220-223

[2] Sawin J et al. Renewables 2015 Global Status Report. France: International Nuclear Information System; 2015

[3] Ilmi M. Enzymatic Biodiesel Synthesis Using Novel Process Intensification Principles. Groningen: Chemical Reaction Engineering; 2017

[4] Ong HC, Mahlia TMI, Masjuki HH. A review on energy scenario and sustainable energy in Malaysia. Renewable and Sustainable Energy Reviews. 2011;**15**(1):639-647

[5] Saito S. Role of nuclear energy to a future society of shortage of energy resources and global warming. Journal of Nuclear Materials. 2010;**398**(1):1-9

[6] EIA J. International Energy Outlook. 2013, DOE/EIA-0484, ed. Washington, DC: US Energy Information Administration; 2013

[7] Avhad MR, Marchetti JM. A review on recent advancement in catalytic materials for biodiesel production. Renewable and Sustainable Energy Reviews. 2015;**50**:696-718

[8] Bereczky A, Torok A. International literature review on the possibilities of biodiesel production. Periodica Polytechnica Transportation Engineering. 2011;**39**:31-37

[9] Sajjadi B, Raman AAA, Arandiyan H. A comprehensive review on properties of edible and non-edible vegetable oil-based biodiesel: Composition, specifications and prediction models. Renewable and Sustainable Energy Reviews. 2016;**63**:62-92

[10] Gielen D, Boshell F, Saygin D, Bazilian MD, Wagner N, Gorini R. The role of renewable energy in the global energy transformation. Energy Strategy Reviews. 2019;**24**:38-50

[11] Wang A, Sudarsanam P, Xu Y, Zhang H, Li H, Yang S. Functionalized magnetic nanosized materials for efficient biodiesel synthesis via acidbase/enzyme catalysis. Green Chemistry. 2020;**22**:2977-3012

[12] Robles-Medina A, González-Moreno PA, Esteban-Cerdán L, Molina-Grima E. Biocatalysis: Towards ever greener biodiesel production. Biotechnology Advances. 2009;**27**(4):398-408

[13] Ma F, Hanna MA. Biodiesel production: A review1 Journal Series #12109, Agricultural Research Division, Institute of Agriculture and Natural Resources, University of NebraskaLincoln. 1. Bioresource Technology. 1999;**70**(1):1-15

[14] Bilgin A, Durgun O, Sahin Z. The effects of diesel-ethanol blends on diesel engine performance. Energy Sources. 2002;**24**(5):431-440

[15] Yusuf NNAN, Kamarudin SK, Yaakub Z. Overview on the current trends in biodiesel production. Energy Conversion and Management. 2011;**52**(7):2741-2751

[16] Mishra V, Goswami R. A review of production, properties and advantages of biodiesel. Biofuels. 2017;**9**(2):273-289

[17] Demirbas A. Comparison of transesterification methods for production of biodiesel from vegetable

oils and fats. Energy Conversion and Management. 2008;**49**(1):125-130

[18] Sivaprakasam S, Saravanan CG. Optimization of the transesterification process for biodiesel production and use of biodiesel in a compression ignition engine. Energy & Fuels. 2007;**21**(5):2998-3003

[19] Soltani S et al. Synthesis of biodiesel through catalytic transesterification of various feedstocks using fast solvothermal technology: A critical review. Catalysis Reviews. 2015;**57**(4):407-435

[20] Chen C, Cai L, Shangguan X, Li L, Hong Y, Wu G. Heterogeneous and efficient transesterification of Jatropha curcas L. seed oil to produce biodiesel catalysed by nano-sized $SO_4^{2-}$/$TiO_2$. Royal Society Open Science. 2018;**5**:181331

[21] Leung DYC, Guo Y. Transesterification of neat and used frying oil: Optimization for biodiesel production. Fuel Processing Technology. 2006;**87**(10):883-890

[22] Hiwot T. Mango (Magnifera indica) seed oil grown in Dilla town as potential raw material for biodiesel production using NaOH-a homogeneous catalyst. Chemistry International. 2018;**4**(4):198-205

[23] Helwani Z, Othman MR, Aziz N, Kim J, Fernando WJN. Solid heterogeneous catalysts for transesterification of triglycerides with methanol: A review. Applied Catalysis A: General. 2009;**363**(1):1-10

[24] Pinzi S, Garcia IL, Lopez-Gimenez FJ, Luque de Castro MD, Dorado G, et al. The ideal vegetable oil-based biodiesel composition: A review of social, economical and technical implications. Energy & Fuels. 2009;**23**(5):2325-2341

[25] Aransiola EF, Ojumu TV, Oyekola OO, Madzimbamuto TF, Ikhu-Omoregbe. A review of current technology for biodiesel production: State of the art. Biomass and Bioenergy. 2014;**61**:76-297

[26] Bernardes OL, Bevilaqua JV, Leal MCMR, Freire DMG, Langone MAP. Biodiesel fuel production by the transesterification reaction of soybean oil using immobilized lipase. Applied Biochemistry and Biotechnology. 2007;**137**(1):105-114

[27] Shah S, Gupta MN. Lipase catalyzed preparation of biodiesel from Jatropha oil in a solvent free system. Process Biochemistry. 2007;**42**(3):409-414

[28] Schuchardt U, Sercheli R, Matheus V. Transesterification of vegetable oils: A review. Journal of the Brazilian Chemical. Society. 1998;**9**:199-210

[29] Arumugam A, Ponnusami V. Production of biodiesel by enzymatic transesterification of waste sardine oil and evaluation of its engine performance. Heliyon. 2017;**3**(12):e00486

[30] Thangaraj B et al. Catalysis in biodiesel production: A review. Clean Energy. 2018;**3**(1):2-23

[31] Navas MB et al. A sustainable process for biodiesel production using Zn/Mg oxidic species as active, selective and reusable heterogeneous catalysts. Bioresources and Bioprocessing. 2020;**7**(1):4

[32] Gawande MB, Pandey RK, Jayaram RV. Role of mixed metal oxides in catalysis scienceversatile applications

in organic synthesis. Catalysis Science & Technology. 2012;**2**(6):1113-1125

[33] Joergensen KA. Transition-metal-catalyzed epoxidations. Chemical Reviews. 1989;**89**(3):431-458

[34] dos Reis SCM, Lachter ER, Nascimento RSV, Rodrigues JA, Reid MG. Transesterification of brazilian vegetable oils with methanol over ion-exchange resins. Journal of the American Oil Chemists' Society. 2005;**82**(9):661-665

[35] Corma A, Iborra S. Optimization of alkaline earth metal oxide and hydroxide catalysts for base-catalyzed reactions. Advances in Catalysis. 2006;**49**:239-302

[36] Ilgen O, Akin A. Development of alumina supported alkaline catalysts used for biodiesel production. Turkish Journal of Chemistry. 2009;**33**:281-287

[37] Liu X et al. Calcium ethoxide as a solid base catalyst for the transesterification of soybean oil to biodiesel. Energy & Fuels. 2008;**22**(2):1313-1317

[38] Calero J et al. Development of a new biodiesel that integrates glycerol, by using CaO as heterogeneous catalyst, in the partial methanolysis of sunflower oil. Fuel. 2014;**122**:94-102

[39] Zabeti M, Wan Daud WMA, Aroua MK. Activity of solid catalysts for biodiesel production: A review. Fuel Processing Technology. 2009;**90**(6):770-777

[40] Demirbas A. Biodiesel from sunflower oil in supercritical methanol with calcium oxide. Energy Conversion and Management. 2007;**48**(3):937-941

[41] Ruhul AM, Kalam MA, Masjuki HH, Fattah IMR, Reham SS, Rashed MM. State of the art of biodiesel production processes: A review of the heterogeneous catalyst. RSC Advances. 2015;**5**(122):101023-101044

[42] Kawashima A, Matsubara K, Honda K. Development of heterogeneous base catalysts for biodiesel production. Bioresource Technology. 2008;**99**(9):3439-3443

[43] Pugnet V et al. Stability, activity and selectivity study of a zinc aluminate heterogeneous catalyst for the transesterification of vegetable oil in batch reactor. Applied Catalysis A: General. 2010;**374**(1):71-78

[44] Xie W, Zhao L. Aminopropylsilica as an environmentally friendly and reusable catalyst for biodiesel production from soybean oil. Fuel. 2013;**103**:1106-1110

[45] Nomanbhay S, Ong MY. A review of microwave-assisted reactions for biodiesel production. Bioengineering (Basel, Switzerland). 2017;**4**(2):57

[46] Alhassan FH, Rashid U, Taufiq-Yap YH. Synthesis of waste cooking oil based biodiesel via effectual recyclable bi-functional $Fe_2O_3MnOSO_4^{2-}/ZrO_2$ nanoparticle solid catalyst. Fuel. 2015;**142**:38-45

[47] Sakthivel A, Komura K, Sugi Y. MCM-48 Supported tungstophosphoric acid: An efficient catalyst for the esterification of long-chain fatty acids and alcohols in supercritical carbon dioxide. Industrial & Engineering Chemistry Research. 2008;**47**(8):2538-2544

[48] Borges ME, Díaz L. Recent developments on heterogeneous catalysts for biodiesel production by oil esterification and transesterification reactions: A review. Renewable and Sustainable Energy Reviews. 2012;**16**(5):2839-2849

[49] Hayyan A, Alam MZ, Mirghani MES, Kabbashi NA, Hakimi NINM, Siran YM, et al. Sludge palm oil as a renewable raw material for biodiesel production by two-step processes. Bioresource Technology. 2010;**101**(20):7804-7811

[50] Charoenchaitrakool M, Thienmethangkoon J. Statistical optimization for biodiesel production from waste frying oil through two-step catalyzed process. Fuel Processing Technology. 2011;**92**(1):112-118

[51] Kondamudi N, Mohapatra S, Misra M. Quintinite as a bifunctional heterogeneous catalyst for biodiesel synthesis. Applied Catalysis A-General. 2011;**393**:36-43

[52] Salinas D, Guerrero S, Araya P. Transesterification of canola oil on potassium-supported $TiO_2$ catalysts. Catalysis Communications. 2010;**11**(8): 773-777

[53] Abdullah RF, Rashid U, Ibrahim ML, Hazmi B, Alharthi FA, Nehdi IA. Bifunctional nano-catalyst produced from palm kernel shell via hydrothermal-assisted carbonization for biodiesel production from waste cooking oil. Renewable and Sustainable Energy Reviews. 2021;**137**:110638

[54] Ramli A, Farooq M, Naeem A, Khan S, Hummayun M, Iqbal A, et al. Bifunctional Heterogeneous Catalysts for Biodiesel Production Using Low Cost Feedstocks: A Future Perspective 2017. IntechOpen; 2017;**285**:285-299

[55] Wan Omar WNN, Amin NAS. Biodiesel production from waste cooking oil over alkaline modified zirconia catalyst. Fuel Processing Technology. 2011;**92**:2397-2405

[56] Faruque MO, Razzak SA, Hossain MM. Application of heterogeneous catalysts for biodiesel production from microalgal oil: A review. Catalysts. 2020;**10**(9):1025

[57] Cannilla C, Bonura G, Rombi E, Arena F, Frusteri F. Highly effective MnCeOx catalysts for biodiesel production by transesterification of vegetable oils with methanol. Applied Catalysis A: General. 2010;**382**(2):158-166

[58] Furuta S, Matsuhashi H, Arata K. Biodiesel fuel production with solid amorphous-zirconia catalysis in fixed bed reactor. Biomass and Bioenergy. 2006;**30**(10):870-873

[59] Atabani AE et al. A review on global fuel economy standards, labels and technologies in the transportation sector. Renewable and Sustainable Energy Reviews. 2011;**15**(9):4586-4610

[60] Ahmad AL, Yasin NHM, Derek CJC, Lim JK. Microalgae as a sustainable energy source for biodiesel production: A review. Renewable and Sustainable Energy Reviews. 2011;**15**(1):584-593

[61] Rezania S, Oryani B, Park J, Hashemi B, Yadav KK, Kwon EE, et al. Review on transesterification of non-edible sources for biodiesel production with a focus on economic aspects, fuel properties and by-product applications. Energy Conversion and Management. 2019;**201**:112155

[62] Jazie AA, Pramanik H, Sinha ASK. Transesterification of peanut and rapeseed oils using waste of animal bone as cost effective catalyst. Materials for Renewable and Sustainable Energy. 2013;**2**(2):11

[63] Guo W, Li H, Ji G, Zhang G. Ultrasound-assisted production of biodiesel from soybean oil using Brønsted acidic ionic liquid as catalyst. Bioresource Technology. 2012;**125**:332-334

[64] Agarwal AK. Biofuels (alcohols and biodiesel) applications as fuels for internal combustion engines. Progress in Energy and Combustion Science. 2007;**33**(3):233-271

[65] Long Y-D et al. Co-production of biodiesel and hydrogen from rapeseed and Jatropha oils with sodium silicate and Ni catalysts. Applied Energy. 2014;**113**:1819-1825

[66] Deng X et al. Production of biodiesel from Jatropha oil catalyzed by nanosized solid basic catalyst. Energy. 2011;**36**(2): 777-784

[67] Gui MM, Lee KT, Bhatia S. Feasibility of edible oil vs. non-edible oil vs. waste edible oil as biodiesel feedstock. Energy. 2008;**33**(11):1646-1653

[68] Khan J. Economics of Mazri Making as Source of Livelihood in Nomadic Lifestyle of Balochistan, Pakistan. IOSR Journal of Agriculture and Veterinary Science. 2013;**4**:47-52

[69] Latif A, Shinwari Z. Sustainable Market Development for Non Timber Forest Products in Pakistan. Pakistan: Ethnobotanical Leaflets; 2005

[70] Adnan M, Khan A. Challenges faced to Mazri Palm (Nannorhops ritchieana): A case study of Jhandey, District Mardan-Pakistan. Vol. 2004. Ethno-botanical Leaflets Pakistan, 2004

[71] Latif A, Begum S, Adnan M, Hussian K, Waseem M. Challenges faced to Mazri Palm (Nannorhops ritchieana): A case study of Jhandey, District Mardan-Pakistan. Ethnobotanical Leaflets. 2004;**2004**:10

[72] Rashid R, Mukhtar F, Khan A. Antifungal and cytotoxic activities of Nannorrhops Ritchiana roots extract. Acta Poloniae Pharmaceutica. 2014;**71**:789-793

[73] Mahmood A. Phytochemical analysis and comprehensive evaluation of antimicrobial activity of nannorhops ritchiana leaves (mazari palm). World Journal of Pharmacy and Pharmaceutical Sciences. 2017;**6**:173-189

[74] Mohibbe Azam M, Waris A, Nahar NM. Prospects and potential of fatty acid methyl esters of some non-traditional seed oils for use as biodiesel in India. Biomass and Bioenergy. 2005;**29**(4):293-302

[75] Tunio M, Samo S, Ali Z, Chand K. Investigation of Jatropha Biodiesel Production on Experimental Scale. University of Engineering and Technology Taxila. Technical Journal. 2016;**21**(2):14

[76] Kumar A, Sharma S. Potential non-edible oil resources as biodiesel feedstock: An Indian perspective. Renewable and Sustainable Energy Reviews. 2011;**15**(4):1791-1800

[77] Kumar A, Sharma S. An evaluation of multipurpose oil seed crop for industrial uses (Jatropha curcas L.): A review. Industrial Crops and Products. 2008;**28**(1):1-10

[78] Shah SH, Raja IA, Rizwan M, Rashid N, Mahmood Q, Shah FA, et al. Potential of microalgal biodiesel production and its sustainability perspectives in Pakistan. Renewable and Sustainable Energy Reviews. 2018;**81**:76-92

[79] Chakrabarti MH, Ali M, Usmani JN, Khan NA, Hasan DUB, Islam MS, et al. Status of biodiesel research and development in Pakistan. Renewable and Sustainable Energy Reviews. 2012;**16**(7):4396-4405

[80] Khan HM, Ali CH, Iqbal T, Yasin S, Sulaiman M, Mahmood H, et al. Current

scenario and potential of biodiesel production from waste cooking oil in Pakistan: An overview. Chinese Journal of Chemical Engineering. 2019;**27**(10):2238-2250

[81] Khan NA, Usmani JN. Status of jatropha cultivation for biodiesel production in Pakistan. Science Technology and Development. 2010;**29**(3):1-9

[82] Khan N, Ahmed M, Shaukat SS, Wahab M, Siddiqui MF. Structure, diversity, and regeneration potential of Monotheca buxifolia (Falc.) A. DC. dominated forests of Lower Dir District, Pakistan. Frontiers of Agriculture in China. 2011;**5**(1):106-121

[83] Jan S, Khan MR. Protective effects of Monotheca buxifolia fruit on renal toxicity induced by CCl4 in rats. BMC Complementary and Alternative Medicine. 2016;**16**(1):289-289

[84] Haq ZU, Rashid A, Khan SM, Razzaq A, Al-Yahyai RA, Kamran S, et al. A In vitro and in vivo propagation of Monotheca buxifolia (Falc.) A. DC. An economical medicinal plant. Acta Ecologica Sinica. 2019;**39**(6):425-430

[85] Khan I, Ali JS, Ul-Haq I, Zia M. Biological and Phytochemicals Properties of Monotheca buxifolia: An Unexplored Medicinal Plant. Pharmaceutical Chemistry Journal. 2020;**54**:293-301

[86] Ali M, Naqvi B, Watson IA. Possibility of converting indigenous Salvadora persica L. seed oil into biodiesel in Pakistan. International Journal of Green Energy. 2018;**15**(7): 427-435

[87] Kamil M, Ramadan K, Olabi AG, Ghenai C, Inayat A, Rajab MH, et al. Desert palm date seeds as a biodiesel feedstock: Extraction, characterization, and engine testing. Energies. 2019;**12**:3147

[88] Amani M, Davoudi M, Tahvildari K, Nabavi S, Davoudi M. Biodiesel Production from Phoenix dactylifera as a New Feedstock. Industrial Crops and Products. 2013;**43**:40-43

[89] Kamil M, Ramadan K, Olabi AG, Ghenai C, Inayat A, Rajab MH. Desert palm date seeds as a biodiesel feedstock: Extraction, characterization, and engine testing. Energies. 2019;**12**(16):173-189

[90] Azeem MW, Hanif MA, Al-Sabahi JN, Khan AA, Naz S, Ijaz A. Production of biodiesel from low priced, renewable and abundant date seed oil. Renewable Energy. 2016;**86**:124-132

[91] Sulaiman A-Z et al. Biodiesel production from oils extracted from Date pits. Green and Sustainable Chemistry. 2017;**7**:48

[92] Abul Soad A, Mahdi S, Markhand G. Date Palm Status and Perspective in Pakistan. Date Palm Genetic Resources and Utilization. Vol. 2. Asia and Europe; 2015. pp. 153-205

[93] Halder PK, Paul N, Beg MRA. Prospect of *Pongamia pinnata* (Karanja) in Bangladesh: A sustainable source of liquid fuel. Journal of Renewable Energy. 2014;**2014**:647324

[94] Padhi S, Singh RK. Non-edible oils as the potential source for the production of biodiesel in India: A review. Journal of Chemical and Pharmaceutical Research. 2011;**3**:39-49

[95] Naik M et al. Production of biodiesel from high free fatty acid Karanja (Pongamia pinnata) oil. Biomass and Bioenergy. 2008;**32**(4):354-357

[96] Dubey S, Kashyap P. Azadirachta indica: A plant with versatile potential.

RGUHS. Journal of Pharmaceutical Sciences. 2014;**4**(2):39-46

[97] Muthu H, SathyaSelvabala V, Varathachary TK, Kirupha Selvaraj D, Nandagopal J, Subramanian S. Synthesis of biodiesel from Neem oil using sulfated zirconia via tranesterification. Brazilian Journal of Chemical Engineering. 2010;**27**:601-608

[98] Li Y, Horsman M, Wang B, Wu N, Lan CQ. Effects of nitrogen sources on cell growth and lipid accumulation of green alga Neochloris oleoabundans. Applied Microbiology and Biotechnology. 2008;**81**(4):629-636

[99] Pimentel D, Marklein A, Toth MA, Karpoff MN, Paul GS, McCormack R, et al. Food Versus Biofuels: Environmental and Economic Costs. Human Ecology. 2009;**37**(1):1

[100] Manzoor M et al. Lucrative future of microalgal biofuels in Pakistan: A review. International Journal of Energy and Environmental Engineering. 2015;**6**(4):393-403

[101] Ali M et al. Prospects of microalgal biodiesel production in Pakistan: A review. Renewable and Sustainable Energy Reviews. 2017;**80**:1588-1596

[102] Bahadar A, Khan MB, Willmann JC. Accelerated production and analysis of biofuel derived from photobioreactor engineered microalgae using super critical fluid extraction. Energy Sources, Part A: Recovery, Utilization, and Environmental Effects. 2016;**38**:1132-1139

[103] Szmigielski M, Maniak B, Piekarski W. Evaluation of chosen quality parameters of used frying rape oil as fuel biocomponent. International Agrophysics. 2008;**22**:243-248

[104] Srinivasan S. The food v. fuel debate: A nuanced view of incentive structures. Renewable Energy. 2009;**34**(4):950-954

[105] Tsai W-T. Mandatory Recycling of Waste Cooking Oil from Residential and Commercial Sectors in Taiwan. Resources. 2019;**8**(1):38

[106] Banković-Ilić IB, Stojković IJ, Stamenković OS, Veljkovic VB, Hung YT. Waste animal fats as feedstocks for biodiesel production. Renewable and Sustainable Energy Reviews. 2014;**32**:238-254

[107] Janaun J, Ellis N. Perspectives on biodiesel as a sustainable fuel. Renewable and Sustainable Energy Reviews. 2010;**14**(4):1312-1320

[108] Irfan M, Zhao Z-Y, Panjwani MK, Mangi FH, Li H, Jan A, et al. Assessing the energy dynamics of Pakistan: Prospects of biomass energy. Energy Reports. 2020;**6**:80-93

[109] Asif M, Muneer T. Life cycle assessment of built-in-storage solar water heaters in Pakistan. Building Services Engineering Research and Technology. 2006;**27**(1):63-69

[110] Rafique MM, Rehman S. National energy scenario of Pakistan Current status, future alternatives, and institutional infrastructure: An overview. Renewable and Sustainable Energy Reviews. 2017;**69**:156-167

[111] Abdullah A, Ahmed A, Akhter P, Razzaq A, Hussain M, Hossain N, et al. Potential for sustainable utilisation of agricultural residues for bioenergy production in Pakistan: An overview. Journal of Cleaner Production. 2021;**287**:125047

[112] Abdullah FB. A Model for Strategizing Energy Security Dimensions and Indicators Selection for Pakistan.

International Journal of Renewable Energy Research. 2020;**10**(2):558-569

[113] Nguyen QA, Smith WA, Wahlen BD, Wendt LM. Total and sustainable utilization of biomass resources: A perspective. Frontiers in Bioengineering and Biotechnology. 2020;**8**:546-546

[114] Tareen WU, Dilbar MT, Farhan M, Ali Nawaz M, Durrani AW, Memon KA, et al. Present status and potential of biomass energy in Pakistan based on existing and future renewable resources. Sustainability. 2020;**12**(1):249

Chapter 8

# Zero Emission Hydrogen Fuelled Fuel Cell Vehicle and Advanced Strategy on Internal Combustion Engine: A Review

*Babu Dharmalingam, Ramakrishna Reddy Ramireddy, Santhoshkumar Annamalai, Malinee Sriariyanun, Deepakkumar Rajagopal and Venkata Ramana Katla*

**Abstract**

Global energy consumption has gradually increased as a result of population growth, industrialization, economic development, and rising living standards. Furthermore, as global warming and pollution worsen, the development of renewable energy sources is becoming more essential. Hydrogen is one of the most promising clean and sustainable energy carriers because it emits only water as a byproduct without carbon emission and has the highest energy efficiency. Hydrogen can be produced from a variety of raw resources, including water and biomass. Water electrolysis is one of many hydrogen production technologies that is highly recommended due to its eco-friendliness, high hydrogen generation rate, and high purity. However, in terms of long-term viability and environmental effect, Polymer Electrolyte Membrane water electrolysis has been identified as a potential approach for producing high-purity, high-efficiency hydrogen from renewable energy sources. Furthermore, the hydrogen ($H_2$) and oxygen ($O_2$) produced are directly employed in fuel cells and other industrial uses. As a result, an attempt has been made in this work to investigate hydrogen synthesis and utilization in fuel cell vehicles. Low-temperature combustion technology has recently been applied in engine technology to reduce smoke and NOx emissions at the same time. The advantages and limitations of homogeneous charge compression ignition, partially premixed charge compression ignition, premixed charge compression ignition, and reactivity regulated compression ignition are described separately in low-temperature combustion strategy.

**Keywords:** hydrogen, fuel cell vehicle, hybrid vehicle, low temperature combustion

## 1. Introduction

Global energy consumption has increased gradually in recent years due to population growth, and economic development and industrialization. Also, global warming and environmental pollution worsened everyday too much of automobile vehicles and industrialization. Hence, the development of renewable energy sources became increasingly important. Hydrogen is one the most promising clean and sustainable energy sources because it emits only water as a byproduct and generates no carbon emissions [1]. Hydrogen has a quality of high energy carrier including high energy density that is more than ordinary petroleum and diesel fuel [2]. At the moment, global hydrogen production is estimated to over 500 billion cube meters per year [3]. It can be used in much industrial application including fertilizer, petroleum refining operation, fuel cell, chemical industries [4]. Hydrogen can be generated from variety of renewable and non-renewable sources like water and fossil fuels [5], oil reforming [6], coal gasification [7], biomass [8], water electrolysis [9].

Many approaches for manufacturing hydrogen are currently available however water electrolysis is one of the most capable methods for producing hydrogen as a product and oxygen as a by-product. At the moment, only 4% of hydrogen can be obtained by electrolysis of water [10]. Water electrolysis also provides a number of advantages, such as high cell efficiency and a greater hydrogen generation rate with excellent purity, making it a better method for converting water to electrical energy via low-temperature fuel cells. The water molecule is the reactant in the electrolysis process, and under the influence of electricity, it is split into hydrogen ($H_2$) and oxygen ($O_2$). Based on the electrolyte, operating conditions, and ionic agents ($OH^-$, $H^+$, $O_2^-$), water electrolysis is separated into four categories: alkaline water electrolysis (ii), solid oxide electrolysis (SOE), microbial electrolysis cells (MEC), and PEM electrolysis of water [11]. The phenomenon was first described by Troostwijk and Diemann in 1789 [12], and it is a well-established technique for commercial hydrogen production up to the megawatt range in the world.

The hydroxyl ions ($OH^-$) flow through the porous diaphragm to the anode under the effect of the electrical circuit between anode and cathode, where they are discharged to 12 molecules of oxygen ($O_2$) and one molecule of water ($H_2O$). Alkaline electrolysis is performed at lower temperatures, such as 30–80°C, with an aqueous solution (KOH/NaOH) as the electrolyte and a 20–30% concentration. Alkaline water electrolysis uses an asbestos diaphragm and nickel materials as electrodes [13]. In the 1980s, Donitz and Erdle proposed solid oxide electrolysis (SOE). Solid oxide electrolysis has attracted a lot of interest since it converts electrical energy into chemical energy while also producing ultra-pure hydrogen with a higher efficiency. Solid oxide electrolysis runs at high pressures and temperatures of 500–850°C and consumes water in the form of steam. Nickel/zirconia is commonly utilized as an $O_2$ conductor in solid oxide electrolysis [14].

Microbial electrolysis cell (MEC) technology may produce hydrogen from organic matter such as renewable biomass and wastewaters. MEC technology is similar to microbial fuel cells (MFCs), however the operational concept is reversed [15]. In 2005, two independent research institutions, Penn State University and Wageningen University in the Netherlands, established the first Microbial electrolysis cell (MEC) method. Electrical energy is turned into chemical energy in microbial electrolysis cells (MECs). Under the influence of an electric current, MECs created hydrogen from organic molecules. Microbes oxidize the substrate at the anode side of the microbial electrolysis process, producing $CO_2$, protons, and electrons. The electrons move to the cathode side via the external circuit, while the protons travel to the cathode via a proton conducting membrane (electrolyte), where the protons and electrons combine to form hydrogen [15]. However, this MEC technology is still in the early stages of development, and various issues like as high internal resistance, electrode materials, and intricate design must be addressed before the technology can be commercialized [16].

In the early 1950s, Grubb achieved the first PEM water electrolysis, and General Electric Co. was created in 1966 to overcome the drawbacks of alkaline water electrolysis. PEM water electrolysis technique, which is similar to PEM fuel cell technology [17], used solid poly sulfated membranes (Nafion®, fumapem®) as an electrolyte (proton conductor). Lower gas permeability, strong proton conductivity (0.1 0.02 S $cm^{-1}$), thinness (20–300 m), and high-pressure functionality are all advantages of these proton exchange membranes. In terms of sustainability and environmental impact, PEM water electrolysis is one of the most environmentally benign methods for converting renewable energy to high purity hydrogen. Another prospective PEM water electrolysis device has a small footprint, high current density (over 2 A $cm^{-2}$), high efficiency, fast responsiveness, and operates at lower temperatures (20–80°C) while producing ultrapure hydrogen as a byproduct [17].

## 2. Fuel cell vehicles

Fuel cell technology is gaining popularity in the automotive industry due to its ease of use, quiet operation, high efficiency, and modular structure. According to Mustafa et al., recent investigations have showed that the usage of fuel cells in vehicles has expanded rapidly, causing a revolution, and will be an alternative to conventional vehicles in the future (2021). Configuration, system components, control/management, technical obstacles, marketing, and future aspects are all categories for fuel cell cars. Based on chemical characteristics and operating temperature, fuel cells are classed as proton exchange membrane FCs, solid oxide FCs, direct methanol FCs, alkaline FCs, molten carbonate FCs, and phosphoric acid FCs. FCs are used in both commercial and research & development applications. Common stack size, theoretical cell voltage, operating temperature, electrical efficiency, benefits, and downsides are used to classify FC features [18]. In this environment, FCs are used in distributed generation, mobile power, backup power, military, space, and vehicle applications. Low temperature and pressure PEMFCs are the most used FCs in vehicle applications because of their high power density, lower working temperature (60–80°C), and reduced corrosion than other FCs [18].

In the construction of fuel cell hybrid electrical vehicles (FCEVs), fuel cell vehicles (FCs) are coupled to electric motors via controlled electronic interfacing components [19]. The basic components of traditional FCEVs are a voltage regulation converter, motor drive, electric motor, and auxiliary energy generation units [20]. For interfacing components and energy management algorithms, FCEVs vehicles have a variety of configuration topologies [21]. The powertrain structures, voltage regulation topologies, motor drive converters, and energy management technologies can all be used to classify FCEVs. In the operation of FCEVs, the FC stack feeds energy to the dc-bus and maintains the required DC bus voltage [22]. The FC is then connected directly to the Unidirectional DC-DC converter (UDC) as a system element to maintain the dc-bus voltage and send the energy generated for vehicle propulsion to the motor drive converter. A DC-AC converter checks the motor speed and torque for safe operation. Finally, the drive controller is in charge of monitoring the electric motors as they convert electrical energy into kinetic energy [23].

FCs have a higher energy density and efficiency than other power sources such as photovoltaics, batteries, ultra capacitors, and super conducting magnetic energy storage. Because of its modular design, FCs are also suitable for electric vehicle applications. Furthermore, FC has a 20–30 year lifespan [24]. As a portable/rechargeable energy storage system, the battery is also a preferred power source for FCEV hybridization. However, it has a short lifespan and is only useful for a short length of time [25]. Ultra capacitors (UC) are a type of storage element that can be used in FCEV applications to increase the dynamic response of the system. Photovoltaic (PV) is a gadget that generates energy, however it is too large to carry. The output of super conductive magnetic energy storage (SMES) generates a lot of power, however it has a low energy density. Short-duration energy storage is also included in SMES, albeit at a high expense [26]. Based on this, several hybridization topologies are recognized in the literature. Full FC, partial FC, and hybrid FC cars are classified as FC + battery hybridization, FC + UC hybridization, FC + battery + UC hybridization, FC + battery + PV, FC + flywheel hybridization, and FC + SMES [18]. FC + battery + PV, FC + battery + PV, FC + flywheel hybridization, and FC + SMES are all examples of FC-powered cars.

The FCEV scheme clearly shows that this topology's energy generation is exclusively dependent on the FC stack. It simple construction includes a fuel tank, FC stack, DC-DC power converter, inverter, and electric motor [27]. These cars feature a long driving range, a fast charging time, high efficiency, cold start capabilities, silent operation due to the lack of mechanical components, energy supply continuity, and low emissions [27]. Full FCEVs are a suitable fit for low-speed vehicles including forklifts, busses, airline vehicles, trams, and marine vehicles. The combination of FC + battery units is the most common topology in FCEV hybridization [18]. A unidirectional DC-DC converter (UDC) connects FC to the DC bus, while a bidirectional DC-DC converter connects the battery to the DC bus. In the operating procedure of FC + battery hybridization, an initial start-up with the battery is provided to avoid the FC running in the low-efficiency zone. As a result, a huge amount of current is generated to start the electric motor [25]. When the car is turned on for the first time, the FC is activated to keep the electric motor going. After then, the battery is charged according to the charge status criteria. The UC only allows FC to be utilized in emergency situations to meet transient power demands. UC, on the other hand, has a low energy density and is not used to give energy on a long-term basis [28].

In contrast to earlier hybridization topologies, FC + battery + UC hybridization has a primary energy source (FC) and two secondary energy sources (battery and UC) (battery and ultra capacitor). In this design, the FC is connected to the DC bus through a one-way DC-DC converter. The energy storage units, battery and UC, are connected to the DC bus by bidirectional DC-DC converters (BDCs). This architecture combines the advantages of FC + battery and FC + UC systems to provide continuous energy while also boosting FC dynamic response during transient events [29]. In recent years, PV panels have been incorporated with FC-based electric vehicles for hybridization. In FC + battery + PV hybridization, PV panels generate DC voltage that is coupled to the DC bus via a unidirectional converter. The FC is the primary energy source in an FC + battery + PV system, with the PV panel acting as a backup. Both the FC and PV busses are connected to the DC bus by unidirectional converters. PV panels generate varying amounts of power based on the intensity of solar radiation, the temperature, and the sun's direction. As a result, the PV electricity generated is fed directly into the electric motor or is used to charge the battery [30].

FC+ flywheel hybridization is similar to the preceding approach in that the FC serves as the major energy source and the flywheel, rather than batteries, serves as an energy storage method. Flywheels and generators are connected to store energy mechanically with a high rotating speed and transform that mechanical energy into electricity when EM requires a lot of it. Flywheels have a faster charging capability, higher efficiency, and higher power rating than batteries [30]. Flywheels are also environmentally friendly, as they operate over a wide temperature range, have a big energy storage capacity, and have a long lifespan [66]. There are three types of static FC models accessible in the literature. Chamberlin-Kim and Amphlett, Larminie, and Dicks models [31] are examples. The most common static model published in the literature is the Amphlett model, which is based on Nernst and Tafel equations. This model takes into account physical parameters like as pressure, temperature, and concentration. The other static model is the Larmine and Dicks model. This model calculates the FC voltage–current characteristic using empirical equations. This model yields the FC voltage versus current amplitude curve. Three zones can be found in this curve. The three zones are electrochemical activation, linear part, and gas diffusion kinetics [32]. The third static FC model is the Chamberlin-Kim model. In this approach, the FC voltage is described in terms of current density. In addition, the fuel-oxidant rate, local temperature, and humidity all affect five factors in this model [32].

Dynamic modeling of FC is described in the literature such as the impedance model, Becherif-Hissel model, and Dicks-Larminie model have been reported [33]. Layer capacitance, diffusion impedance, and ions transport, membrane, and contact resistances are all included in the impedance model [34]. The Nernst voltage, ohmic polarization, concentration, and activation are all modeled in the Dicks-Larminie model. A voltage supply, two resistances, and a capacitor make up this model. The Nernst voltage is demonstrated via the voltage source. The resistances represent electron-hydrogen flow and activation-concentration losses. The charge layers are represented by the capacitance. The pneumatic feature is taken into account in the Becherif-Hissel model to obtain the comparable model for electrical components. The conservation of mass, energy, and charge is taken into account in pneumatic properties [35].

## 3. Introduction to low temperature combustion techniques

Conventional diesel engine running on petroleum and diesel fuel emits more oxides of nitrogen ($NO_x$), oxides of carbon ($CO_x$) and particulate matter (PM) around the world. Low-temperature combustion (LTC) technology in engine development has dropped the environmental effects by providing better combustion efficiency, and increased the engine efficiency and fuel economy. Several low-temperature combustion strategies are available such as homogeneous charged compression ignition (HCCI), premixed charged compression ignition (PCCI), and reactive controlled compression ignition (RCCI). Before combustion, the entire air and fuel is premixed in the LTC combustion mode. The combustion is controlled by a predetermined equivalency ratio and cylinder temperature which leads to reduce the soot formation, PM, and NOx emissions. In LTC mode, the combustion temperature could be maintained between 1800 and 2200 K, which means no NOx emissions are produced in the rich mixing region and no soot is formed below 1800 K in the lean mixing by Hoekman and Robbins.

### 3.1 Homogeneous charge compression ignition (HCCI)

The homogeneous charge compression ignition (HCCI) engine combines the combustion characteristics of both SI and CI engines in an IC engine. The fuel is premixed in the HCCI engine in the same in SI engines, and the fuel is auto-ignited to start the combustion in the same way in CI engines. Before combustion begins, the fuel is vaporized and homogeneously premixed with air. Due to lean-burn combustion, the HCCI has the ability to reduce $NO_x$ emissions and increased the brake thermal efficiency. The in-cylinder temperature is reduced via lean-burn combustion, resulting in decreased NOx emissions as observed by Komninos and Rakopoulos [36]. In addition, due to the increased displacement capacity, HCCI combustion improves brake thermal efficiency by 50%, while emitting less smoke than conventional diesel combustion. The HCCI engine's compression ratio and premixed fuel combustion has improved the brake thermal efficiency of engine and lower the smoke emissions as noticed by Desantes et al. [37]. The multi-zone auto ignition and spontaneous combustion of the entire mixture is promoted by the homogenous mixture and uniform equivalence ratio in the cylinder. Furthermore, flame propagation has little effect on combustion in the HCCI mode [38].

The unanticipated pressure rise and cycle to cycle variation are exacerbated by multi-zone combustion and unexpected ignition location. Also, knocking is caused by high oscillation frequency and unanticipated pressure surge as noticed by Ganesh and Nagarajan [39]. Contino et al. [40] reported that some of the techniques such as early direct injection, early multiple injection, water injection, port fuel injection, external cold EGR, variable valve timing, variable compression ratio, air preheating, and alcohol injection are commonly employed in HCCI to control combustion and emission. The biofuel auto ignition temperature and viscosity are higher than diesel hence a higher compression ratio was used in HCCI engine. The compression ratio for the various loads can be adjusted to enhance the combustion efficiency as noticed by Zhang, et al. [41]. By modifying the spark timing and spark plug placement, the spark aided HCCI engine was able to achieve combustion phasing and emission reduction [42]. The key factors that have been employed to detect the combustion phenomena in the HCCI engine are the pressure increase rate, combustion noise, and ringing intensity. In a real-time combustion application, the ringing intensity is primarily employed to detect the combustion noise for the needed cylinder pressure [43].

Because of the increased stroke volume, the higher compression ratio HCCI engine improves brake thermal efficiency by achieving the auto-ignition temperature of the fuel. High to low octane fuels can be utilized as a port fuel to solve knocking and NOx formation. In HCCI engine, keeping the inlet charge temperature is critical. Similarly, the HCCI engine's compression ratio could be maintained effectively between 10:1 and 28:1. Compression ratios of 10:1 were favored for higher cetane fuels like n-heptane, and 28:1 were preferred for high octane fuels like iso-octane. For biodiesel, the intermediate compression ratio was favored [44]. Alternative method for achieving lower emission in HCCI engine includes use of alcoholic fuels such as ethanol, n-butonal, and methanol. Due to oxygen enrichment, alcohol fuel accelerated premixed burning and complete oxidization of fuel. Also, because of the latent heat of vaporization is higher, it lowers the combustion temperature, enhancing the quenching effect [41]. The HCCI combustion's power output is mostly determined by the equivalency ratio and fuel intake. For the higher power production, the equivalence ratio should remain at 1 as noticed by Vinod Babu et al. [45].

### 3.2 Premixed charge compression ignition (PCCI)

Too early injection of fuel with a higher injection pressure can result in premixed charge compression ignition. Due to early fuel injection, the time between commencement of injection and start of combustion has been extended, considerably improving the homogeneity of the air-fuel mixture prior to combustion [41]. With a slightly higher intake charge temperature maintained at 170°C, the PCCI engine may operate from a minimal air-fuel ratio of 34:1 to an excessively lean air-fuel ratio of 80:1 [46]. In comparison to a standard SI engine, the PCCI combustion strategy uses lean-burn technology and operates on a higher compression ratio engine. After all of the fuel had been injected, the PCCI began to burn. Also, unlike traditional combustion, the combustion events are primarily identified by chemical kinetics and do not follow the diffusion mixed combustion and speed of burning. As a result, the injection pattern and fuel combustion do not overlap, reducing the odds of direct combustion control [47]. To achieve the premixed charge in the PCCI combustion, a single stage fuel injection pattern with an earlier start of injection was adopted. However, starting the injection too early causes wall impingement and wall wetness, resulting in incomplete combustion and higher HC and CO emissions. The fuel injection pattern has been adjusted with a split and multiple injection method to alleviate these issues. Despite the fact that the period of the many injections is completed before combustion begins. Controlling auto ignition by early injection is also a critical job in PCCI combustion. To manage the auto ignition and lengthen the ignition delay interval, a higher amount of EGR is used. EGR also aids in lowering in-cylinder temperature and NOx generation due to the dilution of a fresh charge mixture [48].

PCCI combustion has performed better than HCCI combustion due to the stability of the combustion by partially premixed charge and controlled auto ignition rage and temperature. The phasing of combustion in the PCCI is mostly determined by chemical kinetics, but it can also be influenced by altering the inlet charge temperature, EGR rate, and fuel injection time and pressure. PCCI combustion has used a variety of fuel patterns, including early single pulse injection, port fuel injection, advanced multiple injections, and advanced injection with a tiny amount of late injection. In the previous section, the effects of early and late injection timings were explored. The modest amount of late injection is mostly used to reduce smoke emissions [49]. The spray angle of 70° was employed to atomize the fuel within the combustion chamber in order to eliminate wall wetness during advanced injection [49]. To avoid the generation of HC and CO emissions, the compression ratio of the PCCI engine was kept at the same level as that of a regular diesel engine. Due to the low volatility and strong flammability of the fuel, PCCI combustion has several limitations, according to a few studies [50].

For low volatile fuels like kerosene, diesel, and biofuels, spark assisted PCCI combustion has been applied. When compared to conventional CI combustion, the use of low-quality cetane fuel in the spark aided PCCI strategy engine enhanced engine performance [51]. The partially premixed combustion mixture is generated in PCCI-DI dual-mode combustion by injecting a large volume of fuel in the intake port or early pilot injection, followed by conventional direct injection of the same or another fuel. Due to the ignition delay interval, the combustion phasing of the PCCI-DI dual-mode combustion is primarily determined by the pilot fuel quantity, and the combustion rate is determined by the pilot fuel ratio [31]. For premixed compression ignition low-temperature combustion, port fuel injection is preferred (PCI-LTC). To create a premixed mixture with a proper air-fuel ratio, single fuel or dual fuel port injection is employed. Dual fuel premixed LTC has a better brake thermal efficiency than single fuel LTC and has achieved a significant reduction in NOx and soot emissions. The single fuel premixed LTC has a higher cycle to cycle variance due to the low temperature and lean air-fuel ratio [52, 53]. Reactivity controlled compression ignition (RCCI) is the name given to the dual-fuel premixed LTC, and a thorough description of the RCCI will be given in the following sections.

### 3.3 Partially premixed charge compression ignition (PPCI)

Partially premixed charge compression ignition is related to the PCCI method, which is a hybrid of traditional diesel and HCCI combustion. However, for low cetane fuels, PPCI combustion is favored. Similar to the PCCI combustion method, a longer ignition delay period and improved air-fuel mixing can be accomplished. Few studies have shown that improved and delayed injection strategies can result in extended ignition delay in PPCI combustion. To achieve a longer ignition delay, low and moderate compression ratio were used, as well as moderate to high EGR dilution. The key benefit of PPCI mode over HCCI mode is that it releases less particulate matter and NOx while providing better combustion phasing. PPCI is divided into two categories: early injection PPCI and late injection PPCI. The fuel is injected at the middle of the compression stroke in early injection PPCI, and at the end of the compression stroke in late injection PPCI. The fuel-injection gases of the early injected PPCI variant are denser and cooler due to partial compression. Similarly, in the late injected PPCI model, the fuel-injected gases are colder and denser due to injection occurring on the expansion cycle, which lowers the temperature in the later stage [54]. Due to incomplete oxidation and non-optimal combustion phasing, the PPCI combustion used slightly more fuel than standard diesel combustion [55].

At low load, a greater EGR rate and a delayed injection time reduce the power output of both low and higher power engines. In EGR assisted PPCI combustion, the advanced injection method was used to avoid a reduction in power output. Another disadvantage of PPCI combustion is that it produces more HC and CO because the amount of non-oxidized fuel in the piston bowl and high-pressure squish region increases [56]. The addition of gasoline to the PPCI is another way to achieve lower NOx and soot emissions without using EGR. The main benefits of adding gasoline to the PPCI are that it reduces HC and CO emissions by reducing residual products in the cylinder [57]. For longer ignition delay times, most of the premixed heat release phase was seen, resulting in higher peak cylinder pressure and noise levels. When the ignition delay periods shorten, the diffusion heat release phase occurs, resulting in a state similar to that of ordinary diesel combustion [58].

### 3.4 Reactive controlled compression ignition (RCCI)

HCCI, PCCI, and RCCI are examples of sophisticated low-temperature combustion technology that have recently been created. RCCI, for example, increases research focus due to its versatility. By achieving low-temperature combustion, HCCI and PCCI improve engine efficiency and reduce pollutants, according to previous studies. These two technologies, however, have considerable limits, and they are not ideal for low and high load settings due to knocking, misfire, and a faster rate of pressure rise. Fuel alteration is required in the HCCI and PCCI combustion to overcome the difficulties [59, 60]. They also stated that combustion quality had improved across a broad range of engine operations Bessonette et al. [61] investigated the effect of a partially mixed gasoline/diesel charge in a CI engine from low to high load. Raw diesel is favored for the lowest load situation, while a higher percentage of gasoline blend is suited for the highest load condition, according to them. In a subsequent stage, this dual fuel PCCI operation is referred to as RCCI combustion [62]. Adjusting the low to high reactive fuel ratio and the injection pattern of the high reactive fuels to achieve the NOx to smoke trade-off and higher efficiency. Reactivity stratification in RCCI combustion can also be influenced by fuel qualities such as viscosity, volatility, and ignite ability.

Biodiesel has been tested in a variety of engines and under a variety of operating circumstances all around the world. Due to the presence of oxygen in the biodiesel fuel, NOx emissions were higher for the engine [63, 64]. The RCCI engine driven by gasoline/biodiesel was mathematically analyzed by Li et al. [65]. When comparing raw biodiesel to gasoline/biodiesel, the study found decreased NOx emissions in the gasoline/biodiesel operation. As a result, using biodiesel under the RCCI method may be a better alternative for reducing NOx pollution than using biodiesel-powered diesel engines. Hanson et al. [58] study the RCCI combustion utilizing direct-injected diesel and biodiesel mixture (B20) as a direct-injected fuel and gasoline, E85 (85% ethanol and 15% diesel blend), and E20 as a port fuel. In the RCCI combustion, the findings of the E20/diesel mixture show that maximum pressure and HRR dropped, allowing the peak load to increase by 2 bar (from 8 bar to 10 bar BMEP). The usage of E20 improves combustion efficiency while lowering the heat release rate and exhaust leakage. The combustion efficiency of gasoline/B20 RCCI operation was also increased by lowering the UHC, albeit with a greater CO. Fuel efficiency also improved, resulting in a 1.68 percent increase in BTE. In comparison to the RCCI gasoline/diesel operation, E85/B20 allowed the RCCI operation to increase the BTE from 40 to 43%. The use of biodiesel as a pilot fuel has improved the stability of the cyclic operation of RCCI engine powered by natural gas/biodiesel, according to Gharehghani et al. [66]. This is due to the fact that biodiesel contains oxygen, which raises the cetane number. In comparison to natural gas/diesel, the mixture of natural gas/biodiesel produced 1.6% higher BTE as noticed by Gharehghani et al. [66].

### 3.5 Low-temperature combustion advantages and challenges

The combustion temperature in the LTC mode was always lower than the combustion temperature in a regular diesel engine. There are primarily two strategies to achieve low-temperature combustion: one is to operate the engine with higher EGR, and the other is to operate with an excess air ratio 0 greater than 1 [67]. Fuel combusted and oxidized at higher temperatures under stoichiometric operating conditions, resulting in more NOx production. Also, due to a reduction in oxygen availability in the fuel spray periphery, maximum soot emission was observed under the stoichiometric condition compared to normal diesel combustion [68]. Higher fuel injection pressure is usually a viable approach for overcoming the aforementioned concerns. Higher fuel injection pressure promotes atomization, mixing, and vaporization. However, the key duty to be remedied in modern injection technology in low-temperature combustion is the wall impingement of fuel caused by spray tip penetration at increased fuel injection pressure [69]. Furthermore, improved injection strategies such as high-pressure injection and CRDI approaches reduce the ignition delay period and boost premixed phase combustion, resulting in increased $NO_x$ emissions. The ignition delay and combustion phasing will be lengthened by using a higher level of cold EGR, lowering the compression ratio, and using variable valve timing control to advance the exhaust valve opening. Increased ignition delay enhances air-fuel mixing, resulting in increased homogeneity in the air-fuel combination. Higher EGR rate and lower compression ratio reduce the cylinder peak pressure and temperature, which has a major impact on engine performance and higher fuel consumption.

Getting LTC mode to work in real-time settings with heavy engine load is difficult. It is impossible to manufacture engines with a larger amount of EGR. In addition, the engine's higher BTE should compensate for the increased EGR. In the LTC condition, an external charge booster is necessary to produce higher BTE [70]. When the engine is running at a higher RPM, moderate EGR with an intake charge booster raises the cylinder peak pressure. The combustion process changes depending on the engine load, and it is influenced by the different equivalency ratio and fuel mixing zone, making the engine demanding and difficult to modify the operating state for each load [71]. The real-time modern diesel engine employs dual fuel technology, multiple injection method, and negative valve overlapping. However, these technologies are costly and difficult to implement across the board. By increasing the premixed charge quantity while lowering peak pressure and temperature, these innovations reduce the fuel-rich zone [72].

## 4. Conclusions

This study provides a comprehensive overview of hydrogen production techniques and fuel cell vehicle also described about the low-temperature combustion (LTC) techniques and how it is improve the reliability and fuel efficiency of the CI engine combustion cycle with low emissions and noise. The important findings are presented in this review can be summarized below:

- Even though fuel cells have demonstrated and shown to be a very promising fuel, there are still a number of limitations that prohibit them from being used on a bigger scale than other fuels. The following are some of the most pressing issues that must be addressed right now: Compared to other kinds of energy, the FC has lower overall efficiency. The material and fabrication of the FC have high production costs.

- One of the most pressing concerns is the cost of hydrogen, as well as its storage. Because hydrogen is a relatively light and dangerous gas, it must be stored in special containers. Thermal management in the case of high-temperature fuel cell like solid oxide fuel cell is a type of fuel cell in which the temperature is higher than the ambient temperature.

- The size and weight of current fuel cell systems must be further reduced to meet the packaging requirements for automobiles. This applies not only to the fuel cell stack, but also to the ancillary components and major subsystems.

- PCCI combustion efficiently decrease the CO and HC emission as compared to the HCCI engine, but $NO_x$ and soot emissions were significantly increased with increase in premixed charge percentage. However, the smoke and $NO_x$ emissions were identified as minimum level when compared with conventional diesel engine combustion.

- Higher cycle-to-cycle variation, unpredictable pressure rise, combustion noise and knocking were occurred in the HCCI mode of combustion due to higher homogeneity and unpredictable auto-ignition zone.

- RCCI combustion is preferable for higher load condition due to combustion phase control and higher brake thermal efficiency than PCCI and HCCI modes. The use of natural gas as a reactive fuel was extending the load limit and attained the efficient, clean combustion which significantly decreases the $NO_x$ and soot emission as compared to other techniques.

- The double injection of high reactive fuel in the RCCI combustion decreases the peak pressure and ringing intensity which efficiently decrease the smoke and $NO_x$ emission. The advanced second injection in the RCCI increases the reactive controlled combustion and late second injection increase the mixed controlled combustion.

- The combustion efficiency was increased while using the B20 as the high reactive fuel. Due to the oxygen availability in the biodiesel promotes the oxidization process, which decreases the HC and CO emission as compared to the diesel/gasoline RCCI combustion.

Many experiments have extensively demonstrated that there is a wide and unexploited scope for improving low-temperature combustion using different fuel injection parameter and different reactive fuel injection. The overall study infers that depending on the operating condition, engine configuration parameters, fuel injection mechanism and fuel mixing method influenced more on the engine performance and emission characteristics. Hence, further research work will be needed to the trade-off between the NOx and soot emission with improvement in the engine performance.

## Acknowledgement

The authors would like to thank King Mongkut's University of Technology North Bangkok (Grant Contract No. KMUTNB-KNOW63-28, KMUTNB-Post-65-09, KMUTNB-Post-65-05) for financial support during this work.

## Author details

Babu Dharmalingam[1*], Ramakrishna Reddy Ramireddy[1], Santhoshkumar Annamalai[2], Malinee Sriariyanun[1], Deepakkumar Rajagopal[3] and Venkata Ramana Katla[4]

1 The Sirindhorn Thai-German International Graduate School of Engineering (TGGS) King Mongkut's University of Technology, North Bangkok

2 Mechanical Engineering, Kongu Engineering College, Tamil Nadu, India

3 Mechanical Engineering, Vellore Institute of Technology, Vellore, Tamil Nadu, India

4 Humanities and Sciences, K.S.R.M. College of Engineering, Andhra Pradesh, India

*Address all correspondence to: monsieurbabu@gmail.com

IntechOpen

© 2022 The Author(s). Licensee IntechOpen. This chapter is distributed under the terms of the Creative Commons Attribution License (http://creativecommons.org/licenses/by/3.0), which permits unrestricted use, distribution, and reproduction in any medium, provided the original work is properly cited.

## References


[1] Kazim A, Veziroglu TN. Utilization of solar hydrogen energy in the UAE to maintain its share in the world energy market for the 21st century. Renewable Energy. 2001;**24**:259-274

[2] Chi J, Hongmei Y. Water electrolysis based on renewable energy for hydrogen production. Chinese Journal of Catalysis. 2018;**39**:390-394

[3] Acar C, Dincer I. comparative assessment of hydrogen production methods from renewable and non-renewable methods. International Journal of Hydrogen Energy. 2014;**39**:1-12

[4] Boyano A, Blanco-Marigorta AM, Morosuk T, Tsatsaronis G. Exergoenvironmental analysis of a steam methane reforming process for hydrogen production. Energy. 2011;**36**:2202-2214

[5] Trane R, Dahl S, Skjøth-Rasmussen MS, Jensen AD. Catalytic steam reforming of biooil. International Journal of Hydrogen Energy. 2012;**37**:6447-6472

[6] Huang, Dincer I. Parametric analysis and assessment of a coal gasification plant for hydrogen production. International Journal of Hydrogen Energy. 2014;**39**:3294-3303

[7] Veziroglu TN, Barbir F. Hydrogen Energy Technologies. Vienna: UNIDO; 1998

[8] Levin DB, Pitt L, Love M. Biohydrogen production: prospects and limitations to practical application. International Journal of Hydrogen Energy. 2004;**29**(2):173-185

[9] Barbir F. PEM electrolysis for production of hydrogen from renewable energy sources. Solar Energy. 2005;**78**:661-669

[10] Cipriani G, Di Dio V, Genduso F, La Cascia D, Liga R, Miceli R, et al. Perspective on hydrogen energy carrier and its automotive applications. International Journal of Hydrogen Energy. 2014;**39**:8482-8494

[11] Ni M, Leung MKH, Leung DYC. Technological development of hydrogen production by solid oxide electrolyzer cell (SOEC). International Journal of Hydrogen Energy. 2008;**33**:2337-2354

[12] Trasatti S. Water electrolysis: Who first. Journal of Electroanalytical Chemistry. 1999;**479**:90-91

[13] Shiva Kumar S, Ramakrishna SUB, Srinivasulu Reddy D, Bhagawan D, Himabindu V. Synthesis of polysulfone and zirconium oxide coated asbestos composite separators for alkaline water electrolysis. Chemical Engineering & Process Techniques. 2017;3:1035/1-1035/6

[14] Liang M, Yu B, Wen M, Chen J, Xu J, Zhai Y. Preparation of LSM-YSZ composite powder for anode of solid oxide electrolysis cell and its activation mechanism. Journal of Power Sources. 2009;**190**:341-345

[15] Kadier A, Simayi Y, Abdeshahian P, Azman NF, Chandrasekhar K, Kalil MS. A comprehensive review of microbial electrolysis cells (MEC) reactor designs and configurations for sustainable hydrogen gas production. Alexandria Engineering Journal. 2016;**55**:427-443

[16] Kadier A, Kalil MS, Abdeshahian P, Chandrasekhar K, Mohamed A, Azman NF, et al. Recent advances and emerging challenges in

microbial electrolysis cells (MECs) for microbial production of hydrogen and value-added chemicals. Renewable and Sustainable Energy Reviews. 2016;**61**:501-525

[17] Cheng J, Zhang V, Chen G, Zhang Y. Study of IrxRu1-xO2 oxides as anodic electro catalysts for solid polymer electrolyte water electrolysis. Electrochimica Acta. 2009;**54**:6250-6256

[18] Inci M, Büyük M, Demir MH, Ilbey G. A review and research on fuel cell electric vehicles: Topologies, power electronic converters, energy management methods, technical challenges, marketing and future aspects. Renewable and Sustainable Energy Reviews. 2021;**137**:110

[19] Niu W, Song K, Zhang Y, Xiao Q, Behrendt M, Albers A, et al. Influence and optimization of packet loss on the internet-based geographically distributed test platform for fuel cell electric vehicle powertrain systems. IEEE Access. 2020;**8**:20708-20716

[20] D'epature C, Lhomme W, Sicard P, Bouscayrol A, Boulon L. Real-time backstepping control for fuel cell vehicle using supercapacitors. IEEE Transactions on Vehicular Technology. 2018;**67**(1):306-314

[21] Garcia-Torres F, Vilaplana DG, Bordons C, Roncero-S'anchez P, Ridao MA. Optimal management of microgrids with external agents including battery/fuel cell electric vehicles. IEEE Transactions on Smart Grid. 2019;**10**(4):4299-4308

[22] Mallikarjuna Reddy B, Samuel P. Analysis, modelling and implementation of multi-phase single-leg DC/DC converter for fuel cell hybrid electric vehicles. International Journal of Modelling and Simulation. 2019;**9**:279-290

[23] Ahmadi P, Torabi SH, Afsaneh H, Sadegheih Y, Ganjehsarabi H, Ashjaee M. The effects of driving patterns and PEM fuel cell degradation on the lifecycle assessment of hydrogen fuel cell vehicles. International Journal of Hydrogen Energy. 2020;**45**(5):3595-3608

[24] Trimm DL, Önsan ZI. Onboard fuel conversion for hydrogen-fuel-cell-driven vehicles. Catalysis Reviews. 2001;**43**(1-2):31-84

[25] Saib S, Hamouda Z, Marouani K. Energy management in a fuel cell hybrid electric vehicle using a fuzzy logic approach. In: 5th International Conference on Electrical Engineering-Boumerdes (ICEE-B). Algeria: IEEE; 2017. pp. 1-4

[26] Yang B, Zhu T, Zhang X, Wang J, Shu H, Li S, et al. Design and implementation of battery/SMES hybrid energy storage systems used in electric vehicles: A nonlinear robust fractional-order control approach. Energy. 2020;**191**:116510

[27] Schneider MT, Schade B, Grupp H. Innovation process 'fuel cell vehicle': What strategy promises to be most successful? Technology Analysis & Strategic Management. 2004;**16**(2):147-172

[28] Uzunoglu M, Alam MS. Dynamic modeling, design and simulation of a PEM fuel cell/ultra-capacitor hybrid system for vehicular applications. Energy Conversion and Management. 2007;**48**(5):1544-1553

[29] García P, Fern'andez LM, Torreglosa JP, Jurado F. Operation mode control of a hybrid power system based on fuel cell/battery/ultracapacitor for an electric tramway. Computers and Electrical Engineering. 2013;**39**(7):1993-2004

[30] Boukettaya G, Krichen L. A dynamic power management strategy of a grid connected hybrid generation system using wind, photovoltaic and flywheel energy storage system in residential applications. Energy. 2014;**71**:148-159

[31] Parks, Vitaly Prikhodko, John M.E. Storey, Teresa L. Barone, Samuel A. Lewis Sr., Michael D. Kass, Shean P. Huff. Emissions from premixed charge compression ignition (PCCI) combustion and affect on emission control devices. Catalysis Today. 2010;**151**(3-4):278-284

[32] Ren G, Ma G, Cong N. Review of electrical energy storage system for vehicular applications. Renewable and Sustainable Energy Reviews. 2015;**41**:225-236

[33] Ceraolo M, Miulli C, Pozio A. Modelling static and dynamic behaviour of proton exchange membrane fuel cells on the basis of electro-chemical description. Journal of Power Sources. 2003;**113**(1):131-144

[34] Sadli I, Thounthong P, Martin JP, Raël S, Davat B. Behaviour of a PEMFC supplying a low voltage static converter. Journal of Power Sources. 2006;**156**(1):119-125

[35] Becherif M, Hissel D, Gaagat S, Wack M. Three order state space modeling of proton exchange membrane fuel cell with energy function definition. Journal of Power Sources. 2010;**195**(19):6645-6651

[36] Komninos NP, Rakopoulos CD. Heat transfer in HCCI phenomenological simulation models: A review. Applied Energy. 2016;**181**:179-209

[37] Desantes JM, García-Oliver JM, Vera-Tudela W, López-Pintor D, Schneider B, Boulouchos K. Study of the auto-ignition phenomenon of PRFs under HCCI conditions in a RCEM by means of spectroscopy. Applied Energy. 2016;**179**:389-400

[38] Megaritis A, Yap D, Wyszynski ML. Effect of water blending on bioethanol HCCI combustion with forced induction and residual gas trapping. Energy. 2007;**32**(12):2396-2400

[39] Ganesh D, Nagarajan G. Homogeneous charge compression ignition engine (HCCI) combustion of diesel fuel with external mixture formation. Energy. 2010;**35**:148-157

[40] Francesco C, Masurier J-B, Foucher F, Lucchini T, D'Errico G, Dagautc P. CFD simulations using the TDAC method to model iso-octane combustion for a large range of ozone seeding and temperature conditions in a single cylinder HCCI engine. Fuel. 2014;**137**(1):179-184

[41] Zhang HFL, Yu J, Yao M. Direct numerical simulation of nheptane/air auto-ignition with thermal and charge stratifications under partially-premixed charge compression ignition (PCCI) engine related conditions. Applied Thermal Engineering. 2016;**104**:516-526

[42] Wiemann S, Hegner R, Atakan B, Schulz C, Kaiser SA. Combined production of power and syngas in an internal combustion engine—Experiments and simulations in SI and HCCI mode. Fuel. 2018;**215**:40-45

[43] Maurya RK, Saxena MR. Characterization of ringing intensity in a hydrogen-fueled HCCI engine. International Journal of Hydrogen Energy. 2018;**43**(19):9423-9437

[44] Fukushima N, Katayama M, Naka Y, Oobayashi T, Shimura M, Nada Y, et al. Combustion regime classification of HCCI/PCCI combustion using

Lagrangian fluid particle tracking. Proceedings of the Combustion Institute. 2015;**35**:3009-3017

[45] Vinod Babu VBM, Madhu Murthy MMK, G. Amba Prasad Rao. Butanol and pentanol: The promising biofuels for CI engines–A review. Renewable and Sustainable Energy Reviews. 2017;**78**:1068-1088

[46] Ulaş E, Leermakers N, Somers B, de Goey P. Modeling of PCCI combustion with FGM tabulated chemistry. Fuel. 2014;**118**(15):91-99

[47] Girish BE, Neeraj S, Suryawanshi JG. Investigations on premixed charge compression ignition (PCCI) engines: A review. Fluid Mech Fluid. 2017;**78**:1068-1088

[48] Wang Y, Li H, Longbao Z, Wei L. Effects of DME pilot quantity on the performance of a DME PCCI-DI engine. Energy Conversion and Management. 2010;**51**(4):648-654

[49] Jia M, Xie M, Wang T, Peng Z. The effect of injection timing and intake valve close timing on performance and emissions of diesel PCCI engine with a full engine cycle CFD simulation. Applied Energy. 2011;**88**(9):2967-2975

[50] Pandey SK, Sarma Akella SR, Ravikrishna RV. Novel fuel injection strategies for PCCI operation of a heavy-duty turbocharged diesel engine. Applied Thermal Engineering. 2018;**143**:883-898

[51] Verma G, Sharma H, Thipse SS, Agarwal AK. Spark assisted premixed charge compression ignition engine prototype development. Fuel Processing Technology. 2016;**152**:413-420

[52] Kocher L, Van Alstine D, Magee M, Shaver G. A nonlinear model-based controller for premixed charge compression ignition combustion timing in diesel engines. In: Proceedings of the American Control Conference. Washington, DC, USA: IEEE; 2013

[53] Li T, Moriwaki R, Ogawa H, Kakizaki R, Murase M. Dependence of premixed low-temperature diesel combustion on fuel ignitability and volatility. International Journal of Engine Research. 2012;**13**:14-27

[54] Singh AP, Ayush J, Agarwal AK. Fuel injection strategy for PCCI engine fueled by mineral diesel and biodiesel blends. Energy Fuel. 2017;**31**(8):8594-8607

[55] Liu H, Ma S, Zheng Z, Zheng Z, Yao M. Study of the control strategies on soot reduction under early injection conditions on a diesel engine. Fuel. 2015;**139**:472-481

[56] Fridriksson H, Sundén B, Tunér M, Andersson Ö. Heat transfer in diesel and partially premixed combustion engines; A computational fluid dynamics study. Heat Transfer Engineering. 2017;**38**:1481-1495

[57] Mina M, Moghiman M. Effects of nanoadditives on pollutants emission and engine performance in a urea-SCR equipped diesel engine fueled with blended-biodiesel. Fuel. 2018;**222**(15):402-406

[58] Hanson CB, Choi SB. Effects of operating parameters on mode transition between low temperature combustion and conventional combustion in a light duty diesel engine. International Journal of Engine Research. 2012;**14**(3):231-246

[59] Epping K, Aceves S, Bechtold R, Dec J. The potential of HCCI combustion for high efficiency and low emissions. In: SAE Technical Papers. SAE International in United States; 2002

[60] Kanda T, Hakozaki T, Uchimoto T, Hatano J, Kitayama N, Sono H. PCCI operation with early injection of conventional diesel fuel. In: SAE Technical Papers. 2005

[61] Bessonette PW, Schleyer CH, Duffy KP, Hardy WL, Liechty MP. Effects of fuel property changes on heavy-duty HCCI combustion. In: SAE Technical Papers. SAE International in United States; 2007

[62] Li Y, Jia M, Chang Y, Xu Z, Xu G, Hong L, Wang T. Principle of determining the optimal operating parameters based on fuel properties and initial conditions for RCCI engines. Fuel. 2018;**216**:284-295

[63] Vallinayagam R, Vedharaj S, Yang WM, Lee PS, Chua KJE, Chou SK. Combustion performance and emission characteristics study of pine oil in a diesel engine. Energy. 2013;**57**:344-351

[64] Vedharaj S, Vallinayagam R, Yang WM, Chou SK, Chua KJE, Lee PS. Experimental investigation of kapok (Ceiba pentandra) oil biodiesel as an alternate fuel for diesel engine. Energy Conversion and Management. 2013;**75**:773-779

[65] Li J, Ling X, Liu D, Yang W, Zhou D. Numerical study on double injection techniques in a gasoline and biodiesel fueled RCCI (reactivity controlled compression ignition) engine. Applied Energy. 2018;**211**:382-392

[66] Gharehghani A. Load limits of an HCCI engine fueled with natural gas, ethanol and methanol. Fuel. 2019;**239**:1001-1014

[67] Huang H, Teng W, Liu Q, Zhou C, Wang Q, Wang X. Combustion performance and emission characteristics of a diesel engine under low-temperature combustion of pine oil–diesel blends. Energy Conversion and Management. 2016;**128**:317-326

[68] Goldsborough SS, Hochgreb S, Vanhove G, Wooldridge MS, Curran HJ, Sung CJ. Advances in rapid compression machine studies of low- and intermediate-temperature autoignition phenomena. Progress in Energy and Combustion Science. 2017;**63**:1-78

[69] Wu B, Zhan Q, Yu X, Lv G, Nie X, Liu S. Effects of miller cycle and variable geometry turbocharger on combustion and emissions in steady and transient cold process. Applied Thermal Engineering. 2017;**118**:621-629

[70] Thangaraja J, Kannan C. Effect of exhaust gas recirculation on advanced diesel combustion and alternate fuels—A review. Applied Energy. 2016;**180**:169-184

[71] Imtenan S, Varman M, Masjuki HH, Kalam MA, Sajjad H, Arbab MI, et al. Impact of low temperature combustion attaining strategies on diesel engine emissions for diesel and biodiesels: A review. Energy Conversion and Management. 2014, 2014;**80**:329-356

[72] Girish BE, Neeraj S, Suryawanshi JG. Investigations on premixed charge compression ignition (PCCI) engines: A review. Journal of Mechanical Science and Technology. 2016;**30**(11):5269-5274

Chapter 9

# Performance and Emission Characteristics of Hydrogenation Derived Renewable Diesel as Diesel Engine Fuel

*Omojola Awogbemi, Daramy Vandi Von Kallon and Josiah Pelemo*


## Abstract

Growing anxieties about the continued depletion of fossil fuel reserves, improving the performance of diesel engines, and mandates to reduce greenhouse gas emissions have made the search for alternative fuels for diesel engines more imperative. Hydrogenation Derived Renewable Diesel (HDRD) is recognized as a sustainable, reliable, and cost-effective alternative to petroleum-based diesel (PBD) fuel for compression ignition (CI) engines. This may be because the physicochemical properties of HDRD are similar to that of PBD fuel. The current effort examines the performance and emission characteristics of HDRD in unmodified CI engines. Performance emissions characteristics such as power, torque, brake specific fuel consumption, thermal efficiency, nitrogen oxides, carbon monoxide, carbon dioxide, particulate matter, and exhaust gas temperature were interrogated and compared with that of PBD fuel in a CI engine. The outcome of the study shows that HDRD is better than biodiesel and a sustainable replacement for PDB fuel to achieve improved performance and reduced emissions of CI engines. Going forward, more investigations are needed to further simplify the preparation and democratize the utilization of HDRD as CI fuels for various applications.


**Keywords:** compression ignition engine, HDRD, performance, emission, fuel, renewable

## 1. Introduction

Fossil fuels, which originated from the anaerobic decomposition of carbon-rich dead plants and animals, have continued to dominate the energy source and drive the industrialized world. About 70–80% of the global energy consumption is gotten from fossil fuels [1]. Fossil fuels, comprising coal, oil, and gas, are non-renewable and the main contributor to global warming and climate change. Extraction, refining, and utilization of fossil fuels have caused unimaginable degradation of the environment. Also, going by the rate of consumption, the global oil reserves estimated to be 1.65



IntechOpen

trillion barrels may be fully depleted within the next five decades [2]. Also, increased population, accelerated industrial revolution, and increased mechanized farming has continued to cause an increased utilization of fossil fuels and consequently increased emission. The global consumption of fossil fuels was recorded as 121, 531 Terawatt-hour (TWh), 129,855 TWh, and 136,131 TWh for 2010, 2015, and 2019 respectively. On the other hand, the total carbon dioxide ($CO_2$) emissions were documented to be 31.49 Billion Tonnes, 33.39 Billion Tonnes and 34.35 Billion Tonnes respectively (**Figure 1**). However, fuel consumption and $CO_2$ emission plummeted in 2020 due to the Covid 19-imposed lockdown. With the relaxation of various travel restrictions and increased commercial and industrial activities, fuel consumption and emissions are expected to increase substantially. This is expected to escalate environmental degradation and climate change.

The use of biofuels is one of the panaceas for the unfavorable effects of fossil fuels in diesel engine applications. Biofuels are renewable fuels generated from fresh and living organisms. They usually occur in solid, liquid, or gaseous forms. Biofuels enjoy several benefits like renewability, ecological friendliness, feedstock accessibility, the elasticity of the production methods, and their amenability to existing fossil fuels pipeline infrastructure. Also, biofuels demonstrate matchless capability for the sustenance of the ecosystem [5, 6]. However, the high cost of production, increased NOx emission, and increased engine wear are major setbacks to the use of some biofuels. Also, the conflict between some of the feedstocks with the food chain, undeveloped production technologies, and unfavorable government policies have continued to militate against the wide production and utilization of biofuels in many jurisdictions. Notwithstanding these impediments and complications, biofuels remain a clean, safe, and sustainable replacement for fossil fuels and a strategic resource for $CO_2$ reduction and carbon mitigation to avert the ominous environmental catastrophe [5, 7, 8].

The transport sector utilizes more than 90% of the total fossil fuel products and about 28% of the total global energy and is a major contributor to the emission of dangerous gases [9]. Solid biofuels (wood chips, briquettes, sawdust), liquid biofuels (biodiesel, renewable diesel, bioethanol), and gaseous biofuels (biogas, biomethane,

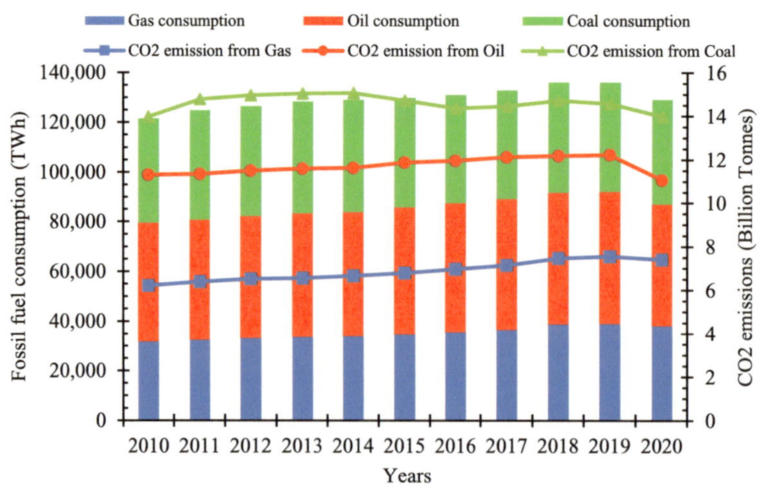


**Figure 1.**
*Global consumption (TWh) and $CO_2$ emission (billion Tonnes) from coal, oil, and gas 2010–2020. Adapted from [3, 4].*

syngas) have been used as reliable and environmentally benign candidates for fossil-based fuels. The overall energy consumed in the transportation sector was 110 million terra joule (TJ) in 2015 while 129 billion liters of liquid biofuel were utilized in 2016 and the quantity is predicted to increase to 180 billion liters by 2050 [10]. The number of global on-road vehicles which was about 1.2 billion is projected to increase to 2 billion and 2.5 billion vehicles in 2035 and 2050 respectively [11]. Compression ignition (CI) engines because of their versatility, strength, and multi-faceted usage, have continued to be used as passenger vehicles, construction machinery, agricultural equipment as well as rail and heavy-duty trucks. Fueling these engines with petroleum-based diesel (PBD) fuel will exacerbate the detrimental effects on the health and environment.

To increase the share of renewable fuels in the transportation sector energy mix, renewable energy sources and other less polluting fuels such as electricity, natural gas, bioethanol, propane, biodiesel, jet fuel, and biomethane have been tested. These renewable and less polluting energy sources have been found to meet the huge demand and requirements for bioenergy and secure the energy supply. For example, the deployment of electric vehicles has been plagued with the high cost, infrastructural deficit, and long duration of charging of the battery in many jurisdictions. The liquid biofuels have the advantage of being produced for wastes and other renewable sources with a low carbon footprint, thereby making them a more economically viable option [12]. Globally, more concerted efforts geared at increasing the production and utilization of renewable fuels are needed to achieve Sustainable Development Goals and ensure environmental sustainability. Also, more public awareness and education, targeted policy, and research and development (R & D) aimed at increasing the production and utilization of liquid biofuels should be intensified.

### 1.1 Motivation, aim, and scope

Concerns over the environmental, social, economic, and supply of world energy have been addressed by governments in various jurisdictions. Possible solutions include the introduction of biofuel into the energy mix by encouraging and incentivizing the production and utilization of biofuels. The desire to popularize the application of these biofuels, particularly for CI engines applications, has gained considerable attention in recent years. A lot of studies have been carried out and reported on the production and utilization of biodiesel and bioethanol as CI engine fuels. In previous research, Saravanan et al. [13], Khan et al. [14], Krishna et al. [15], and Shirneshan et al. [16], among several others investigated the performance and emission characteristic of biodiesel, ethanol, and biodiesel-ethanol blends on CI engines. The outcomes of their studies showed the benefits and shortcomings of the deployment of these renewable fuels in CI engines with particular attention to Hydrogenated Derived Renewable Diesel (HDRD). In their various studies, they confirmed the superiority of HDRD over biodiesel and PBD fuels for CI engine transport applications. Recently Chia et al. [12] and Kumar et al. [17] demonstrated their preference for HDRD over biodiesel, ethanol, and other liquid biofuels. They cited the superior heating value, excellent transport and storage stability, and non-corrosive nature of HDRD as some of the reasons.

Bearing in mind the ongoing efforts at finding more sustainable renewable fuels to power CI engines, and the various challenges encountered with the usage of biodiesel and bioethanol, the relevant question to ask is how has HDRD performed as an alternative fuel for CI engines? . How effective is HDRD as CI engine fuel from the standpoint of performance and emission characteristics? The motivation for this study is the desire to improve the quantum and quality of information and awareness on HDRD as a transportation fuel to assist consumers, fuel refiners, and engine manufacturers in making informed decisions in fuel selection. The current effort aims to investigate the performance and emission characteristics of CI engines fueled with HDRD.

Overall, the outcomes of this work will equip governments, policy formulating agencies, industry experts, researchers, and the general public with the requisite information on the application of HDRD in CI engines. It is also hoped that research funding bodies will be encouraged to provide more funds for future R & D to stimulate investigation into novel strategies for production and utilization of the HDRD. To achieve this, the article will be divided into subheadings to discuss HDRD as a renewable fuel, performance of HDRD in CI engine, emission characteristics of HDRD as CI engine fuel, implications of HDRD as CI engine fuel, and conclusion. The current effort is, however, limited to a desktop review of published literature on the performance and emission behavior of HDRD in diesel engines.

## 2. HRDD as a renewable fuel

HDRD, otherwise called renewable diesel, green diesel, and hydrotreated vegetable oil, is a second-generation liquid biofuel. HDRD is chemically identical to PBD fuel but not the same as biodiesel. Biodiesel, also referred to as Fatty Acid Methyl Ester (FAME), is a mono-alkyl ester mostly generated by the catalytic transesterification process, HDRD is a blend of straight-chain and branched paraffin hydrocarbons within the $C_{15}$–$C_{18}$ range. The similarities in properties of petroleum diesel and HDRD allow it to meet the automotive fuel specifications, seamless application of HDRD in CI engines, and use of the same transport infrastructure [18, 19]. The global production of HDRD grew from 1.5 billion liters in 2011 to 9.5 billion liters in 2017 and is projected to become 15 billion liters in 2024 [20, 21]. Also, due to attractive properties and advantageous utilization of HDRD, the production capacity and the share of

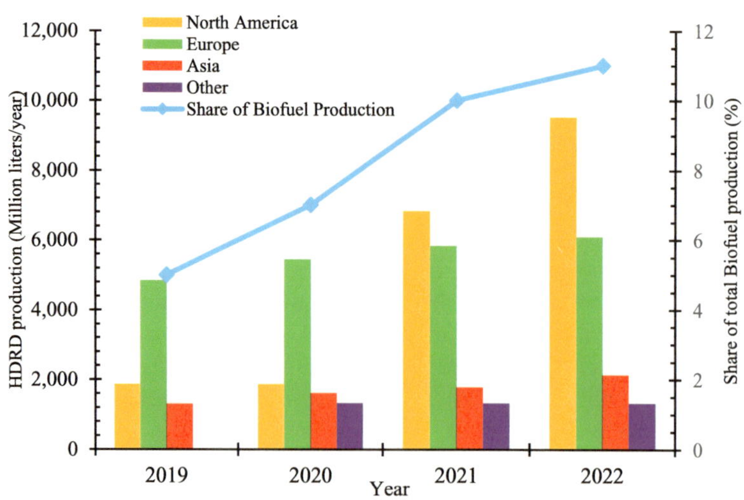


**Figure 2.**
*Global HDRD production capacity and share in biodiesel production 2019–2022. Adapted from [22].*

biofuel production have been increasing since 2019, globally (**Figure 2**). This trend is expected to continue.

To meet up with the growing demand and utilization of HDRD, many commercial production plants have been installed and commissioned using advanced technologies (**Table 1**). **Figure 3** shows the producer, capacity/year, and country of location of HDRD plants, worldwide. The HDRD is usually produced through catalytic

| Company | Location | Capacity (tonnes/year) | Technology/process |
|---|---|---|---|
| Neste | The Netherlands | 1,000,000 | NExBTL |
| Neste | Singapore | 1,000,000 | NExBTL |
| Diamond Green Diesel | USA | 900,000 | Ecofining™ |
| UOP/Eni | Italy | 780,000 | Ecofining™ |
| Total | France | 500,000 | Vegan® by Axens |
| Petro oil & Gas | UAE | 500,000 | UOP Renewable jet fuel process |
| Neste | Finland | 380,000 | NExBTL |
| REG Inc | USA | 250,000 | Dynamic Fuels LLC |
| AltAir Fuels | USA | 130,000 | Ecofining™ |
| UPM Biofuels | Finland | 100,000 | UPM BioVerno |
| Petro oil & Gas | UAE | 500,000 | UOP Renewable jet fuel process |

*REG = Renewable Energy Group, UAE = United Arab Emirate, USA = United States of America.*

**Table 1.**
*HDRD production plants [20, 21, 23].*

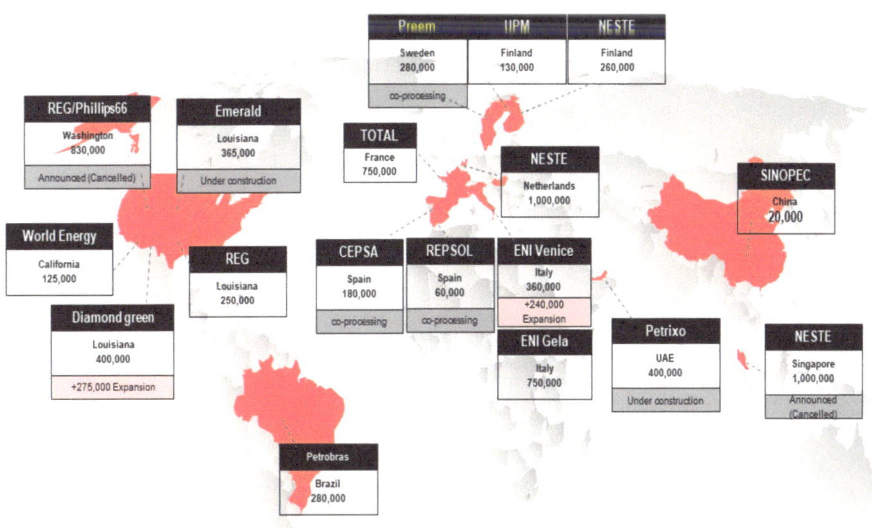


**Figure 3.**
*Locations, company, and capacity/year of major HDRD plants [21].*

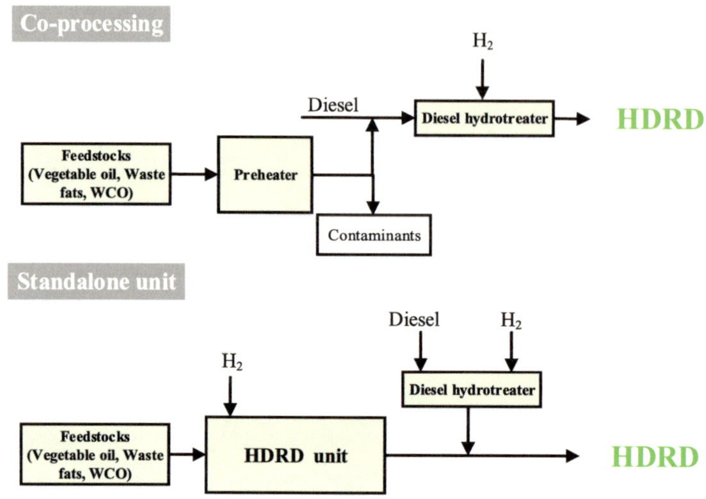


**Figure 4.**
*Schematic diagram of HDRD production by hydroprocessing. Adapted from [25].*

hydroprocessing, decarboxylation, and/or decarbonylation of triacylglycerol. During hydroprocessing, hydrogen is applied for the removal of oxygen from the triglyceride molecules through decarboxylation and hydrodeoxygenation, depending on the catalyst selection and process conditions [24]. This can be accomplished either through a co-processing arrangement of a distillate hydroprocessing unit or by building a standalone unit as shown in **Figure 4**. **Figure 5** shows the reaction pathways for HDRD production.

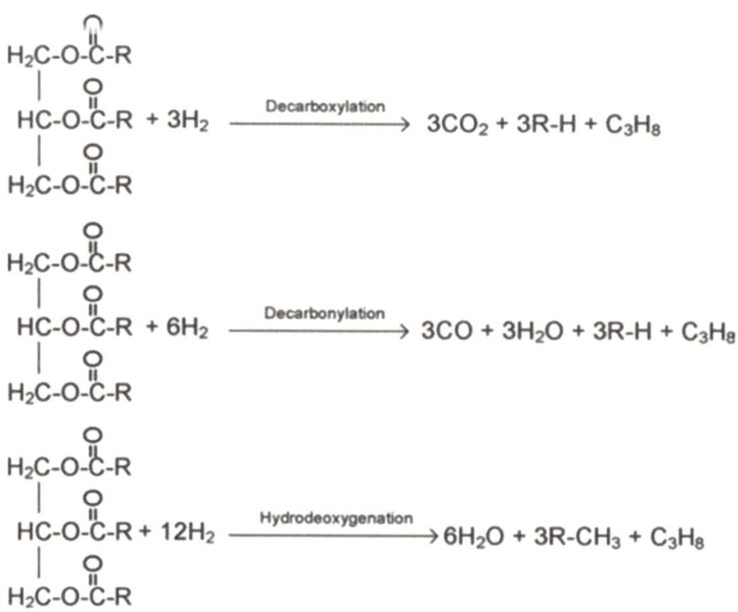


**Figure 5.**
*Reaction pathways for HDRD production [26].*

| Feedstock | Production process | Yield (%) | Remark |
|---|---|---|---|
| Waste cooking oil | Deoxygenation | >95 | Cheap and readily available feedstock |
| Waste cooking oil | Hydrodeoxygenation | 43.8 | Low-cost and non-edible feedstock |
| Waste cooking oil | Hydrotreatment | 100 | High product yield<br>Waste to fuel |
| Karanja oil | hydroprocessing | 80 | Nonedible oil |
| Karanja oil | hydrogenation | 100 | High product yield |
| Karanja oil | Hydrotreating | 82.6 | Non-edible oil |
| Algae oil | Hydroprocessing | 80 | Non-edible feedstock |
| Palm oil | Hydrodeoxygenation | 100 | High product yield |
| Palm oil | Deoxygenation | >95 | Readily available feedstock<br>Easy conversion method |
| Palm oil | Hydrodeoxygenation | 100 | High product yield |
| Palm oil | Hydrodeoxygenation | >89 | Readily available feedstock |
| Animal Fats | Deoxygenation | 90 | Waste to fuel, cheap feedstock |
| Animal Fats | Deoxygenation | 100 | High product yield |
| Animal Fats | Deoxygenation | 94.2 | High product yield, non-edible feedstock |
| Jatropha oil | Hydroprocessing | 98.5 | High product yield |

**Table 2.**
*Performance of some renewable feedstocks for HDRD production [12].*

Generally, HDRD can be synthesized from feedstocks such as sugar, starch, or cellulosic materials through various techniques like catalytic conversion, Biomass to Liquid, and pyrolysis. Also, vegetable oil, waste cooking oil, waste animal fats, recovered fats, and other triglycerides-bearing oils are converted into HDRD by pyrolysis and hydroprocessing. The outcome of the use of some renewable feedstocks such as waste cooking oil, animal fats, algae oil, jatropha oil, and Karanja oil have shown high product yield under moderate production conditions (**Table 2**). The conversion of triglycerides to HDRD through hydroprocessing entails chemical reactions such as hydrogenation, decarboxylation, decarbonylation, and hydrodeoxygenation reactions [12]. HDRD is produced in line with the methods and specifications of the American Society for Testing and Materials (ASTM) D975 and the European Committee for Standardization EN 590 [27]. **Table 3** shows the specifications and International Standards Organizations (ISO) test method for HDRD.

The cetane number of HDRD, a measure of the ignition quality of diesel fuel in CI engines, is usually between 820 $Kg/m^3$ and 845 $Kg/m^3$ and higher than PBD fuel and biodiesel. The high value of cetane number allows a CI engine fueled with HDRD to operate with higher thermal efficiency and at a lower fuel consumption [12]. The lower value of density, compared with biodiesel or PBD fuel indicates reduced volumetric heating value and increased fuel consumption. The high lubricity of HDRD ensures minimum engine wear, noiseless running, and smooth engine operation [12, 20].

CI engines are a form of an internal combustion engine. As heat engines, CI engines convert the chemical energy in the fuel into mechanical work [30]. The diesel fuel is passed into the engine through a fuel injector into the cylinder and mixed with preheated air where the mixture auto ignites due to the movement of the piston.

| Property | Unit | EN 590 | ASTM D975 | Test method |
|---|---|---|---|---|
| Density @15°C | $kg/m^3$ | 820–845 | — | EN ISO 3675, EN ISO 12185 |
| Kinematic viscosity @ 40°C | $mm^2/s$ | 2.0–4.5 | 1.9–4.1 | EN ISO 3104 |
| Flashpoint (Closed cup) | °C | 55 | 52 | EN ISO 2719 |
| Cloud point | °C | — | W: −5 °C<br>S: 3 °C | — |
| Cold filter plugging point | °C | — | W: −15 °C<br>S: −5°C | — |
| Cetane number | — | 51 | 40 | EN ISO5165 |
| Cetane index | — | 46 | 40 | EN ISO 4264 |
| Water and sediment | % vol | 0.02w/w | 0.06 | EN ISO 12937 |
| Total contamination | ppm | 24 | — | EN ISO 12662 |
| Carbon residue | wt % | 0.3 | 0.36 | EN ISO 10370 |
| Total ash | wt % | 0.01 | 0.01 | EN ISO 6245 |
| Total sulfur | mg/kg | 10 | 15 | EN ISO 20846, EN ISO 20847, EN ISO 2088 |
| Lubricity @ 60°C | WSD, microns | 460 | 520 | EN ISO 12156-1 |
| Copper strip corrosion | 3 h @ 50 °C | No. 1 | No. 3 | EN ISO 2160 |

*W = winter, S = summer.*

**Table 3.**
*Specifications and testing methods of HDRD [28, 29].*

The piston reciprocates between the Bottom Dead Center (BDC) and the Top Dead Center (TDC). The application of HDRD in CI engines makes the engine behave in a certain way and the efficacy of the fuel is measured in line with some set performance criteria and emission characteristics.

## 3. Performance of HDRD in CI engines

Desirous to find solutions to the obvious inadequacies in the utilization of PBD fuel in CI engines, HDRD has been used by various researchers. However, the major performance criteria used in measuring the performance characteristics include power, torque, fuel consumption, thermal efficiency, and mean effective pressure. For example, the brake specific fuel consumption (BSFC) is an important performance metric that measures the conversion of the fuel to useful work while mechanical efficiency calculates the effectiveness of the engine as the ratio of the brake power to the indicated power. On the other hand, the brake thermal efficiency (BTE) measures the ability of the engine to efficiently convert the chemical energy in the fuel to useful work.

Using these performance criteria, the engine metrics of HDRD is compared with that of PBD fuel when used in a CI engine are compiled in **Table 4**. When HDRD was used to power a 6.5 liters, indirect-injection, water-cooled military diesel engine, it was reported that an increase in load led to increased fuel consumption and improved brake mean effective pressure (BMEP). Also, at a given fuel consumption threshold, an increment in engine speed caused a reduction in the brake torque. It was also reported that the best BSFC was achieved at high loads and low engine speed. This is because at low speed, engine friction is reduced and fuel consumption is minimized. When compared with PBD fuel, the application of HDRD resulted in better engine performance in all the engine metrics measured [31]. Also, Ogunkoya et al. [32], Mangus et al. [33], and Kim et al. [34] reported that their respective tested CI engines fueled with HDRD presented lower BSFC when compared with PBD fuel.

| Fuel tested | Engine parameters | Test conditions | Result of the test | Remark | Ref. |
|---|---|---|---|---|---|
| HRDR and PBD | 6.5 L, WC, indirect injection | Varying engine speed and load | ↑ BMEP<br>↑ BSFC at high load and low engine speed | HDRD was found better than PBD fuel | [31] |
| HDRD and PBD | 1C, 1S, DI, AC | Constant engine speed at varying load | Higher mechanical efficiency and BTE with increased load<br>BSFC decreased with increased engine load | HDRD performed better than PBD fuel | [32] |
| HDRD and PBD | 1S, NA, 0.435 L, 6.2 kW, common rail | Varying engine load and speeds | Reduced BSFC at all load and speeds | HDRD is better than PBD fuel | [33] |
| HDRD, biodiesel, and PBD | 1.5 L passenger car, intercooler | Varying engine load and speeds | Reduced BSFC at all load and speeds | Lower fuel consumption than biodiesel and PBD fuel | [34] |
| HDRD and biodiesel | 1C, 4S, DI, 4.3 kW | Varying loading | ↑BSFC at higher loading<br>↑ BTE as loading increases | HDRD performed better than biodiesel | [35] |
| HDRD, biodiesel, PBD, and their blends | 1C, 4S, DI, water-cooled | Varying engine loads | Higher BSFC, BTE, and EGT than biodiesel and PBD | HDRD is preferred over biodiesel and PBD | [36] |
| HDRD and PBD fuel blends | 1C, 4S, common rail AVL 501 heavy duty engine | Varying engine loading conditions | ↑BSFC increased by 2.8% compared with PBD<br>↑HDRD displayed better BTE | HDRD performed better than PBD in heavy duty CI engines across engine loads | [37] |
| HDRD and PBD blends | 1C, 4S, common rail Ricardo Hydra light duty engine | Varying engine loading conditions | ↑Better BSFC compared with PBD<br>↑HDRD displayed better BTE | HDRD was adjudged a better fuel than PBD in light duty CI engines across engine loads | [37] |
| HDRD and biodiesel blends | 4C, DI, WC, 1.9 TDI diesel engine | Varying engine loads | ↑ Improved BTE and BSFC across the loading condition | HDRD performed better than biodiesel and the blends | [38] |

*↑ = increased, ↓ = decreased, L = liters, C = Cylinder, S = Stroke, DI = Direct injection, NA = Naturally aspirated, AC = air-cooled, WC = water-cooled.*

**Table 4.**
*Performance of CI engine fueled with HDRD.*

They attributed the lower BSFC to the lower viscosity and the impact of high heating value which allows for better fuel atomization.

In research, Janarthanam et al. [35] compared the engine performance of HDRD with that of biodiesel in a vertical single-cylinder, four strokes, and 4.3 kW Kirloskar engine across engine loads. They reported that HDRD showed higher BSFC and BTE, particularly at higher engine loads. They attributed these results to the kinematic viscosity and calorific value of HDRD. Similarly, Singh et al. [36], compared the performance of HDRD with biodiesel and PBD blends in a single cylinder, four strokes 3.5 kW direct injection water-cooled test rig at various engine loads. They reported a higher BSFC, BTE, and EGT with HDRD than with biodiesel and PBD blends, as shown in **Figure 6**. According to them, higher calorific value and cetane index of HDRD compared with biodiesel and PBD accounted for these results. Though HDRD has not been widely used in CI engines, a few reported cases show that HDRD is a better alternative to PBD fuel when compared with biodiesel. The properties of HDRD are a major factor propelling the application of HDRD as a viable and effective substitute for PBD fuel.

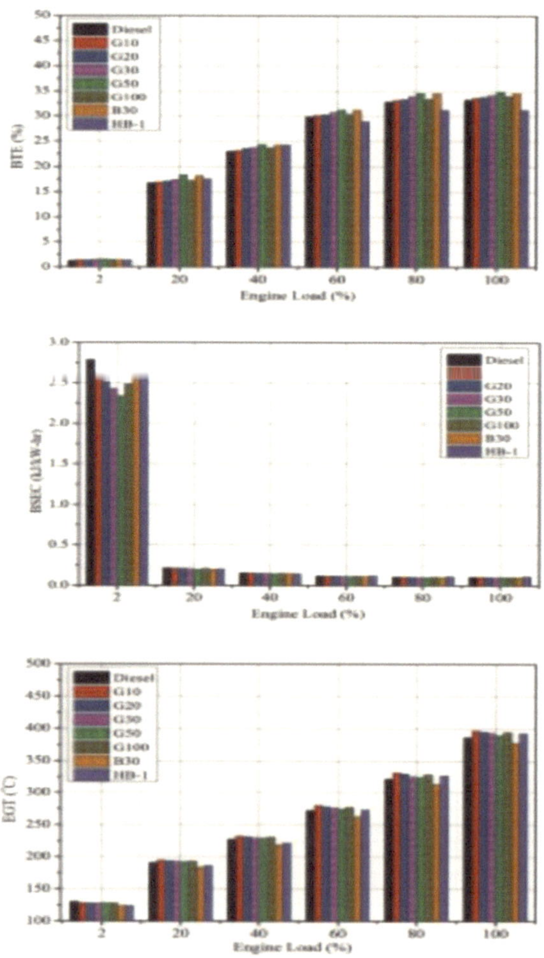


**Figure 6.**
*BTE, BSEC, and EGT of HDRD at various engine loads [36].*

Similarly, Preuß et al. [37] tested HDRD and its blends on both light and heavy duty single cylinder CI engines and compared the results with PBD fuel under various operating conditions. The heavy duty research engine was equipped with an AVL 501 single cylinder engine while the light duty research engine had a Ricardo Hydra engine equipped with a Volvo NED4 cylinder head. The authors reported that the use of HDRD in both light and heavy duties engines led to improved BTE and BSFC for all the engine loading conditions. They attributed these results to the high oxygen content and lower heating value of HDRD compared to PBD and their blends. Using HDRD and biodiesel blends in a 4 cylinder 1.9 TDI CI engine text bed, Shepel et al. [38] reported that HDRD generated better BTE and BSFC than biodiesel and its blends. The results are due to the heating value and the specific heat of combustion of HDRD. This result confirms the assertion that HDRD is a better fuel than biodiesel for transportation applications of CI engines.

## 4. Emission characteristics of HDRD in CI engines

Kim et al. [34] reported the outcomes of the exhaust test carried out on a passenger car with an intercooler fueled with unblended HDRD and compared the results with when PBD fuel was used. They reported a reduction in particulate matter (PM), nitrogen oxide (NOx), carbon monoxide (CO), and total hydrocarbon content (THC) emissions. They attributed these results to the properties of HDRD which allows more complete combustion. Mangus et al. [33] also reported the same pattern of results affirming that the CI engine fueled with HDRD emits less NOx, CO, PM, and THC than when the same engine is fueled with PBD fuel under the same engine speed and load. In another research, da Costa et al. [39] reported that a single cylinder power generation CI engine operated with HDRD synthesize from sugarcane emitted less CO, HC, NOx, and PM when compared with PBD fuel. The same pattern of results was reported by Ogunkoya et al. [32], Vojtisek-Lom et al. [40], and Na et al. [41] who, in their separate studies, affirmed that CI engines fueled on HDRD emitted less CO, $CO_2$, HC, NOx, and soot.

However, Karavalakis et al. [42] and Gysel et al. [43] reported a slight increment in the CO, $CO_2$, NOx, and PM emissions in their studies, as shown in **Table 5**. The higher PM was attributed to the higher cetane number of the tested HRDR fuel which promoted the growth of the diffusive combustion. The higher NOx and PM emissions eliminate the benefits of the aromatic-free characteristics associated with using HDRD fuels. The emission of two greenhouse gases, $CO_2$ and $N_2O$ were found to be lower with the use of HDRD. This is one of the benefits of the application of HDRD in CI engines. Janarthanam et al. [35] studied the emission characteristics of compared the engine performance of a vertical single cylinder, four strokes, 4.3 kW Kirloskar fueled with HDRD and biodiesel. They reported lower emissions of CO, HC, NOx, and smoke due to higher methyl esters and oxygen contents of the tested HDRD. Similarly, Singh et al. [36], compared the performance of HDRD with biodiesel and PBD blends in a single cylinder, four strokes 3.5 kW direct injection water-cooled test rig at various engine loads. They reported that HDRD generates lower CO, UHC, and smoke but higher NOx emission compared with biodiesel and PBD and their blends (**Figure 7**). Reduction in CO, $CO_2$, and smoke emissions were due to higher oxygen content and cetane index of HDRD while the increment in NOx emission was attributed to the higher cetane index, ignition delay, higher cylinder temperature, and pressure as compared to biodiesel and PBD [36]. When HDRD and biodiesel blends

| Fuel tested | Engine parameters | Test conditions | Result of the test | Remark | Ref. |
|---|---|---|---|---|---|
| HDRD, biodiesel, and PBD | 1.5 L passenger car, intercooler | Varying engine load and speeds | ↓reduced CO, NOx, PM, and THC emission | HDRD emits less dangerous gases | [34] |
| HDRD and PBD | 1S, NA, 6.2 kW, common-rail | Varying engine load and speeds | Emission of less NOx, CO, PM, and THC | Emission of less dangerous gases | [33] |
| HDRD and PBD | 1C, 4S, NA, AC, | Varying engine loads | Less CO, HC, NOx, and PM | Emission of fewer pollutants | [39] |
| HDRD and PBD | 1C, 1S, DI, AC | Varying load | Lower emission of CO, $CO_2$, HC, NOx, and soot | HDRD generates fewer pollutants than PBD fuel | [32] |
| HDRD and PBD | 6C, turbocharged, WC, common rail | Varying engine loads and speeds | Lower HC, CO, CO2, NOx | HDRD produces less toxic emissions | [40] |
| HDRD, biodiesel, and PBD | Freightliner truck with 2000 C15 Caterpillar engine | Test cycles | Reduced CO, THC, PM, and NOx | HDRD generates fewer exhaust gases than biodiesel and PBD fuel | [41] |
| HDRD and PBD blends | 6C, 2014 Cummins ISX15 400ST diesel engine | Engine load and blends | Lower, CO, $CO_2$, THC<br>Higher NOx, PM | The use of HDRD provides some emission benefits | [42] |
| HDRD and PBD | 12C, 4S, Caterpillar D398 engine | Engine load | Reduced NOx<br>A slight increment in CO, $CO_2$, and PM | The use of HDRD provides some emission benefits | [43] |
| HDRD and biodiesel | 1C, 4S, DI, 4.3 kW | Varying loading | Reduced CO, HC NOx, and smoke emissions | HDRD generates lesser exhaust gases than biodiesel | [35] |
| HDRD, biodiesel, PBD and their blends | 1C, 4S, DI, water cooled | Varying engine loads | ↓CO, UHC, and smoke opacity emissions<br>↑NOx emission | HDRD generates lower CO, UHC, smoke but higher NOx emission compared with biodiesel and PBD | [36] |
| HDRD, PBD, and biodiesel | 6C, 2014 model year Cummins ISX15 400ST diesel engine | Varying engine loads | ↓CO, $CO_2$, and smoke emissions<br>↑ NOx emission | HDRD generates lower CO, $CO_2$, smoke but higher NOx | [40] |
| HDRD and biodiesel blends | 6C, 6.37 L, Mercedes-Benz engine turbocharger and intercooler | Varying engine loads | ↓ NOx emission<br>↑ PM emission | Lower NOx but higher PM emission than PBD and biodiesel<br>Emissions within Euro III limits | [44] |
| HDRD and PBD fuel blends | 1C, 4S, common rail AVL 501 heavy duty engine | Varying engine loading conditions | ↑ Slight increment in NOx emission<br>↓ About 50% reduction in PM and soot emissions | The use of HDRD contributed to improved air quality | [37] |
| HDRD and PBD blends | 1C, 4S, common rail Ricardo Hydra light duty engine | Varying engine loading conditions | ↑ Slight increment in NOx emission<br>↓ About 50% reduction in PM and soot emissions | HDRD fuel ensured lower soot and improved air quality | [37] |

| Fuel tested | Engine parameters | Test conditions | Result of the test | Remark | Ref. |
|---|---|---|---|---|---|
| HDRD and biodiesel blends | 4C, DI, WC, 1.9 TDI diesel engine | Varying engine loads | ↑ 8% $CO_2$ emission<br>↓ 15% CO emission<br>↓ 18% smoke emission<br>↓ 14% HC emission<br>↓ 19% NOx emission | HDRD was more ecologically beneficial than biodiesel fuel | [38] |
| HDRD and PBD fuel | Euro 3, 51 kW Fiat Panda vehicle | Varying engine loads | ↓27% HC<br>↓ 30% NOx<br>↓18% CO<br>↓ 3% $CO_2$<br>↓5% PM | HDRD will contribute to the attainment of air quality and environmental sustainability | [45] |

*↑ = increased, ↓ = reduced, L = liters, C = Cylinder, S = Stroke, DI = Direct injection, NA = Naturally aspirated, AC = air-cooled, WC = water-cooled.*

**Table 5.**
*Emission characteristics of CI engine fueled with HDRD.*

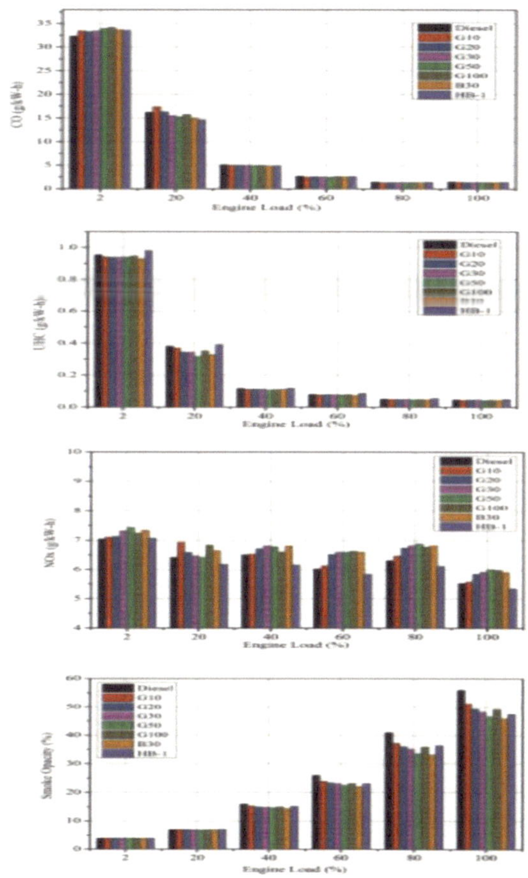

**Figure 7.**
*Emission characteristics of HDRD at various engine loads [36].*

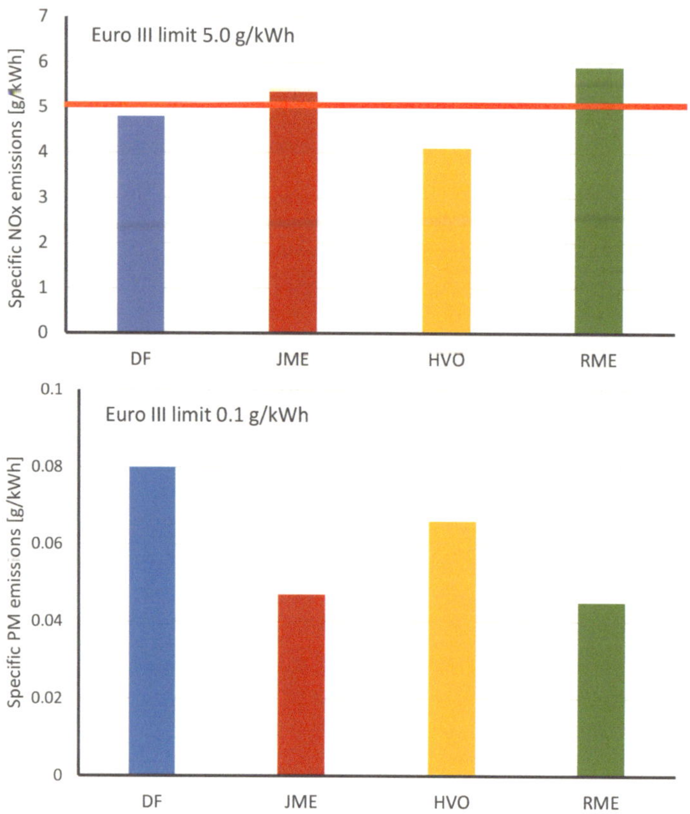


**Figure 8.**
*NOx and PM emissions of HDRD [44].*

were tested a six-cylinder, 6.37 L Mercedes-Benz CI engine equipped with a turbocharger and intercooler, HDRD generated less NOx but more PM emissions. The NOx and PM emissions generated from HDRD were found to be lesser than that from PBD and within the acceptable Euro III limit, as shown in **Figure 8** [44].

Similarly, light and heavy duty CI engines were fueled with HDRD and PBD fuels blends across various engine loads. The light duty engine was fixed a single cylinder, common rail, Ricardo Hydra, and Volvo NED4 cylinder head engine while the heavy duty engine consisted of a single cylinder, common rail, AVL 501, and Volvo D13 cylinder head. The outcome of the emission characteristics showed HDRD a slight increment in NOx emission and reduction in PM and soot emissions for both engine types fueled with HDRD across tested engine loads [37]. Shepel et al. [38] reported a reduction in CO, smoke, HC, and NOx emissions when HDRD was tested in a 4 cylinder, direct injection, water cooled, 66 kW, 1.9 TDI diesel engine test blend and the results compared with biodiesel fuel. There was, however an increment of 8% in $CO_2$ emission which was a result of the higher oxygen content of HDRD compared to other tested fuels. Similar results were obtained when Dobrzyńska et al. [45] tested both HDRD and PBD fuels on a Euro 3, 51 kW Fiat Panda vehicle. They recorded a 27% reduction in HC, 30% in NOx, 18%, in CO, 3% in $CO_2$, and 5% in PM emissions. They concluded that the adoption of HDRD as fuel for CI engines, particularly in the transport sector will reduce the emission of environmentally hazardous gasses, ensure cleaner air quality, and ultimately improve human health.

## 5. Implications and justifications

The increased utilization of HDRD as transport engine fuel has triggered renewed interest in R & D and funding of the production infrastructure across the globe. Also, the share of HDRD in global biofuel moved from about 5% in 2019 to about 10% in 2021 is pointed to its increased global production capacity. This trend, which is expected to continue, typifies the concerted efforts by countries to increase their share of renewable fuel in their energy mix. The simple production method, low cost feedstock, ecofriendly nature, improved performance, and moderate emission generated from HDRD, in comparison with PBD and biodiesel makes HDRD a fuel of the future.

| Parameter | PBD | Biodiesel | HDRD | Ref. |
|---|---|---|---|---|
| Engine performance | Poor performance in ICEs | • High BSFC at high load and low engine speed<br>• Good thermal efficiency | • BTE improved by 20–23%<br>• Better mechanical efficiency<br>• Lower BSFC | [32, 33] |
| Emission Characteristics | High emission of CO, $CO_2$, smoke, and PM | • Low emission of CO<br>• High emission of NOx | • HC reduced by 30%<br>• CO was 35% lower<br>• NOx reduced by 15%<br>• Smoke reduced by 75% | [33, 41] |
| Renewability | Nonrenewable | Renewable though can affect food chain | Renewable | [36] |
| Sustainability | Not sustainable | • Ecologically and environmentally sustainable<br>• High water consumption | • Highly sustainable<br>• Improved sustainability by using waste oil and fats as feedstock<br>• Biooil and biodiesel can be used as feedstock.<br>• Lower water footprint | [46] |
| Cost of production | • High cost of crude oil exploration<br>• High cost of refining | Moderately high | Reduced cost of production | [35] |
| Production infrastructure | Complex and expensive | Costly | Can be upgraded | [36, 46] |
| Application | • Used in trucks and stationary engines<br>• Used for earth moving equipment and agricultural machinery<br>• Used for house heating | • Alternative fuel for ICEs<br>• Viable liquid biofuel turbine. | • Useful in ICEs<br>• Viable bio-jet fuel for aviation gas turbine.<br>• Supplemental fuel for turbines. | [47] |

**Table 6.**
*Comparison of PBD and biodiesel with HDRD.*

The outcome of most research showed that HDRD increased the BTE of the tested CI engine by more than 20%. For example, Kumar et al. [17] reported an increase in BTE from 21 to 23% at lower engine loads. However, at higher loads, a 34%, 36% and 32% increment were recorded at 40%, 80%, and 100% engine loads, respectively. In terms of emission, most studies reported a reduction of about 15%, 30%, 35%, and 75% reductions in NOx, HC, CO, and smoke emission at full load conditions. However, some authors reported that the NOx is unchanged while some reported an increased NOx emission from HDRD fueled CI engine. The aggregate of opinions suggests that HDRD performs better and generates fewer emissions than biodiesel, and PBD. This is very significant because it justifies increased investment in the production and utilization of HDRD.

When compared with HDRD with PBD and biodiesel, available information shows the preference for HDRD by most researchers and consumers. For example, using the major performance criteria, HDRD performed better as ICEs fuel than PDF and biodiesel. Also, the cost of production of HDRD is comparably lower than that of biodiesel. Just like biodiesel, HDRD is generated from lignocellulosic biomass, waste oils, and animal fats. HDRD is not only ecofriendly, and cost effective but also safeguards the environment by emitting fewer toxic gases. **Table 6** compares the performance, emission, production, and application of PBD, biodiesel, and HDRD.

## 6. Conclusions

The utilization of HDRD as CI fuel is to assist in energy security and provide sustainable and environmentally friendly alternatives to the use of PBD fuel. Though the use of biodiesel, bioethanol, and biogas have been well established with notable advantages, HDRD is to help fill the performance gap created by these renewable fuels. The use of HDRD ensures better engine performance creating more options for running a CI engine. One of the disadvantages of using biodiesel in an unretrofitted CI engine is the emission of NOx. The use of HDRD emits less NOx in most cases. In other to reduce the emission of NOx, the concentration of hexadecane and dodecane in the fuel should be increased. This can be achieved during production by altering the feedstock after production by the addition of additives. This however negates the idea of the carbon chain length effect on NOx emission [48].

In the current effort, the performance and emission characteristics of using HDRD in a CI engine have been presented. HDRD is a sustainable replacement for PBD fuels and a more effective renewable fuel than biodiesel. The application of HDRD in CI engines allows improved mechanical efficiency, BTE, and reduced fuel consumption across all engine loads and speeds. CI engines fueled with HDRD are reported to generate less CO, $CO_2$, NOx, and PM when compared with PBD fuel. Though the production process for HDRD is more complex and expensive than biodiesel due to the high temperature and pressure involved, the overall advantage of using HDRD surpasses that of biodiesel.

Going forward, more investigations are needed to simplify the production process of HDRD to domesticate the procedure. More awareness is needed to popularize the production and utilization of HDRD among the population. There should be tax holidays and other incentives for the producers of HDRD as a way to encourage its production and utilization. Governments, across jurisdictions, should provide more funds for R & D in the feedstock, production techniques, standardization, and utilization of HDRD for various applications.

## Conflict of interest

The authors declare no conflict of interest.

## Author details

Omojola Awogbemi[1*], Daramy Vandi Von Kallon[1] and Josiah Pelemo[2]

1 Department of Mechanical and Industrial Engineering Technology, University of Johannesburg, South Africa

2 Discipline of Mechanical Engineering, University of KwaZulu-Natal, South Africa

*Address all correspondence to: jolawogbemi2015@gmail.com

IntechOpen

© 2022 The Author(s). Licensee IntechOpen. This chapter is distributed under the terms of the Creative Commons Attribution License (http://creativecommons.org/licenses/by/3.0), which permits unrestricted use, distribution, and reproduction in any medium, provided the original work is properly cited.

## References


[1] Fossil Fuels. Available from: https://www.statista.com/markets/410/topic/444/fossil-fuels/#overview. [Accessed: January 13, 2022]

[2] World oil reserves. Available from: https://www.worldometers.info/oil/. [Accessed: January 13, 2022]

[3] Fossil Fuels. Available from: https://ourworldindata.org/fossil-fuels. [Accessed: January 13, 2022]

[4] CO2 emissions by fuel. Available from: https://ourworldindata.org/emissions-by-fuel.[Accessed: January 13, 2022]

[5] Awogbemi O, Kallon DVV, Onuh EI, Aigbodion VS. An overview of the classification, production, and utilization of biofuels for internal combustion engine applications. Energies. 2021;**14**:5687. DOI: 10.3390/en14185687

[6] Awogbemi O, Kallon DVV, Aigbodion VS, Mzozoyana V. Property determination, FA composition and NMR characterization of palm oil, used palm oil and their methyl esters. PRO. 2022;**10**:11. DOI: 10.3390/pr10010011

[7] Adewuyi A. Challenges and prospects of renewable energy in Nigeria: A case of bioethanol and biodiesel production. Energy Reports. 2020;**6**:77-88. DOI: 10.1016/j.egyr.2019.12.002

[8] Bhan S, Gautam R, Singh P, Sharma A. A comprehensive review of performance, combustion, and emission characteristics of biodiesel-fueled diesel engines. In: Das LM, Sharma A, Hagos F, Tiwari S, editors. Recent Trends in Thermal Engineering. Singapore: Lecture Notes in Mechanical Engineering. Springer; 2022. pp. 27-41. DOI: 10.1007/978-981-16-3428-4_7

[9] IEA. Key World Energy Statistics 2018. Available from: https://webstore.iea.org/key-world-energy-statistics-2018. [Accessed: January 13, 2022]

[10] IRENA. Global Energy Transformation: The REmap Transition Pathway (Background Report to 2019 Edition), International Renewable Energy Agency, Abu Dhabi, 2019. Available from: https://www.irena.org/publications/2019/Apr/Global-energy-transformation-The-REmap-transition-pathway. [Accessed: January 13, 2022].

[11] Gis W. Electromobility and hydrogenization of the motor transport in Poland now and in the future. Journal of KONES. 2018;**25**:95-101. DOI: 10.5604/01.3001.0012.4780

[12] Chia SR, Nomanbhay S, Ong MY, Shamsuddin AHB, Chew KW, Show PL. Renewable diesel as fossil fuel substitution in Malaysia: A review. Fuel. 2022;**314**:123137. DOI: 10.1016/j.fuel.2022.123137

[13] Saravanan A, Murugan M, Reddy MS, Parida S. Performance and emission characteristics of variable compression ratio CI engine fueled with dual biodiesel blends of rapeseed and Mahua. Fuel. 2020;**263**:16751. DOI: 10.1016/j.fuel.2019.116751

[14] Khan O, Yadav AK, Khan ME, Parvez M. Characterization of bioethanol obtained from Eichhornia Crassipes plant; its emission and performance analysis on CI engine. Energy Sources, Part A: Recovery, Utilization, and Environmental Effects. 2021;**43**:1793-1803. DOI: 10.1080/15567036.2019.1648600

[15] Krishna SM, Salam PA, Tongroon M, Chollacoop N. Performance and emission

assessment of optimally blended biodiesel-diesel-ethanol in diesel engine generator. Applied Thermal Engineering. 2019;**155**:525-533. DOI: 10.1016/j.applthermaleng.2019.04.012

[16] Shirneshan A, Bagherzadeh SA, Najafi G, Mamat R, Mazlan M. Optimization and investigation of the effects of using biodiesel-ethanol blends on the performance and emission characteristics of a diesel engine by genetic algorithm. Fuel. 2021;**289**:119753. DOI: 10.1016/j.fuel.2020.119753

[17] Kumar N, Koul R, Singh RC. Comparative analysis of ternary blends of renewable diesel, diesel and ethanol with diesel. Sustainable Energy Technologies and Assessments. 2022;**50**:101828. DOI: 10.1016/j.seta.2021.101828

[18] Tuli D, Kasture S. Biodiesel and green diesel. In: Tuli D, Kasture S, Kuila A, editors. Advanced Biofuel Technologies: Present Status, Challenges, and Future Prospects. Oxford, United Kingdom: Elsevier; 2022. pp. 119-133. DOI: 10.1016/B978-0-323-88427-3.00010-6

[19] Schütze A. Alternative biofuel options–diesel. In: Elvers B, Schütze A, editors. Handbook of Fuels: Energy Sources for Transportation. Boschst: Wiley; 2021. pp. 315-371. DOI: 10.1002/9783527813490.ch12

[20] Douvartzides SL, Charisiou ND, Papageridis KN, Goula MA. Green diesel: Biomass feedstocks, production technologies, catalytic research, fuel properties and performance in compression ignition internal combustion engines. Energies. 2019;**12**:809. DOI: 10.3390/en12050809

[21] Renewable Diesel: The Fuel of the Future. Available from: https://www.futurebridge.com/industry/perspectives-energy/renewable-diesel-the-fuel-of-the-future/. [Accessed: January 20, 2022]

[22] Global hydrotreated vegetable oil production capacity, 2019-2022. Available from: https://www.iea.org/reports/renewable-energy-market-update-2021/transport-biofuels. [Accessed: January 20, 2022]

[23] Zhang B, Wu J, Yang C, Qiu Q, Yan Q, Li R, et al. Recent developments in commercial processes for refining bio-feedstocks to renewable diesel. BioEnergy Research. 2018;**11**:689-702. DOI: 10.1007/s12155-018-9927-y

[24] Hongloi N, Prapainainar P, Prapainainar C. Review of green diesel production from fatty acid deoxygenation over Ni-based catalysts. Molecular. Catalysis. 2021;**111696**. DOI: 10.1016/j.mcat.2021.111696

[25] Green diesel. Available from: http://www.oil-gasportal.com/green-diesel/. [Accessed: January 14, 2022]

[26] Vonortas A, Papayannakos N. Comparative analysis of biodiesel versus green diesel. Wiley Interdisciplinary Reviews: Energy and Environment. 2014;**3**:3-23. DOI: 10.1002/wene.78

[27] Green Diesel. Green Diesel. Available from: https://www.businesswire.com/news/home/20191022005713/en/Green-Diesel-Developments-Alternative-Fuel-Space-2019. [Accessed: January 14, 2022]

[28] Lambert N. Study of Hydrogenation Derived Renewable Diesel as a Renewable Fuel Option in North America. Available from: https://www.nrcan.gc.ca/sites/www.nrcan.gc.ca/files/oee/files/pdf/transportation. [Accessed: January 19, 2022]

[29] Fuels: Reference Diesel Fuel. Available from: https://dieselnet.com/

standards/eu/fuel_reference.php. [Accessed: January 19, 2022]

[30] Rao KVS, Kurbet S, Kuppast VVA. Review on performance of the IC engine using alternative fuels. Materials Today: Proceedings. 2018;**5**:1989-1996. DOI: 10.1016/j.matpr.2017.11.303

[31] Caton PA, Williams SA, Kamin RA, Luning-Prak D, Hamilton LJ, Cowart JS. Hydrotreated algae renewable fuel performance in a military diesel engine. In: Proceedings of the ASME 2012 Internal Combustion Engine Division Spring Technical Conference; 6-9 May 2012. Vol. 44663. Torino, Piemonte, Italy: ASME; 2012. pp. 121-132. DOI: 10.1115/ICES2012-81048

[32] Ogunkoya D, Roberts WL, Fang T, Thapaliya N. Investigation of the effects of renewable diesel fuels on engine performance, combustion, and emissions. Fuel. 2015;**140**:541-554. DOI: 10.1016/j.fuel.2014.09.061

[33] Mangus M, Mattson J, Depcik C. Performance and emissions characteristics of hydroprocessed renewable jet fuel blends in a single-cylinder compression ignition engine with electronically controlled fuel injection. Combustion Science and Technology. 2015;**187**:857-873. DOI: 10.1080/00102202.2014.982794

[34] Kim D, Kim S, Oh S, No SY. Engine performance and emission characteristics of hydrotreated vegetable oil in light duty diesel engines. Fuel. 2014;**125**:36-43. DOI: 10.1016/j.fuel.2014.01.089

[35] Janarthanam H, Ponnappan VS, Subbiah G, Mani P, Suman D, Rajesh M. Performance and emission analysis of waste cooking oil as green diesel in 4S diesel engine. AIP Conference Proceedings. 2020;**2311**:020022. DOI: 10.1063/5.0034194

[36] Singh D, Sarma AK, Sandhu SS. A comprehensive experimental investigation of green diesel as a fuel for CI engines. International Journal of Green Energy. 2019;**16**:1152-1164. DOI: 10.1080/15435075.2019.1653882

[37] Preuß J, Munch K, Denbratt I. Performance and emissions of renewable blends with OME3-5 and HVO in heavy duty and light duty compression ignition engines. Fuel. 2021;**303**:121275. DOI: 10.1016/j.fuel.2021.121275

[38] Shepel O, Matijošius J, Rimkus A, Duda K, Mikulski M. Research of parameters of a compression ignition engine using various fuel mixtures of hydrotreated vegetable oil (Hvo) and fatty acid esters (fae). Energies. 2021;**14**:3077. DOI: 10.3390/en14113077

[39] da Costa RBR, Coronado CJ, Hernández JJ, Malaquias ACT, Flores LFV, de Carvalho JA. Experimental assessment of power generation using a compression ignition engine fueled by Farnesane–a renewable diesel from sugarcane. Energy. 2021;**121187**. DOI: 10.1016/j.energy.2021.121187

[40] Vojtisek-Lom M, Beránek V, Mikuška P, Křůmal K, Coufalík P, Sikorová J, et al. Blends of butanol and hydrotreated vegetable oils as drop-in replacement for diesel engines: Effects on combustion and emissions. Fuel. 2017;**197**:407-421. DOI: 10.1016/j.fuel.2017.02.039

[41] Na K, Biswas S, Robertson W, Sahay K, Okamoto R, Mitchell A, et al. Impact of biodiesel and renewable diesel on emissions of regulated pollutants and greenhouse gases on a 2000 heavy duty diesel truck. Atmospheric Environment. 2015;**107**:307-314. DOI: 10.1016/j.atmosenv.2015.02.054

[42] Karavalakis G, Jiang Y, Yang J, Durbin T, Nuottimäki J, Lehto K.

Emissions and fuel economy evaluation from two current technology heavy-duty trucks operated on HVO and FAME blends. SAE International Journal of Fuels and Lubricants. 2016;**9**:177-190. DOI: 10.4271/2016-01-0876

[43] Gysel NR, Russell RL, Welch WA, Cocker DR, Ghosh S. Impact of sugarcane renewable fuel on in-use gaseous and particulate matter emissions from a marine vessel. Energy & Fuels. 2014;**28**:4177-4182. DOI: 10.1021/ef500457x

[44] Westphal GA, Krahl J, Munack A, Rosenkranz N, Schröder O, Schaak J, et al. Combustion of hydrotreated vegetable oil and jatropha methyl ester in a heavy duty engine: Emissions and bacterial mutagenicity. Environmental Science & Technology. 2013;**47**:6038-6046. DOI: 10.1021/es400518d

[45] Dobrzyńska E, Szewczyńska M, Pośniak M, Szczotka A, Puchałka B, Woodburn J. Exhaust emissions from diesel engines fueled by different blends with the addition of nanomodifiers and hydrotreated vegetable oil HVO. Environmental Pollution. 2020;**259**:113772. DOI: 10.1016/j.envpol.2019.113772

[46] Wong A, Zhang H, Kumar A. Life cycle water footprint of hydrogenation-derived renewable diesel production from lignocellulosic biomass. Water Research. 2016;**102**:330-345. DOI: 10.1016/j.watres.2016.06.045

[47] Miller P, Kumar A. Techno-economic assessment of hydrogenation-derived renewable diesel production from canola and camelina. Sustainable Energy Technologies and Assessments. 2014;**6**:105-115. DOI: 10.1016/j.seta.2014.01.008

[48] Knothe G, Sharp CA, Ryan TW. Exhaust emissions of biodiesel, petrodiesel, neat methyl esters, and alkanes in a new technology engine. Energy & Fuels. 2006;**20**:403-408. DOI: 10.1021/ef0502711

Chapter 10

# Characteristics Analysis of Performance as Well as Emission of Elaeocarpus Ganitrus Additive Based Pumpkin and Juliflora Mixed Biodiesel Blend in CI Engine

*Vinoth Kannan Viswanathan and Pushparaj Thomai*


## Abstract

Up-to-the-minute researches of different countries have used conformist seed oils such as jatropha oil, coconut oil for the synthesis of biodiesel. In the present investigation, (pumpkin) *Cucurbita pepo. L* along *Prosopis juliflora* seed oil was used for the synthesis of mixed biodiesel with 5 ml Elaeocarpus Ganitrus (Rudraksha) as additive. Performance tests were conducted using biodiesel blend in water cooled CI engine and the emissions were analyzed using a five-gas analyzer. Pumpkin and Juliflora biodiesels were blended in equal ratio to form mixed biodiesel denoted by PJB. It was observed that there was 51% amplify in Brake Thermal efficiency and 33% diminution in Brake Specific Fuel consumption at the maximum load when compared to diesel for PJB20 blend with 5 ml additive. It was also noted that emission of CO bargain by 75%. $CO_2$ bargain by 16.95%, HC compact by 49.2% and NO compact by 34.2% for PJB20 blend with 5 ml Rudraksha additive than that of diesel. The smoke opacity with respect to diesel was also noted to be concentrated by 6% for PJB20 with additive used as fuel in CI engine without any engine modification.


**Keywords:** *Cucurbita pepo*. L (pumpkin), *Prosopis Juliflora*, B20 biodieseel blend, Elaeocarpus Ganitrus, Rudraksha, Emission characteristics, 5-gas analyzer

## 1. Introduction

Liquid fuels of agricultural origin are being increasingly considered as alternatives to gasoline and gas oil as sources of energy. Specifically, biodiesel has a substantial potential to reduce oil imports and ensure continuity in the energy supply [1]. The main alternative fuels operated so far are oxygenates (alcohol, ether etc.), vegetable oils and their esters, gaseous fuel (hydrogen, liquefied petroleum gas etc.), gas to liquids (GTL) and coal imitatives. Ethanol has attracted attention wide-reaching because of its potential use as an alternative fuel [2]. New cheap oil crops are wanted to produce economical oils appropriate for biodiesel production. One of the possible substitute oil crops for the Mediterranean area is pumpkin seed (*C. pepo* L.). To the authors' best knowledge, studies on pumpkin seed oil as the feedstock for methyl ester making were never conducted. In the search for substitute oils for biodiesel production, pumpkin seed oil here a promising choice; however, this cannot be regarded as a massive raw material for biodiesel production on a large scale. The percentage increase of biodiesel in the blends increases emissions due to improper combustion process [3]. However, the selecting of oil is an important measure for biodiesel performance. In biodiesel production, it is needed to make the low cost oil crops. Pumpkin oil and juliflora oil are few of the good oil crops for biodiesel production. The pumpkin seed oil contains an oil intensity of 42–54% depending on numerous factors such as plant area, climate and state of ripeness [3]. The biodiesel is extracted from digestible or non-digestible vegetables by using transestrification production method [4, 5]. The biodiesel have most widely used five types of methyl esters like methyl linolenate, methyl stearate, and methyl oleate and methyl palmitate [6]. During the combustion phase of the biodiesel, engine emitted 9% carbon composition which is less than high speed diesel [7]. In any type of biodiesel preparations FFA oil content of less than 3% has been easily converted by using a catalyst [8, 9]. Similarly, the transesterification process of base catalyst cannot access the high amount of FFA content in vegetable oils [10–12]. EG biodiesel can be blended with other biodiesel as additives for improving the performance and reducing the emission at effective cost [13]. mixture of two biodiesels in equal weight ratio namely *C. pepo*.L (pumpkin) and *Tectona grandis* (teak) seed oil was used for the synthesis of biodiesel with 5-ml Diethyl ether as additive [14]. Present work investigates the biodiesel obtained from transesterification process of Pumpkin and Maize is used as an alternative fuel to diesel [15]. Present investigation, (pumpkin) *C. pepo*.L and *prosopis juliflora* seed oil was used for the synthesis of biodiesel [16].

In this experiment, pumpkin seed oil, juliflora seed oil are converted to fatty acid, by adding 15 g potassium hydroxide (KOH) as catalyst for transesterification [17]. Rudraksha was added as additive to the exceeding mixed biodiesel blend and the engine was driven with minimum load to maximum load and at excess load condition. From the results, it was noted that here was an enhancement in BTE and diminish in Specific fuel consumption. The advantage of *C. pepo* oil along with *prosopis juliflora* oil over other edible oil stretch out in the oil price. In this regard the proposed research work is aimed to focus on the performance and emission of the pumpkin and juliflora mixed biodiesel blend with Rudraksha additive.

5 ml Rudraksha biodiesel as additive was added to mixed biodiesel blend Pumpkin+ Juliflora biodiesel called PJB were then blended with diesel. Take 900 ml of diesel in 1000 ml measuring jar first and add 100 ml PJB biodiesel and add 5 ml Rudraksha additive named PJB10 biodiesel. Take 800 ml of diesel in 1000 ml measuring jar first and add 200 ml PJB biodiesel and add 5 ml Rudraksha additive named PJB20 biodiesel. Take 700 ml of diesel in 1000 ml measuring jar first and add 300 ml PJB biodiesel and add 5 ml Rudraksha additive named PJB30 biodiesel. Take 600 ml of diesel in 1000 ml measuring jar first and add 400 ml PJB biodiesel and add 5 ml Rudraksha additive named PJB40 biodiesel. Take 500 ml of diesel in 1000 ml measuring jar first and add 500 ml PJB biodiesel and add 5 ml Rudraksha additive named PJB50 biodiesel. In India Rudraksha is available in abundance from the Gangetic plane in foothills of the Himalayas and hilly regions, which can be converted into biodiesel. Rudraksha trees can be cultivated in larger scale which may cause severe impact on expenditure of EG biodiesel production. The EG biodiesel as additive are blended in minimal quantity of 5 ml with diesel fuel consumption which led to better performance and clean environment.

Make manual mixing of both fuels, and then take this blend into a mechanical agitator and mixing thoroughly for 15–25 min. The blend is under observation for assuring that there was no separation and settling. This blend was poured to diesel tank in the experiment engine and run the experiments. The volume of addition of Rudraksha biodiesel is very little amount and there is no need to remove the same amount from the blended mixed biodiesel. Rudraksha biodiesel is used as additive to other biodiesel blends. The optimum Rudraksha biodiesel as additive with mixed biodiesel blends were also used to conduct the performance and emission tests at varying loads and compared diesel. Tests were carried out for analyzing various parameters such as brake thermal efficiency, brake specific fuel consumption and exhaust gas temperature. Besides these, the other tests were carried out to measure the emissions of CO, $CO_2$, HC, NO and smoke in the exhaust [18].

## 2. Methodology and raw materials

### 2.1 Pumpkin seed and Juliflora seed oil

Both the oils are not gladly accessible in the marketplace as there is no more viable making. The seeds of pumpkin and juliflora were procured and the oil drawing out was prepared in a laboratory.

**Figure 1a** and **b** represents the seeds of *Cucurbita pepo* and *Prosopis Juliflora*. The preparation of biodiesel from the pumpkin oil and juliflora oil were done unconnectedly using catalytic transesterification process. 15 g of KOH (6,1 ratio) [19] was added to pumpkin and juliflora oil followed by 200 ml of methanol. The mixture was maintained at 65–70°C for 1 h and then residues were allowed to settle down for 2 h in a titration setup [20]. After few hours the glycerin were separated from the biodiesel by

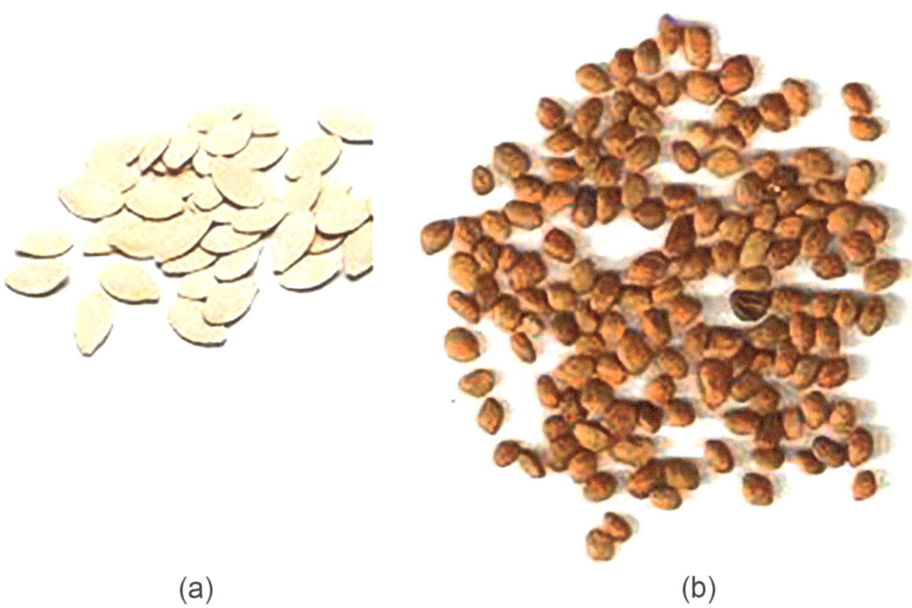


**Figure 1.**
*Seeds of Raw materials (a) Cucurbita pepo seed (b) prosopis juliflora seed.*

| Fuels | Density (kg m$^{-3}$) | Calorific value (kJ/kg) | Kinematic viscosity (poise) | Fire point (°C) | Flash point (°C) |
|---|---|---|---|---|---|
| Diesel | 825 | 42,000 | 2.870 | 0.78 | 0.65 |
| PJ Biodiesel | 766 | 39,846 | 4.960 | 1.06 | 0.98 |
| B20 blend with Rudraksha additive | 828 | 42,400 | 2.950 | 0.82 | 0.68 |

**Table 1.**
*Physical properties of fuels and its blend.*

titration. The two different biodiesel namely pumpkin biodiesel and juliflora biodiesel, half a liter both were assorted to variety the mixed biodiesel. It was followed by stimulated well using magnetic stirrer at the range of 60–80°C temperature. Both the biodiesel are mixed in alike ratio of 50:50 vol. each, for the point of matching the calorific value of mixed biodiesel by way of diesel and also to meet up the average flash point of the diesel fuel. Pumpkin biodiesel has privileged flash point which guides to delayed firing of fuel during ignition, while juliflora biodiesel has minor flash point nearer to diesel and increase the possibility of easy and fast ignition of air fuel mixture. 50:50 combination of pumpkin and juliflora biodiesel blend is identified as PJ biodiesel. 5 ml Rudraksha bio additive was added to each sample to facilitate the combustion process as well as to reduce the emissions from the burnt fuel. It was denoted by R5. The, following blends with additives were prepared with the diesel fuel and mixed biodiesel fuels. 900 ml diesel and 100 ml PJ-biodiesel with 5 ml additive labeled PJB10 + R5, 800 ml diesel and 200 ml PJ-biodiesel with 5 ml additive labeled PJB20 + R5, 700 ml diesel and 300 ml PJ-biodiesel with 5 ml additive labeled PJB30 + R5, 600 ml diesel and 400 ml PJ-biodiesel with 5 ml additive labeled PJB40 + R5 and 500 ml diesel and 500 ml PJ-biodiesel with 5 ml additive labeled PJB50 + R5. **Table 1** represents the physical properties of diesel, PJ biodiesel and B20 blend with Rudraksha additive.

The properties of pumpkin biodiesel, juliflora biodiesel and the blends such as viscosity, density, calorific value, flash point and fire point were measured in the laboratory scale. The hydrometer was used to determine the density of the fuel samples. Viscosity was measured with red wood viscometer; bomb calorimeter was used to measure the calorific value of sample fuels and flash point, fire point apparatus was used to find the flash point and fire point for the sample fuels.

## 2.2 Experimental setup and experimental uncertainty analysis

A single cylinder, 4-strokes, constant speed (1500 rpm) and water cooled CI engine whose compression ratio 17.5:1 with utmost power output of 5.2 kW was used to examine the performance and emission characteristics of mixed biodiesel. An eddy current dynamometer associates the load to the motor.

**Table 2** shows the test engine specifications in detail. The specific fuel consumption was determined using solenoid controller. The flywheel speed was measured using a non-contact type of sensor mounted up on the engine. The cooling water transmits the heat to the surrounding that was generated during the engine operation. The engine load were applied at different percentage such as 20%, 40%, 60%, 80%, 100% (maximum load) and 120% (over load) by eddy current dynamometer. **Table 3** shows the specification of exhaust gas analyzer and smoke meter. **Figure 2** shows the schematic arrangement of KIRLOSKAR TV-1 test engine.

| Item | Specifications |
|---|---|
| Engine power | 5.2 kW (7 hp) |
| No. of cylinder | 1 |
| No. of strokes | 4 |
| Fuel | H.S Diesel |
| Cylinder bore | 0.0875 m |
| Stroke length | 0.11 m |
| Engine speed | 1500 rpm |
| Compression ratio | 17.5:1 |
| Orifice dia. | 0.02 m |
| Loading | Eddy current dynamometer |
| Dynamometer arm length | 0.195 m |
| Inlet valve open | 5° before TDC |
| Inlet valve close | 36° after BDC |
| Exhaust valve open | 36° before BDC |
| Exhaust valve close | 5°after TDC |
| Fuel injection | 23° before TDC |
| Injection pressure | 220 bar |

**Table 2.**
*Specifications of the engine.*

| Equipment name | Model | Measurement range |
|---|---|---|
| Exhaust gas analyzer | Make Model and AVL 444 di-gas analyzer | CO,HC, NO, $CO_2$<br>CO: 0–10 (% Volume), HC: 0–20,000 ppm, $CO_2$: 0–10 (% Volume), NO: (0–5000 ppm) |
| Smoke meter | AVL 437C smoke meter | smoke density<br>0–100 (Opacity in %) |

**Table 3.**
*Specification of exhaust emission measuring equipment.*

Error and uncertainties occurs in the experimentation can come up from gadget selection, state, calibration, examination, environment, reading as well as test planning. Errors will crawl into all experimentations regardless of the care which is put forward. Uncertainty analysis is wanted to prove the accuracy of the untried results. Uncertainty analysis is conceded out using the procedure given by Holman (1994) and Moffat (1988). The uncertainty in brake power is 0.21% brake specific fuel consumption is 2.22% and brake thermal efficiency is 2.56%. The instruments used in the investigational study and their accuracy and uncertainty proportions are given in **Table 4**. The uncertainty of the entity measurements has been taken from the manufacturer's data sheet. Since the equipment is within the calibration validity period, it is predictable that the uncertainties of entity measurements are in agreement with the manufacturer's claim.

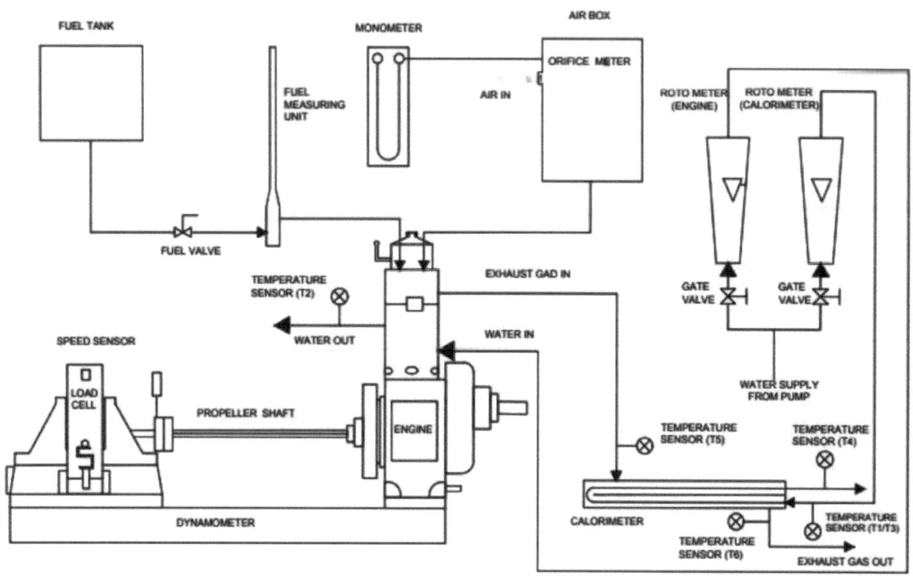


**Figure 2.**
*Schematic diagram of experimental setup.*

| Instruments | Accuracy | Uncertainty |
|---|---|---|
| Burette for fuel measurement | ±0.2 cm$^3$ | ±1.5 |
| Speed measuring unit | ±1 rpm | ±1.0 |
| Manometer | ±1 mm | ±1.0 |
| Crank angle encoder | ±1° | ±0.2 |
| Gas analyzer | | |
| NO | ±20 ppm | ±0.2 |
| HC | ±15 ppm | ±0.2 |
| $CO_2$ | ±0.03% | ±1.0 |
| Smoke meter | ±0.2 | ±1.0 |

**Table 4.**
*Instruments and uncertainties.*

## 3. Result and discussions

The trial engine was examined for the performance and emission individuality using PJ biodiesel blend with additive and the same was judge against the diesel. 5 ml Rudraksha bio additive was put in to the PJ biodiesel blends for civilizing the performance individuality of biodiesel. The emissions of blended biodiesel such as CO, $CO_2$, HC, and NO was hardnosed using five-gas analyzer. The smoke opacity proportion was studied with AVL 437C Free accelerometer Smoke meter. The performance of the engine was also investigated for brake specific fuel consumption and brake thermal efficiency.

### 3.1 Performance characteristics

**Figure 3** shows the comparison of Brake Thermal Efficiency (BTE) for mixed biodiesel used in the test engine. At maximum load (100%) the BTE was observed to be 27.83% for diesel, 42.02% for PJB20 + R5 blend that shows, PJ biodiesel with Rudraksha additive results in higher BTE and good thermal performance with B20 blend compared to other blends. The BTE improvement of about 51% for PJB20 blend with Rudraksha additive compared to diesel was noted. When the load exceeds the maximum limit, the efficiency tends to decrease for both diesel as well as biodiesel blends.

**Figure 4** shows the comparison of Specific Fuel Consumption (SFC) for mixed biodiesel used in the test engine. At maximum load (100%) the SFC was observed to be 0.307 kg/kW-h for diesel, 0.206 kg/kW-h for PJB20 + R5 blend that shows, PJ biodiesel with Rudraksha additive results in lower SFC with B20 blend compared to other

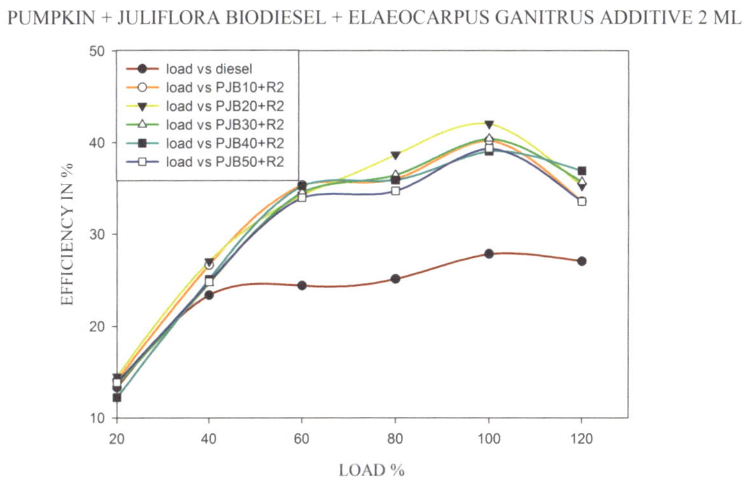


**Figure 3.**
*BTE vs. load variations for PJB + R5 biodiesel blend.*

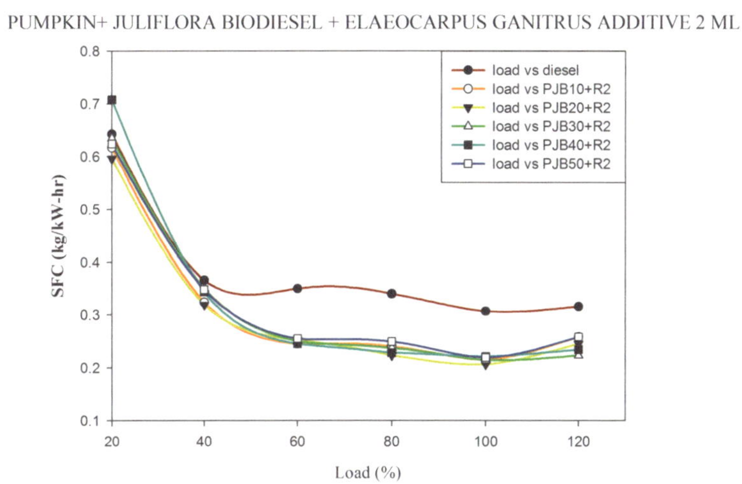


**Figure 4.**
*SFC vs. load variations for PJB + R5 biodiesel blend.*

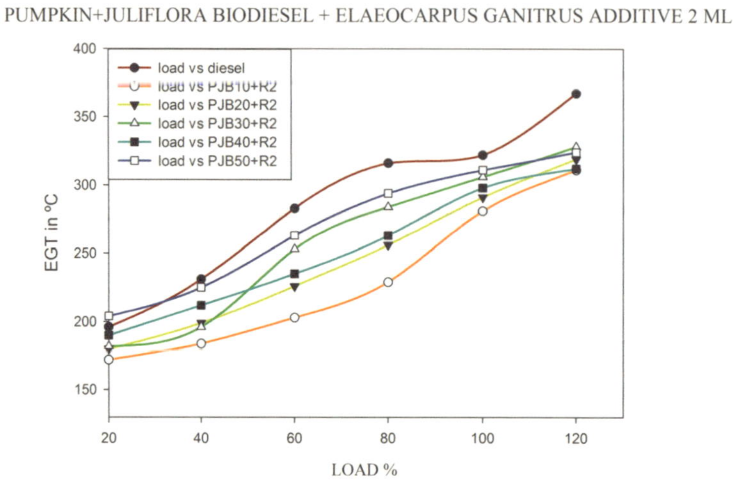


**Figure 5.**
*EGT vs. load variations for PJB + R5 biodiesel blend.*

blends. The SFC decrement of about 33% for PJB20 blend with Rudraksha additive compared to diesel was noted. When the load exceeds the maximum limit, the fuel consumption was observed to increase for both diesel as well as biodiesel blends.

**Figure 5** shows the comparison of Exhaust Gas Temperature (EGT) for mixed biodiesel used in the test engine. Lower engine temperature was observed when PJ biodiesel is used with Rudraksha and this improves the combustion process due to excess amount of oxygen present in additive. At maximum load (100%) the EGT was observed to be 322°C for diesel, 291°C for PJB20 + R5 blend that shows, PJ biodiesel with Rudraksha additive results in lower EGT with B20 blend compared to other blends. The EGT decrement of about 9.63% for PJB20 blend with Rudraksha additive compared to diesel was noted. When the load exceeds the maximum limit, the exhaust gas temperature was observed to increase for both diesel as well as biodiesel blends.

### 3.2 Emission characteristics

**Figure 6** shows the emission of carbon monoxide (CO) at different loads from the test engine by using diesel, PJ biodiesel blend with Rudraksha additive. The features affecting CO emission were air-fuel mix and oxygen. CO emission was because of the inadequate burning of fuel, where the oxidation has not occurred properly [21]. This is due to inadequate air quantity and inability of carbon conversion to $CO_2$ at exhaust manifold. At maximum load the CO emission was observed to be 0.32% by volume for diesel, 0.08% by volume for PJB20 blend with Rudraksha additive. The CO emission was observed to be reduced by 75% for PJB20 + R5 blend. It was observed that the blended biodiesel with additive has comparatively lower emission. This decrease in CO output was because of increase in burning chamber temperature and nearness of more oxygen in additive based biodiesel.

**Figure 7** shows the emission of Carbon dioxide ($CO_2$) at different loads as of the test engine by means of diesel and PJ biodiesel blend with additive. This $CO_2$ emission shows absolute combustion process due to the quantity of oxygen there in the biodiesel. At maximum load the $CO_2$ release was observed to be 5.9% by vol. for diesel, 4.9% by vol. for PJB20 blend with additive. $CO_2$ emission was observed to be reduced by 16.95% for B20 blend with additive compared to diesel was noted.

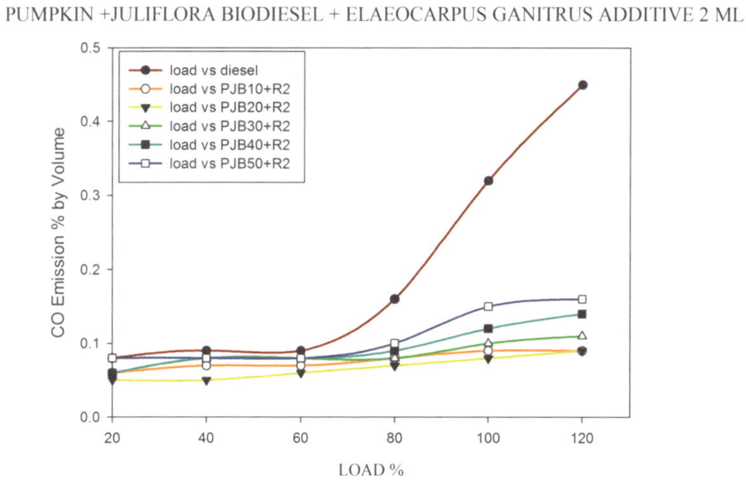


**Figure 6.**
*Load vs. emission characteristics of CO for PJB + R5 blended biodiesel.*

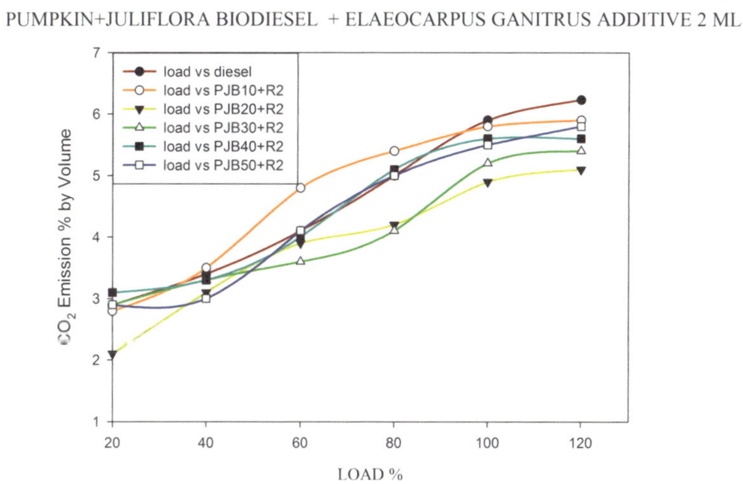


**Figure 7.**
*Load vs. emission characteristics of $CO_2$ for PJB + R5 blended biodiesel.*

**Figure 8** show the emission of hydro carbon (HC) at different loads from the test engine by using diesel and PJ biodiesel blend with additive. HC emission was observed with unburned fuels due to inadequate temperature formation at near the cylindrical walls in the engine [12, 22]. The lesser HC release occurs due to lower heat rejection by high in-cylinder temperature. At maximum load the HC emission was observed to be 118 ppm for diesel and 60 ppm for B20 blend with additive, which is decreased by 49.2% for PJB20 + R5 blend compared to diesel, was noted. It was observed that the blended biodiesel with additive has lower emission than that of diesel. It shows the oxygen content in additive increases the possibility of complete fuel burning [23]. At minor loads, HC emission was noted to be lesser but when blend ratio enlarges, HC emission also elevates compared to diesel. Yet, at greater loads the cutback in HC emission was typically inclined by rising wall temperature in the cylinder towards the exhaust manifold [7]. The researchers found an analogous reduction

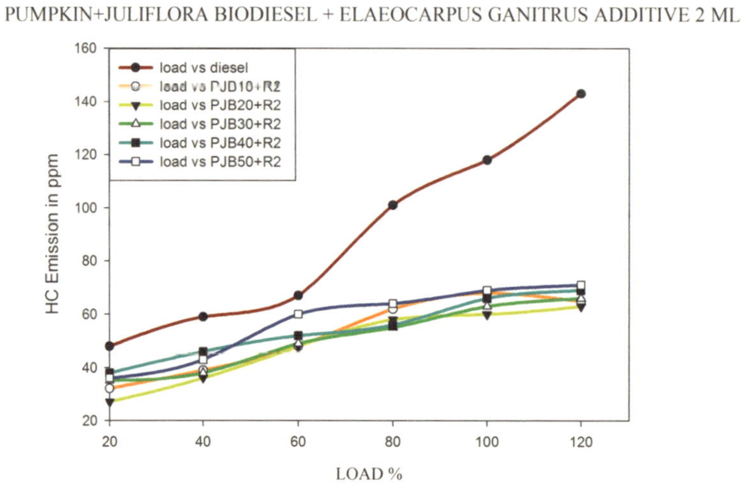


**Figure 8.**
*Load vs. emission characteristics of HC for PJB + R5 blended biodiesel.*

in HC emission by using Rudraksha additive with biodiesel processed with uncoated engines that shows the possibility for further reduction of HC with coated engine.

**Figure 9** shows the emissions of nitrogen oxide (NO) from the exhaust system while using different biodiesel blends with additive. Yilmaz N found reduction in NO, by using di-tertiary-butyl peroxide (DTBP) in coated engine while using biodiesel [24]. NO emission happens due to combustion process at privileged temperature and poorer oxygen attention. While using biodiesel blends, the oxygen intensity were higher and this results in minor NO emission. At maximum load, NO emission was viewed to be 1212 ppm for diesel and 798 ppm for PJB20 blend with additive Rudraksha. The NO emission was observed to be decreased by 34.2% for B20 blend with Rudraksha additive compared to diesel was noted. In case of biodiesel with additive, an optimized concert was achieved at B50 blend of biodiesel due to sufficient heat generation and better oxygen concentration.

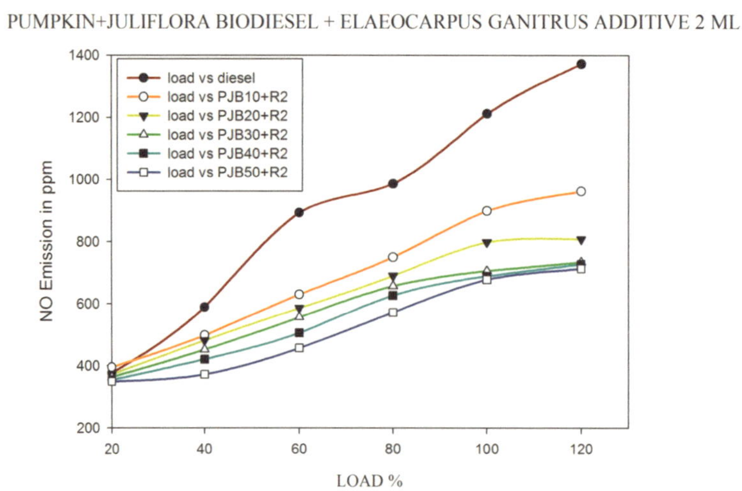


**Figure 9.**
*Load vs. emission characteristics of NO for PJB + R5 blended biodiesel.*

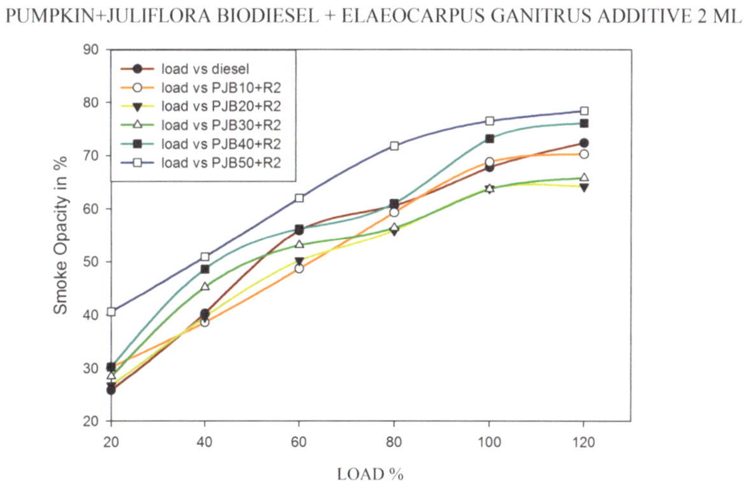


**Figure 10.**
*Load vs. smoke opacity for PJB + R5 blended biodiesel.*

**Figure 10** shows the smoke opacity for diesel as well as biodiesel blends. The enhancement of oxygen plays a key role in smoke emission from the engine at different loading conditions. At maximum load, the smoke opacity was observed to be 67.8% for diesel and 63.8% for PJB20 blend with additive. The smoke emission was observed to be decreased by 6% for PJB20 blend with additive judge against to diesel was noted. Use of biodiesel generates higher smoke levels when match up with the diesel due to the rate of adjourned oxidation process. In this stare use of PJ biodiesel with Rudraksha additive diminish the smoke level in the engine.

The incidence of additive has abridged the smoke opacity than that of diesel and all biodiesel blends. For PJB20 blend with additive the smoke opacity was almost identical to the diesel. It shows sign of the presence of sufficient oxygen substance and non-defective combustion progression. As the result, it was incidental that the exhaust temperature is essential for characterizing the smoke actions. Also, the smoke density was pragmatic to increase with percentage of lift in biodiesel blends with growing loads.

## 4. Conclusion

The experimental work Pumpkin-Juliflora biodiesel with Rudraksha additive was conducted on a single cylinder, four-stroke direct infusion water cooled diesel engine with an eddy current dynamometer. Performance and emission characteristics of PJ biodiesel with additive were observed as follows:

- The diesel and PJ biodiesel blend with additive demonstrated a reduction pattern for BSFC and increasing pattern of BTE. PJB20 + R5 blend shows decreased BSFC by 33% at ceiling load evaluated with the diesel.

- PJB20 + R5 merge with additive shows 51% augmented BTE at maximum load compared with diesel.

- PJB20 merge with Rudraksha additive demonstrated a diminution in $CO_2$ emission by 16.95% at ceiling load in contrast to diesel.
- Alike results were established in the PJB20 merge with additive for the CO release that was abridged by 75% at ceiling load owing the ignition process as of the company of surplus oxygen and superior temperature of the arrangement.
- The HC discharge for PJB20 merge with additive shows a decrease of 49.2% compared to diesel.
- The NO emission for PJB20 merge with additive demonstrates a reduction of 34.2% by volume at maximum load compared with diesel. NO output decreases with increasing biodiesel blends with additive
- Smoke opacity for PJB20 + R5 merge was 6% lower than that of diesel. Smoke will be generally higher for biodiesel when compared with diesel. Whereas in this research the smoke was observed to be decreased due to Rudraksha additive mixed with biodiesel.
- From the results, it is evident that PJB20 mixed biodiesel merge with Rudraksha additive exhibits better performance and decreased emission characteristics and therefore could be considered as an alternative fuel to diesel without any engine modifications.

## Author details

Vinoth Kannan Viswanathan[1]* and Pushparaj Thomai[2]

1 Department of Mechanical Engineering, Parisutham Institute of Technology and Science, Thanjavur, Tamil Nadu, India

2 Department of Mechanical Engineering, Kings College of Engineering, Pudukkottai, Tamil Nadu, India

*Address all correspondence to: vinkan18mech@gmail.com

IntechOpen

© 2022 The Author(s). Licensee IntechOpen. This chapter is distributed under the terms of the Creative Commons Attribution License (http://creativecommons.org/licenses/by/3.0), which permits unrestricted use, distribution, and reproduction in any medium, provided the original work is properly cited.

## References

[1] Berrios M et al. A kinetic study of the esterification of free fatty acids (FFA) in sunflower oil. Fuel. 2007;**86**:2383-2388

[2] Masum BM et al. Effect of ethanol–gasoline blend on NOx emission in SI engine. Renewable and Sustainable Energy Reviews. 2013;**24**:209-222

[3] Schinas P et al. Pumpkin (*Cucurbitapepo* L.) seed oil as an alternative feed stock for the production of biodiesel in Greece. Biomass and Bioenergy. 2009;**33**:44-49

[4] Karthickeyan V, Balamurugan P. Effect of thermal barrier coating with various blends of pumpkin seed oil methyl ester in DI diesel engine. Heat Mass Transfer. 2017;**53**:3141-3154

[5] Karthickeyan V. Effect of cetane enhancer on ceramic coated diesel engine fuelled with neat *Moringa oleifera* methyl ester. Data in BriefFuel. 2019;**24**(103932):1-5

[6] Demirbas A. Biodiesel production from vegetable oils via catalytic and non-catalytic supercritical methanol transesterification methods. Progress in Energy and Combustion Science. 2005;**31**:466-487

[7] Lee CS et al. An experimental study on the atomization and combustion characteristics of biodiesel-blended fuels. Energy Fuels. 2005;**19**:2201-2208

[8] Shailendra S et al. Biodiesel development from rice bran oil: Transesterification process optimization and fuel characterization. Energy Conversion and Management. 2008;**49**:1248-1257

[9] Haas MJ. Improving the economics of biodiesel production through the use of low value lipids as feedstocks: Vegetable oil soapstock. Fuel Processing Technology. 2005;**86**(10):1087-1096

[10] Thangaraj S, Govindan N. Evaluating combustion, performance and emission characteristics of diesel engine using karanja oil methyl ester biodiesel blends enriched with HHO gas. International Journal of Hydrogen Energy. 2018;**xxx**:1-13

[11] Dorado MP et al. An alkali–catalyzed transesterification process for high free fatty acid waste oils. American Society of Agricultural Engineers. 2013;**45**(3):525-529

[12] Centinkaya M, Karaosmanoglu F. Optimization of base-catalyzed transesterification reaction of used cooking oil. Energy and Fuels. 2004;**18**:1888-1895

[13] Viswanathan VK, Thomai P. Performance and emission characteristics analysis of Elaeocarpus Ganitrus biodiesel blend using CI engine. Fuel. 2021;**288**:119611. DOI: 10.1016/j.fuel.2020.119611

[14] Vinothkannan V, Pushparaj T. Performance and emission characteristic analysis of *Cucurbita Pepo* l. and *Tectona Grandis* seed oil biodiesel blends in CI engine with additive. Energy Sources, Part A: Recovery, Utilization, and Environmental Effects. 19 Nov 2020. DOI: 10.1080/15567036.2020.1849453. [Accessed: 01 Nov 2020], [Published online: 19 Nov 2020]

[15] Magesh N, Thomai P, Viswanathan VK. Experimental investigation on performance, combustion and emission characteristics of CI engine fuelled with pumpkin and maize biodiesel blends. Tierärztliche Praxis. 2021;**41**:51-64

[16] Viswanathan VK, Thomai P. Investigation on performance and emission characteristics of CI engine fuelled with *Cucurbita Pepo* L. and Prosopis Juliflora seed oil biodiesel blends. Tierärztliche Praxis. 2020;**40**:203-212

[17] Leung DYC, Guo Y. Transesterification of neat and used frying oil: Optimization for biodiesel production. Fuel Processing Technology. 2006;**87**:883-890

[18] Senthil R et al. Assessment of engine operating parameters on working characteristics of a diesel engine fuelled with 20% proportion of biodiesel diesel blend. Recent Advance Petrochemical Science. 2017;**3**(1):001-008

[19] Emiroğlu AO, Şen M. Combustion, performance and emission characteristics of various alcohol blends in a single cylinder diesel engine. Fuel. 2018;**212**:34-40

[20] Ramadhas AS et al. Biodiesel production from high FFA rubber seed oil. Fuel. 2005;**84**:335-340

[21] Subramani L et al. Novel Garcinia gummi-gutta Methyl Ester (GGME) as a potential alternative feedstock for existing unmodified DI diesel engine. Renewable Energy. 2018;**125**:568-577

[22] Nanthagopal K et al. Influence of exhaust gas recirculation on combustion and emission characteristics of diesel engine fuelled with 100% waste cooking oil methyl ester. Waste Biomass Valorization. 2018;**10**:2001-2014

[23] Herbinet O et al. Detailed chemical kinetic oxidation mechanism for a biodiesel surrogate. Combustion Flame. 2008;**154**:507-528

[24] Yilmaz N, Atmanli A. Experimental assessment of a diesel engine fueled with diesel-biodiesel- 1-pentanol blends. Fuel. 2017;**191**:190-197

Chapter 11

# Bio-Circular Engine: Simultaneous and Successive Use of BioDiesel as Bio-Lubricant and Bio-Fuel in Diesel Engines-(B100) New Bio-Lubricant for all Engines

*Cesar Bautista Sterling*


## Abstract

The scientific literature allows us to demonstrate the characteristics of high lubricity of biodiesel (particularly B100 from palm), which as a bio-fuel, can fulfill the function of bio-lubricant (B100 = 3 Ester); even surpassing motor oils in some respects (Synthetic Base = 2 Ester). Once its characteristics have been reviewed, we can affirm that it is possible to use B100 as a Bio-Lubricant in Diesel internal combustion engines, but also in spark-ignition engines. A comparison is made between commercial synthetic esters and fatty acid methyl esters ("FAME") designated B100. In the same context, we describe a procedure and a device designed to use B100 in diesel engines, not only as Bio-Fuel, but also as Bio-Lubricant, for both functions, successively and simultaneously, called "Bio-Circular Engine"; so: in Stage 1; biodiesel is taken from the fuel tank (B100) to the engine crankcase (previously filtered), where it will fulfill its first function as Bio-Lubricant. In Stage 2; the same B100 is conducted from the same crankcase to the fuel injection system (previously filtered and, if required, cooled) to fulfill the second function of Bio-Fuel. The "Bio-Circular Engine" is "Circular Economy", because it uses a single substance (B100), for two different and simultaneous functions (as Bio-Fuel and as Bio-Lubricant).


**Keywords:** biolubricant, esters, synthetic esters, lubricity, B100, diesel engine, internal combustion engine, B100 double function, bio-Circular engine

## 1. Introduction

The greenhouse gases, the main cause of global warming, emitted mostly by mobile sources that use, petroleum diesel fuel, which emits toxic gas and known as



IntechOpen

carcinogenic. On the other hand, these mobile sources generate large amounts of used lubricating oil, considered as hazardous waste, which affect human health and the environment "We only have 10–12 years to save ourselves from the new limit of 1.5°C of average temperature increase (not 2°C as was believed), which would be the point of no return" IPCC 2018 [1]. "Lyon (France), June 12, 2012 (AIIC/WHO)—After a week of the meeting of international experts, the International Agency for Research on Cancer (AIIC), which is part of the World Health Organization (WHO), announced this on June 12, that the exhaust fumes from diesel engines have been classified as carcinogenic to humans (Group 1), based on sufficient scientific evidence showing that such exposure is associated with an increased risk of lung cancer" [2]. According to the Basel Convention, ratified and adopted in Colombia through Law 253 of 1996, used oils of automotive and industrial origin are classified as a hazardous waste of mineral oils. Their dangerous characteristics vary according to the processes or equipment in which they have been used; Among its possible dangerous components are lead, chlorine, barium, magnesium, zinc, phosphorus, chromium, nickel, aluminum, copper, tin, sulfur, and polynuclear aromatic hydrocarbons, among others; others, which if released or mishandled may have immediate adverse effects or retarded in the environment. This is due to the bioaccumulation and its toxic effects on biotic systems, which effect human health and natural resources [3]. Undoubtedly, the real solution is electrical energy in all its possibilities, from renewable sources, but in the meantime, these energies, as in the case at hand, biodiesel (B100), obliges as soon as possible to break paradigms and take risks with bold, alternative, or unconventional solution proposals that facilitate an efficient and safe transition, from fossil fuels, through biofuels and biolubricants, toward totally clean energies.

It is not unreasonable then to propose a new use for B100, as a biolubricant; its physicochemical characteristics are very close not only to fossil diesel fuel but also to the physicochemical characteristics of commercial mineral and synthetic lubricants.

The review of similar characteristics of B100 with commercial lubricants refers to the direct use of B100 inside the crankcase, performing the integral function of engine lubrication. This procedure can be adapted to virtually any modern diesel engine, as any modern diesel engine would be capable of using B100; but it also applies to all diesel engines capable of using mixtures (B10, B20, B50, etc.), complementing it with an alternate mechanism, arranged in the procedure to make the mixtures (B100 + petroleum diesel), before entering the injection system of fuel, and after the passage of B100 (100% pure), through the crankcase in its role as a lubricant. It is also possible to use B100 as a lubricant for spark-ignition engines.

This new function of B100 as biolubricant means that the use of commercial mineral and/or synthetic oils in diesel engines disappears; motor oil disappears into the maintenance budget for vehicles and transport fleets, which would be limited to the normal change of filters. At the same time, the environmental contamination caused by used oils also disappears. In the procedure, immediately after the function of B100 as biolubricant, the following function of B100 as biofuel, reduces about 80% of greenhouse gas emissions, thus contributing significantly to the reduction of global warming.

The process is based on the physicochemical characteristics of B100, which in principle are key to what is required, such as a high boiling point; low vapor pressure; flashpoint higher than 170°C (for palm biodiesel), much higher than that of diesel (64°C); a density of about 0.88 $g/cm^3$; viscosity at 40°C of 4.5 $mm^2/S$; lubricity of 6000 g BOCLE [4].

The biodiesel produced from pre-esterified Colombian palm oil meets the technical specifications required by European Standard 14214 for the properties evaluated.

Biodiesel produced stands out for its high chemical stability (oxidation stability). The main quality deficiency of biodiesel from palm oil is its low value in determining the cold filter plugging point (POFF. 10–12°C.), which would not greatly affect tropical or during the summer season [5].

## 2. Consideration of biodiesel (B100) as biolubricant

Biolubricants have several advantages compared to mineral-based lubricants:

- High degree of biodegradability.
- Low toxicity to humans and aquatic organisms.
- Good lubricating properties.
- High viscosity index.
- High flashpoint.
- Good adhesion to metal surfaces.

These basic physicochemical characteristics are what we must find in Biodiesel to consider it as a biolubricant. The higher lubricity of biolubricants results in less friction, and a higher viscosity index has more efficient heat transfer. In addition, due to their better adherence to metal surfaces, biolubricants, being polar substances, produce more resistant lubricating films, in any type of lubrication.

The arguments (physical and chemical aspects) to consider B100 as a biolubricant:

1. High Flash or Spark Point of Palma B100. 160°C. Available ASTM tests include ASTM D56, Tag Closed Cup Flash Point, used for viscosities below 5.5 cSt at 40°C, as well as for viscosities below 9.5 cSt at 25°C and flashpoints below 93°C. ASTM D93, Pensky-Martens Closed Cup Flash Point is used for petroleum products with a temperature range from 40 to 360°C and biodiesel with a temperature range from 60 to 190°C. Flashpoint is also very useful in used oil analysis to detect dilution with fuel, thermal degradation of the base oil, and contamination [6]. These three previous aspects would not apply in the case of B100 as a lubricant.

2. Kinematic viscosity of Palma B100: 4.71 cSt at 40°C; 2.25cSt at 100°C.

3. Low Volatility of B100 from Palma: Flat distillation curve; with 309°C initial boiling point, 50% recovered temperature of 321°C, and 338°C final boiling point. This indicates that the fatty acid methyl esters that makeup palm biodiesel do not have very high differences in their boiling points, while petroleum derivatives have a wide variety of hydrocarbons with different volatilities [7]. Volatilization is a term used to describe the “boiling off” of lighter molecules in fluids. It is closely related to oil consumption in car engines. The test simulates the reaction of oil to internal temperatures found in the ring/piston/cylinder area of internal combustion engines. Known as ASTM D5800,

the Noack volatility test reveals the evaporation loss at high temperatures of the oil's lighter molecules and additives. Depending on the method, a quantity of sample is placed in a crucible or reaction flask and heated at 250°C with constant airflow for 60 min. The comparison of the weight before and after the test will determine the loss by evaporation. Evaporation losses can also result in a change in oil properties, as additives can evaporate during the volatilization process. As the lighter molecules "burn" or evaporate, the heavier molecules remain, causing a change in fluid viscosity [6].

4. Low phosphorus content of B100: 1.26 mg/kg. + ASTM 6751 Standard: Max. 10 mg/kg. The Selby-Noack volatility test uses a noble metal heater; eliminates the need for Wood's metal and collects the evaporated material for further analysis. This is particularly useful for identifying elements such as phosphorus, which are known to lead to premature failure of catalytic systems [6].

5. Lubricity; B100 as Newtonian Fluid: The shear rate is proportional to the shear stress, a typical characteristic of lubricating oils [7].

6. Lubricity; B100 as Polar Substance: Excellent adherence to metal surfaces. The metal is considered as an electrostatic surface, where there is an electronic cloud, therefore, it is negatively charged, attracting polar substances such as esters. This attraction occurs through the carbon atom of the ester carbonyl (C = O), which has a charge density δ(+) [8–12]. The molecules are attached to the surface under the mechanism of physical adsorption [10, 11]. The lubricating film is formed by the absorption on the metallic surface of molecules of polar substances. When compatibility of forces occurs, an attraction occurs between the absorbent and the substance to be adsorbed, resulting in the fixation of the molecules of the substance on the surface of the solid. For this type of adsorption, we can have several layers of adsorbed molecules [10, 11].

7. Blow By; harmless phenomenon for B100. The problem of contamination of the lubricant, caused by fuel entering the crankcase through the cylinders and piston rings (Blow-By), is eliminated. Petroleum diesel fuel (fuel dilution), depletes additives and introduces sulfur and aromatic compounds into the motor oil, affecting viscosity. This will never happen with B100 because its runoff to the crankcase due to Blow By will be mixed with the same B100 that will be in its first stage or function as engine lubricant; also, because the B100 is free of the polluting elements.

8. Water content in B100 from Palma: 380 mg/kg. + ASTM 6751 Standard: Max. 500 mg/kg. Water has a devastating effect on oil and lubricated parts because it leads to rust and corrosion, respectively. In oils with anti-wear additives based on Zinc Dithiophosphate (ZDDP), the water reacts with them and gives rise to the formation of sulfuric acid, eliminating the limited lubricating film. Similarly, this will never happen with B100 because it is free of additives.

9. Its prolonged oxidation stability of 26 h (minimum 3 h for the ASTM 6751-S500 standard; minimum 6 h for the EN 14214 standard), maintains the chemical stability of B100 for a longer time in optimal working conditions, compared to fossil mineral oils.

*DOI: http://dx.doi.org/10.5772/intechopen.103663*

10. The thermal stability stands out, with a thermal reflectance value of 99% (above the thermal reflectance of aluminum of 93%), which allows it to carry out very efficient thermal management of cooling inside the engine, as a contour element.

**Figure 1** shows the oil temperature gauge of a truck, for a maximum expected temperature of 300°F (148.89°C), which is below the spark point of palm biodiesel (180°C. Bio-D Colombia).

Thermal requirement for diesel engine oil in normal operation:

- 190–220°F (88–104°C) for Caterpillar.
- 180–225°F (82–107°C) for Cummins.
- 200°230°F (93–110°C) for Detroit Diesel.
- 181–203°F (83–95°C) for Mercedes-Benz

11. Other considerations in favor of B100: The main effects of motor oil contamination, such as dilution with fuel, thermal degradation of the lubricant base, and contamination occur, as has been said, largely as a result of the mentioned blow-bay phenomenon; but the passage of substances through the piston and cylinder rings does not only occur in the direction of the crankcase but also known that no matter how new the engine is, there will always be a passage of lubricating oil from the crankcase to the combustion chamber because the oil will always be circulating (even if it is in a minimum amount) through the spaces between the rings and the grooves, the pistons and cylinder walls which also require lubrication. These minimum amounts of oil grow over kilometers and time, affecting to a greater or lesser extent the ignition process of the mixture, the quality of combustion, and the quality of emissions, as well as generating greater contamination by volatile compounds generated by the combustion of gases and heavy chemical elements from oil additives and their mineral bases.

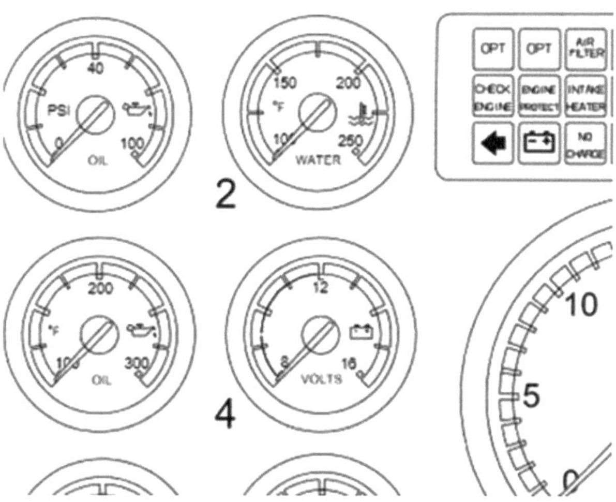


**Figure 1.**
*Temperature data of some brands of commercial motors.*

## 3. Comparison between synthetic ESTER and B100 or FAME (fatty acid methyl Ester)

### 3.1 Synthetic Ester

**Esters** are the result of the chemical reaction of an organic acid and an alcohol. Acid with two carboxyl groups (a functional group characteristic of organic acids) is called a diacid and the product of its reaction with alcohol is called diester. The alcohol that has more than one hydroxyl group (functional group characteristic of alcohols) is called a polyol. The product of the reaction of an organic acid with a polyol is called a polyol ester [13].

**Diester** These are mostly used among synthetic esters. They are more stable to oxidation and heat than hydrocarbons, beginning to decompose at 200°C. They contain two carboxyl, (C = 0) responsible for the characteristic of the polar substance (**Figure 2**).

### 3.2 B100 or Fame (fatty acid methyl Ester)

They contain three carboxyl (C = 0), responsible for the characteristic of the polar substance. These various adsorbed layers of biodiesel molecules (3 esters, 3(c = o)), constitute the key to the performance of B100 as a biolubricant, which allows it to withstand high pressures and high shear rate (HPHS), in any lubrication regimen and engine workload (**Figure 3**).

When machine surfaces interact with higher pressures and temperatures, additives mitigate the typical effects of metal-to-metal contact (wear) by creating initial molecular layers on the machine surface that are more ductile. These friction control layers directly reduce shear resistance during contact and are sacrificed.

The first layers can mitigate friction by allowing the weaker molecular bonds in the lubricant to be released with less force compared to the strong bonds that result from film boundary conditions due to metal-to-metal contact of surface asperities. The formation of low shear strength films is also affected by the type of base oil and the metallurgy of the surfaces.

There are three types of lubricant additives that help reduce this friction and control wear: friction modifiers, anti-wear additives, and extreme pressure additives [14].

$$2ROH + HO-\overset{\overset{O}{\|}}{C}-(CH_2)_n-\overset{\overset{O}{\|}}{C}-OH \rightarrow RO-\overset{\overset{O}{\|}}{C}-(CH_2)_n-\overset{\overset{O}{\|}}{C}-OR + 2H_2O$$

| | | | |
|---|---|---|---|
| ALCOHOL (R=8, 9, 10, 13) | ACIDO DIBASICO (n = 4, 7, 8) | DIESTER | AGUA |
| | ADIPICO | ADIPATOS | |
| | AZELAICO | AZELATO | |
| | SEBASICO | SEBACATOS | |
| | FTALICO | FTALATOS | |

**Figure 2.**
*Diester: Two carbonyl.*

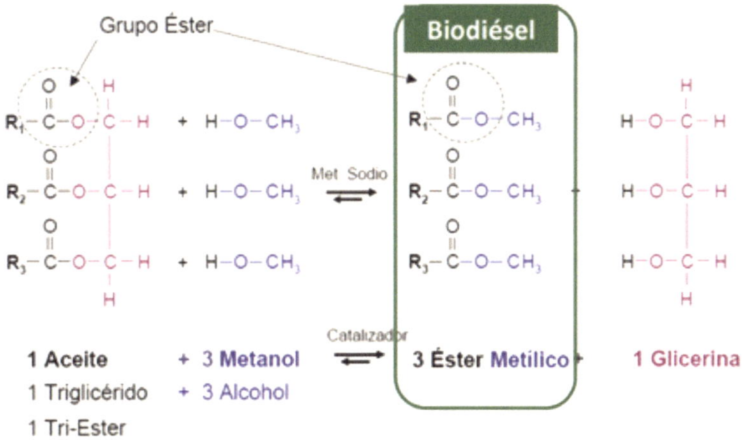


**Figure 3.**
*Triester: (biodiesel). Transesterification process.*

In this tribological scenario, the main characteristic of B100 as biolubricant becomes important (B100 is Tri-Ester). This means that B100 has an adsorption intermolecular force 50% higher than commercial fossil synthetic ester bases (Di-Ester). This condition of intermolecular power superiority translates into a constant presence of the lubricant film, even in extreme lubrication conditions such as those mentioned here.

It is also important to highlight that most additives for motor oils, for the different functions required (detergents, dispersants, anti-wear, anti-rust, high pressure, etc.), are polar substances (eg. ZDDP); the vast majority are also based on sulfur, phosphorus, zinc, and others that are highly polluting the environment, in addition to triggering chemical processes that deteriorate the quality of the oils inside the engines.

### *3.2.1 Fatty acid composition*

**Table 1** shows the composition of fatty acids, in the central column, it is shown in yellow, half corresponds to saturated fatty acids and the other half to unsaturated

| Composicion de acidos grasos | | | | | |
|---|---|---|---|---|---|
| **Acidos Grasps** | | **Aceite de Palma** | | **Aceite de Palmiste** | |
| Caprico | C10:0 | | | 3.7 | |
| La aim | C12:0 | 0.2 | | 48.3 | Acidos |
| Miriam | C14:0 | 1.1 | | 15.8 | la Lin cos |
| Palmfti co | C18:0 | 44 | 50% | 7.8 | |
| Esteasico | C18:0 | 4.5 | Saturados | 2 | |
| 01Aim | Cl | 39.2 | 50% | 15.1 | |
| Lindeico | C18:2 | 10.1 | Insalura dos | 2.7 | |

**Table 1.**
*Fatty acid composition of palm oil.*

fatty acids. This represents the connection key that, from the concept of biofuel, brings us closer to the concepts of biolubricants.

The unsaturated compounds have an iodine value close to 85 g I/100 g of sample, while the saturated ones have an almost zero value. Unsaturated oils have a low viscosity and a lower pour point than their saturated counterparts.

Cold properties are strongly influenced by the degree of unsaturation; unsaturated compounds remain liquid at temperatures below 0°C, while saturated compounds are solid at room temperature. Saturated compounds have higher oxidative stability due to the absence of double bonds, but poor cold properties. Only unsaturated compounds are suitable for use as lubricants, but they can only be used under operating conditions that do not require high oxidative stability [15].

### 3.3 Common characteristics

Some characteristics between diesters (synthetics) and triester (B100). Diesters and triester have natural properties of lubricity and high detergency and dispersance, so they receive the name of clean operating lubricants. Their thermal stability allows them to work up to 180°C. They can operate at low temperatures since its freezing points are between –50 and –60°C (only some B100s). The viscosity index is high, close to 140. They have low volatility, high solvency for both additives and tanks, cleaning the sludge left previously; they tend to dissolve varnishes and lacquers. Soften the elastomers of the seals, therefore, it is recommended to use with these oils, viton seals and medium to high nitrile buna N. They are compatible with mineral oils and are biodegradable [12].

## 4. Bio-lubricant biodiesel (B100) as a new oil for 4-stroke internal combustion engines, 100% biodiesel

This statement includes the formulation of Biolubricants for 4-stroke internal combustion engines, using as bio-lubricant base, biodiesel (B100), hydro-biodiesel, and/or other products for the same function from biomass. As an additive, biodiesel (B100), hydro-biodiesel, and/or other products for the same function from biomass obtained from raw materials other than the biolubricant base, with physical–chemical characteristics to enhance the properties of the base and/or introduce new properties, always obtaining B100 in the end, this being a suitable biolubricant, which guarantees the superior protection and durability required by engine manufacturers, without prejudice to the fact that only the biolubricant base (biodiesel, hydro-biodiesel, others) fulfill the double function of biolubricant and/or biofuel; and notwithstanding that any crude or processed vegetable or animal oil may be used as biolubricant base oil or as an additive. The physicochemical characteristics of B100 as a new biolubricant for 4T engines correspond to those that are representative of high-quality commercial oils, required by manufacturers, and regulated by international regulations (ISO, SAE, EN, ASTM, API, etc.).

In the case of the formulation described here of B100 as a biolubricant, it refers to combining the characteristics of biodiesel obtained from different types of fatty acids, for example:

Saturated fatty acids: such as some used edible oils or animal fats that, when transformed into biodiesel, have good characteristics for high temperatures.

Unsaturated fatty acids: such as rapeseed, soybean, sunflower, castor oil, jatropa curcas, and others, when transformed into biodiesel, have good properties for low temperatures.

Fatty acids with virtually the same proportion of saturated and unsaturated: such as Colombian palm oil, which contains an approximate proportion of 50% saturated and 50% unsaturated fatty acids, when transformed into biodiesel, present virtually a sum of the good properties of the two types of fatty acids above, that is, relatively good characteristics for high temperatures and low temperatures.

In some cases, such as Colombian palm oil, the main deficiency is its cloud point of 10–12°C (which would not be a problem in tropical countries or in the summer season). This deficiency can be corrected by adding (as an additive) biodiesel (B100) obtained from rapeseed, soybean, sunflower, canola, castor, or jatropha oil, which have excellent cold characteristics due to the high degree of unsaturation, with cloud points below between 0°C and −30°C and even lower in some cases.

B100 or 100% biodiesel, as a biolubricant, is the only fluid known and indicated for the total performance of the procedure indicated here for 4T diesel engines, with Bio-Circular technology, in stage 1 as biolubricant and stage 2 as bio-fuel.

Based on the fact that a 4-stroke (compression-ignition) diesel engine is mechanically similar to a 4-stroke (spark-ignition) gasoline engine, except for the fuel delivery and ignition systems, as well as the type of fuel, and then of carrying out driving tests in a passenger vehicle with a 4T gasoline engine, using B100 as a lubricant (only as a lubricant), the optimal operation of the engine has been verified without presenting damage or wear outside the norm. Consequently, B100 can be used as a lubricating oil for 4T gasoline engines (spark ignition) and in general, for engines with a sump or engine oil tank.

In no case can B100 be used as fuel, in a gasoline engine (spark ignition), because B100 require very high pressures inside the combustion chamber for their self-ignition; compression ratio between 17:1 and 24:1 for 4T diesel engines; while the compression ratios for a gasoline engine is usually between 9:1 and 12:1.

This new function of B100 as lubricating oil for 4T engines will allow not only significant savings in scheduled engine oil changes; it means that this item disappears forever within the maintenance budget for vehicles, transport fleets, industrial equipment, limiting it only to the normal change of filters but environmental pollution due to the dumping of used oils into the environment (soil, air, water) also disappears.

All the execution and elements of the Bio-Circular Engine are developed under sufficiently documented and proven procedures and techniques; without prejudice to the fact that it can be executed under the application of new or similar technologies, or even without the application of the technologies described herein.

## 5. B100 and "Bio-circular engine"

The "Bio-Circular Engine" that results from considering the possibility of using B100, not only as a biofuel but also as biolubricant, once its properties have been determined, we can define it as: "Procedure, apparatus, bio-lubricant and bio-fuel for simultaneous use of b100 with dual function in diesel engines". But it can also be expressed as follows in a more descriptive way as: "Procedure and apparatus for simultaneous use of b100, with double function, in diesel engines: stage 1, as bio-lubricant and stage 2, as bio-fuel".

The Bio-Circular Engine is related to the use of biodiesel in 4T diesel engines (vehicular, commercial, industrial, railway, river, sea, aviation, etc.), but also applicable to all kinds of internal combustion engine which run a crankcase (oil inside the engine), from B100 or pure biodiesel to different mixtures (B10, B20, B50, etc.), to reduce polluting emissions from petroleum diesel. It is also related to the use of biodiesel (B100), with or without additives of organic origin, such as biolubricant, to avoid contamination of lubricating fossil oils used in internal combustion engines in general. It is also related to a procedure to make diesel engines work with biodiesel, in such a way that it simultaneously and successively fulfills the two functions: as biolubricant in stage 1, and as biofuel in stage 2; either with B100 or with the required diesel-biodiesel mixture. The application or use of any type of procedure, apparatus, or device that uses biodiesel to lubricate the engine and once this task is carried out, is successively and continuously taken to the injection system to be used as fuel is unknown. Currently, it is very common to use biodiesel obtained from different oilseeds (with different characteristics, depending on the geographical region where they are grown), to be used as fuel, in a pure form (B100 or 100% Biodiesel); but also mixed with petroleum diesel (B10, B20, B50, etc.). The first aspect of the new Bio-Circular Engine, a procedure applicable to diesel engines is provided, certified to use biodiesel B100 (100% pure biodiesel.), but also certified for any percentage of mixture with mineral diesel fuel (B10, B20, B50, etc.); with double function, in two successive and/or simultaneous stages: In stage 1, B100 or pure biodiesel is conducted from the fuel tank to the crankcase or engine lubricant tank (previously filtered), where it will fulfill the first integral function as biolubricant. In stage 2,—successively and/or simultaneously—it is conducted from the same crankcase to the injection system of the diesel engine (previously filtered), where it will fulfill the second function, as biofuel. The complete process will be governed by the Control Unit (ECU), which will control the two successive and simultaneous stages, through the data obtained through specific sensors. This procedure can be adapted to virtually any modern diesel engine, as virtually any modern diesel engine would be capable of using B100. The second aspect of the Bio-Circular Engine consists of an apparatus or device to carry out the 2 stages of the aforementioned procedure, which is made-up of a series of electrical, electronic, and mechanical elements. These items include low- and high-pressure piping for the B100; electric pumps, pressure, temperature, level, flow sensors, etc.; filters; in some cases, radiator for the B100 just removed from the crankcase; electric cables; electronic control unit (ECU) and others, which when installed in 4T diesel engines and then interconnected with each other, will allow the supply of the B100 from the vehicle's fuel tank, passing through the engine crankcase for its first function as biolubricant and from there direct it to the fuel injection pump and later to the injectors located in the cylinder head for its second function as biofuel. This general configuration of the device is designed for use with B100, in 4T diesel engines manufactured from the year 2000 onwards, as they were built with sealing materials and plastic elements (Viton), resistant to the detergent action of B100. It is included in the Bio-Circular Engine, a variation of the device, which in addition to being able to use B100, allows it to be installed in 4T diesel engines manufactured before the year 20,000 that have been refurbished to use biodiesel but for technical reasons or due to poor availability of B100 must add fossil diesel (B10, B20, B50, etc.). This addition of the fossil diesel is made after the B100 has performed its first lubrication function. Once extracted from the engine crankcase, it is taken through the duct corresponding to a new device where B100 duct and the fossil diesel duct converge, which arrives from its own totally independent tank and pipe, located in a

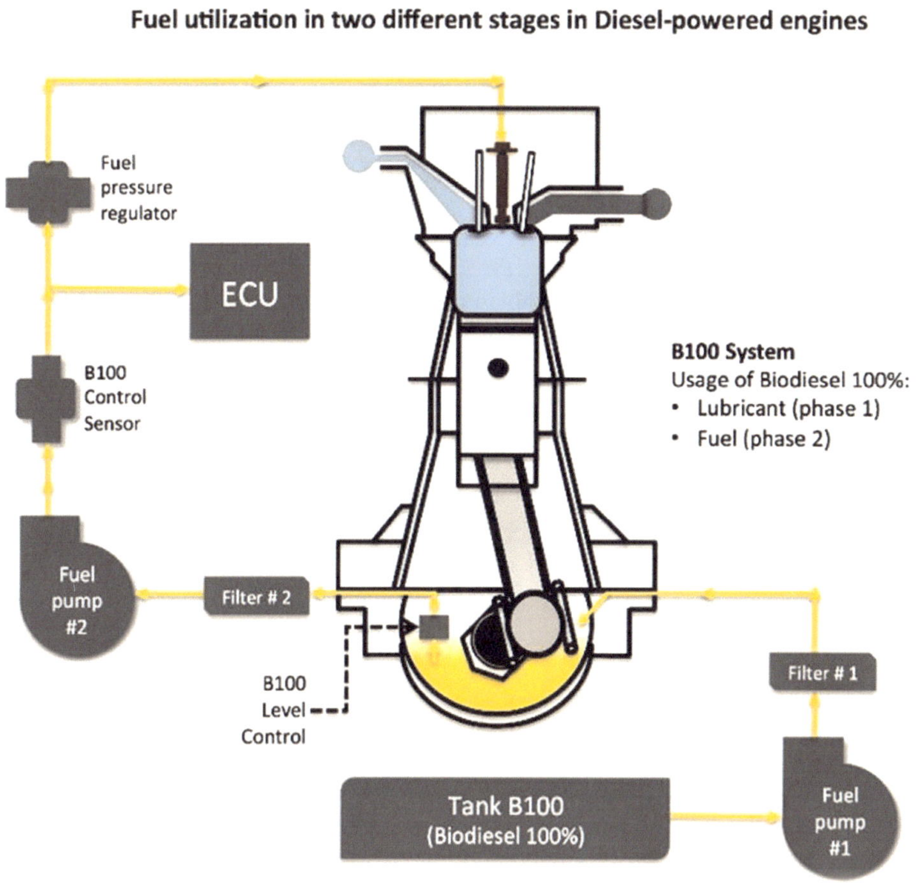


**Figure 4.**
*Bio-circular engine, conceptual diagram.*

place away from the fuel tank B100. For this variation of the original procedure and apparatus, some additional elements must be installed, such as, the mixing station and fuel percentage selection monitor, in addition to its interconnections with the ECU, which allows the use of at least B5–B100 (with this variation, operation with only B100 in the two stages is also planned; it is enough to suspend the supply of fossil diesel), which will guarantee that B100 will always be renewed in its function as a biolubricant inside the engine, to a lesser or greater degree flow rate, which will depend on the amount of B100 chosen on the selector provided for it. As mentioned earlier, the entire process will be governed by the ECU, which will control the two successive and simultaneous stages, using the data obtained through specific sensors. The third aspect of the Bio-Circular Engine, a biolubrican biodiesel (B100) A as a new oil for 4-stroke internal combustion engines, 100% biodiesel, which is the only fluid known and indicated for the total performance of the procedure and device characteristic of the Bio-Circular Engine; with all the required and recognized physicochemical characteristics of modern commercial synthetic oils, for its performance in 4T engines, composed entirely of 100% biodiesel (B100); the base is a B100 obtained from a certain raw material, to which one or more portions of other B100 of the same or different generation (including hydro-biodiesel) and the same or different raw materials are added, to enhance the properties of the base and/or introduce new properties, always obtaining B100 at the end. In exceptional cases, it is also possible to add some type of substance that improves some of its properties (eg cold flow). The main characteristics of B100 as a biolubricant may vary depending on the raw materials (rapeseed, soybean, sunflower, palm, animal fat, used vegetable oil, etc.) from which it is obtained; the most important ones are: B100 is a polar substance (all additives in commercial mineral and synthetic oils are polar substances). It is a triester (a commercial synthetic oil is a diester). It is a Newtonian fluid (shear rate proportional to shear stress). It forms layers of molecules with greater strength of adhesion to metals (3 carbonyls per molecule). The higher the effort and pressure, the higher the viscosity. Palm reduction of $NO_x$ emissions due to saturated triglycerides (B100 from Colombian palm), greater thermal and chemical stability, no rust, no sludge, no deposits. There is no acidification of the lubricant due to the Blow-By phenomenon (passage of fuel through the piston rings towards the crankcase). High polar strength (3 carbonyls). Low viscosity (fuel saving). High detergency (impeccable cleaning). No carbon in carter. Does not require the use of toxic additives. It does not pollute water, air, and land. High flashpoint (>170°C; Colombian palm B100). Volatility >300°. Excellent antiwear protection in limit and elastohydrodynamic lubrication in cases of high pressure, load, and temperature.

### 5.1 B100 and "bio-circular engine" the best way to execute the procedure, for the case of only for B100 in the 2 stages

**Figure 5**, (only for B100 in the 2 stages); shows all the elements for the use of B100 in both stages and throughout the procedure; provides the schematic of an internal combustion engine with the complete schematic drawing of the apparatus or device, with all the components arranged to enhance said procedure, where the conduction of the B100 can be observed, from the fuel tank to the Carter or engine oil tank, so that it performs the first function, as biolubricant (stage 1); Next, and on the opposite side of the engine, we can see the elements designed to extract the B100 from the Carter,

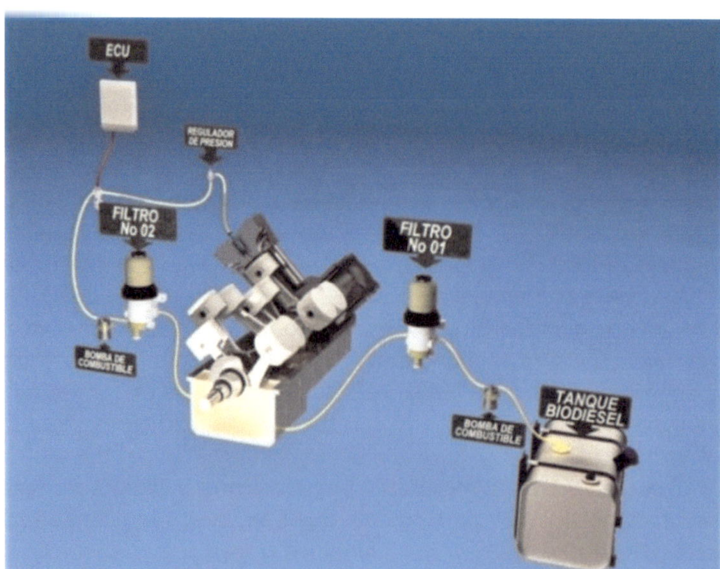


**Figure 5.**
*Bio-circular engine, only for B100 in two stages.*

drive it through the injection system to the injectors located in the cylinder heads, so that it performs the second function, such as biofuel (stage 2). The entire procedure will be governed by the ECU.

In **Figure 6**, (only for B100 in the 2 stages); You can see the conduction of the B100, by means of the electric pump (9) from the fuel tank (8), passing through the filter (7) to the Carter or engine oil tank, so that it performs the first function, such as biolubricant (stage 1); the flow and level controller (6) ensures the constant supply of B100 to the Carter, keeping it at the appropriate level line (horizontal intermittent double line), in such a way that the B100 fulfills its function as biolubricant, for a time t, which is determined by the constant flow of B100 to the fuel injection system (B100). Note: The residence or service time (t) of B100 inside the engine as a lubricant is determined by the fuel consumption of the Bio-Circular engine; for this, it provides that the transit of the B100 from the fuel tank (8), passing through the engine crankcase, until reaching the injectors (10) in the cylinder heads, is continuous; consequently, the volume of B100 that passes through said engine crankcase is exactly that consumed by the same engine in its internal combustion process, either as a function of the operating time and/or the distance traveled and/or the work performed. We will take as an example a vehicle with a 12,000 CC four-stroke (4T) diesel engine; 287 Kw; crankcase capacity of 40 l of lubricating oil and fuel consumption of 5 km/4 l. With the previous data, we know that approximately every 50 km of travel, it consumes 40 l of fuel (B100), which is the capacity of the crankcase, with which, for the case of this example,

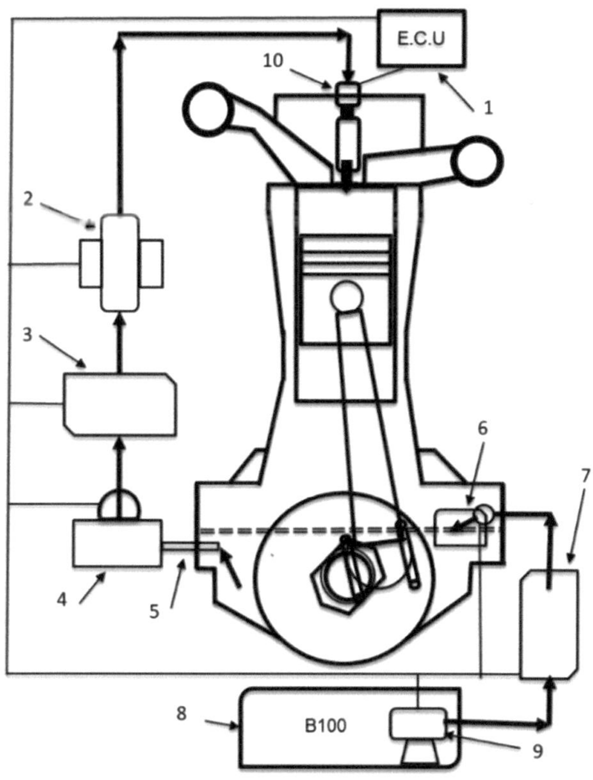


**Figure 6.**
*Bio-circular engine diagram, for the exclusive use of biodiesel (B100) in the 2 stages. Video 1 functional model, bio-circular diesel engine; https://bit.ly/3HgVXJi.*

we can affirm the following: Every 50 km of travel, 40 l of B100 pass through the engine crankcase in its stage 1, as biolubricant. Every 50 km of travel, virtually all the 40 l of lubricating fluid (B100) would be renewed. Every 50 km of travel, virtually a complete change of engine lubricating oil would be done, at no additional cost. (Normally it is done every 15,000 km). Assuming that the same vehicle in the example moves at an average speed of 50 km/h, a complete change of lubricating oil would be done virtually every 1 h. From the conditions of the previous point, it can be deduced that every 6 min, 4 l of B100 would be entering the engine crankcase, that is, 1 l every 1.5 min. With the previous example, which takes real data from driving and normal operating conditions, it can be stated that biodiesel (B100) in its transit through the interior of the engine, in its first function as biolubricant (stage 1), at an average of 1 l every 1.5 min, can keep its physicochemical characteristics unchanged, so that once extracted from inside the engine, to continue its transit to the injection system, it will be in perfect condition for its second function as biofuel. (stage 2). Once B100 has passed through the interior of the engine (crankcase) in its biolubricant function, it is extracted from the crankcase through a duct (5), provided with a pre-filtering element, located at a certain level height, to guarantee the permanence of B100 inside the engine, if the supply of B100 from the fuel tank is suspended. Such level corresponds to the original level of the lubricant preset by the manufacturer; (END of stage 1). (START of stage 2); the conduit (5) is directly connected to the electric pump (4) in charge of sending the B100, already in its second function as biofuel, to the injection pump (2), which previously passed through the filtering station (3) (in some cases also cooled), to be conducted to the injector (10) located in the cylinder head where the B100 will be atomized inside the cylinder and ignited by the high compression of the piston and the help of the high cetane number (68 for the case of palm biodiesel) that will guarantee combustion with low levels of polluting emissions. (END of stage 2). All the execution and the elements of the present invention are interconnected and controlled by the ECU (1), notwithstanding that it can be executed under the application of new technologies or even without the application of the technologies described here; that is to say that, for the case of the example, it could eventually be possible to make use of the described procedure, using the elements installed at the factory, in the vehicles, to execute the same functions of the B100, not only in the conventional engine but also, the same, converted or adapted as a Bio-Circular Engine.

### 5.2 "Bio-circular engine" the best way to execute the procedure, for the case of B100 in stage 1 and B100-diesel blends in stage 2

**Figure 7**, (for B100 in stage 1 and B100-Diesel blends in stage 2); like **Figure 5**, it shows all the elements for the use of B100 in stage 1 as biolubricant; during stage 2, B100 is blended with fossil diesel. Additionally, the necessary elements for the realization of said mixtures are shown (B10, B20, B50, etc.), which includes from a diesel fuel tank, passing through the B100-diesel mixer, once mixed in the indicated proportion, it is carried out through the pressure system to the injectors in the realization of said mixtures are shown (B10, B20, B50, etc.), which includes from a diesel fuel tank, passing through the B100-diesel mixer, once mixed in the indicated proportion, it is carried out through the pressure system to the injectors in the engine cylinders. Before the passage of the B100 (100% pure) through the engine in its first function as a biolubricant, it is subsequently led to the mechanism arranged to make said biodiesel-diesel mixtures and finally to the injectors in the engine cylinders, where it will fulfill the second function as biofuel, in the mixture. The whole procedure will be governed by the ECU.

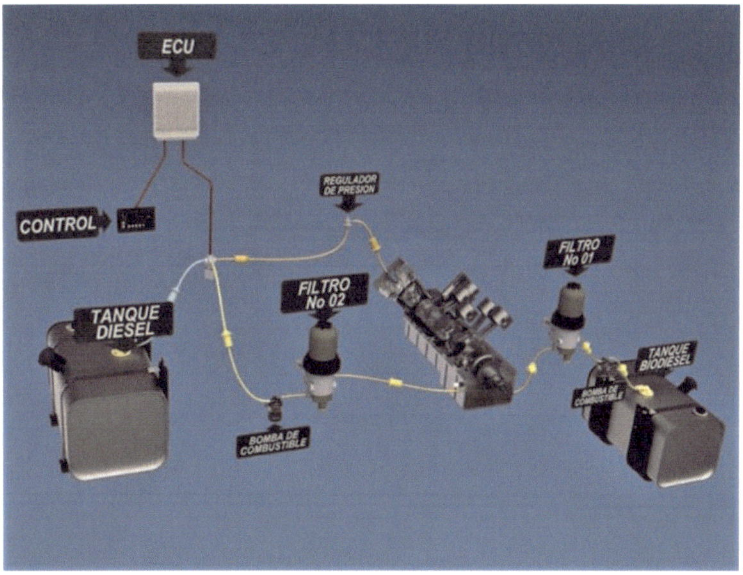


**Figure 7.**
*Bio-circular engine, for B100 in stage 1 and B100-diesel blends in stage 2.*

In **Figure 8**, (for B100 in stage 1 and B100-Diesel mixtures in stage 2), exactly the same operating scheme is executed as in **Figure 6**, until the end of the function of B100 as biolubricant (End of stage 1). For this case, in stage 2 (as biofuel) elements are added that have the task of mixing B100 or 100% pure biodiesel, with fossil diesel, to obtain mixtures technically known as B10, B20, B50, etc.; such mixtures are usually regulated by entities or states, according to environmental, economic, industrial requirements, among others. Returning to the case of **Figure 6**, at the point where the B100 passes through the filtering station (3) (and subsequent refrigeration if necessary), this B100 is led to the device (15), designed to make the mixture (B100 + fossil diesel); the fossil diesel for its part is conducted through the action of the electric pump (13), from the reservoir or tank (12), passing through the filtering station (14), to the aforementioned mixing element (15); proportions of B100 and petroleum diesel already mixed and previously selected on the display (16), is conducted to the high-pressure system or pump (2), to finally be transported to the injector (2), at the head of the cylinder. Also, for this case of **Figure 8**, all the implementation and elements of the present invention are interconnected and controlled by the ECU (1); without prejudice to the fact that it can be executed under the application of new technologies or even without the application of the technologies described here and even without their assistance. Note: It should be noted that both for the case of **Figures 6** and **8**, B100 that enters the Bio-Circular Engine crankcase in stage 1 (as biolubricant), is 100% pure biodiesel. For the case of **Figure 4,** B100 is only mixed with the fossil diesel, in stage 6, in the mixing element (15), consequently, there is no possibility that the fossil diesel enters the interior of the engine (carter and lubrication system), by the action of the process or device described in this new Bio-Circular Engine. For the case of **Figure 8**, the permanence of B100 as biolubricant inside the engine will not depend on the total fuel consumption as in the example of **Figure 6**; yes, it will depend on consumption, but depending on the percentage of B100 contained in the mixture (Biodiesel + fossil diesel). If we take the same conditions and engine data

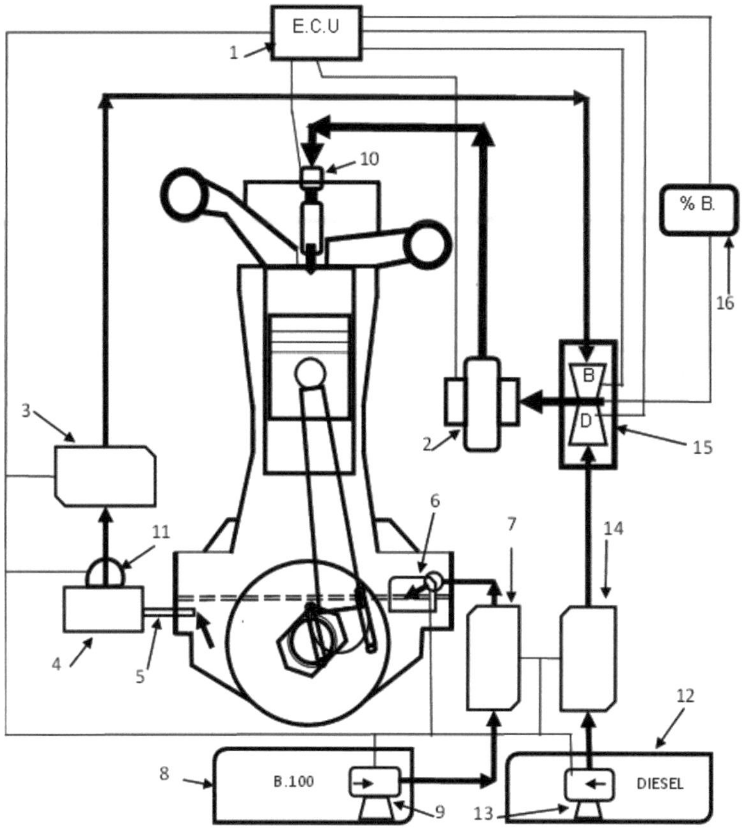


**Figure 8.**
*Bio-circular engine diagram, for B100 in stage 1 and B100-diesel blends in stage 2.*

from the previous example and select a B20 mixture, that is, 20% biodiesel with 80% fossil diesel; with the data of the previous example in which the consumption of the engine is 5 km/4 l and the oil capacity of the crankcase is 40 l of lubricating oil, then every 50 km the entire lubricant would be renewed (40 l of B100). In the case of B20 (mixture of 20% biodiesel +80% fossil diesel) or (20% pure biodiesel in the fuel mixture), this pure biodiesel (B100) would then take five times longer to be completely renewed, that is, the total renewal would take 250 km of travel, to consume 40 l of B100 as a lubricant. If in the same way, we assume that the vehicle in the example moves at 50 km/h, then it would take 5 h for the 40 l of B100 to pass through the interior (crankcase) of the engine. This means that every hour 8 l of B100 would be renewed; that is, 4 l every 30 min; or what is the same, 1 l every 7.5 min. In conclusion, for the example that concerns us, we observe the difference between pure biodiesel B100 and a mixture of B20: With B100 (100% pure Biodiesel), the biolubricant in the crankcase is renewed at a flow rate of 1 l/1.5 min. With B20 (20% pure Biodiesel), the crankcase biolubricant is renewed at a flow rate of 1 l/7.5 min. Bearing in mind that in the case of the B20, the renewal time is five times greater, due to the fact that five times less biolubricant B100 circulates inside the engine (crankcase), it would only take 250 km to complete the total renewal. If we take into account that it is a vehicle with commercial characteristics, those 250 km would normally be traveled in 1 day or less, which will ensure the efficiency and stability of the physical–chemical characteristics of most biodiesels of different materials. Caution: All the execution and the elements of the present invention are interconnected and controlled by the ECU (1), notwithstanding that it can be executed under the application of new technologies or even without the application of the technologies described here; that is to say that, for the case of the example, it could be possible to make use of the described procedure, using the elements installed at the factory, in the vehicles, to execute the same functions of the B100, not only in the conventional engine but also, the same, converted or adapted as a Bio-Circular Engine.

## 6. Conclusions

1. Yes, it is possible to use biodiesel B100 as a biolubricant and once this first function has been fulfilled, it can fulfill its role as a biofuel.

2. The physicochemical characteristics of B100 remain intact due to the rapid transition from its function as a lubricant to its subsequent function as a fuel.

3. For the operation of the proposed procedure and apparatus, it is not necessary to add additive packages, since the natural characteristics of B100 are those that are usually obtained with said additive packages in lubricants of mineral origin; if necessary (low temperatures), it can be mixed (as an additive) with other biodiesel from unsaturated fatty acids (soy, sunflower oil, etc.), for such a requirement.

4. The rapid transit of the B100 guarantees a substantial improvement in internal cooling, as well as efficient cleaning, which translates into greater engine longevity.

5. The problem of contamination of the lubricant, is caused by the entry of fuel into the crankcase through the cylinders and piston rings (Blow-By). Fuel that depletes additives and introduces sulfur and aromatic compounds into the motor oil, affecting viscosity; this will never happen with B100, as it is free of harmful chemicals.

6. Water has a devastating effect on oil and lubricated parts by causing rust and corrosion, respectively. In oils with anti-wear additives based on zinc dithiophosphate (ZDDP), it reacts with them giving rise to the formation of sulfuric acid, which eliminates the boundary film. It will never happen with B100 because it is free of these additives.

7. Savings in maintenance and oil changes by the described procedure; do not repurchase motor oil.

8. B100 does not need to be added with any combustible substance (neither fossil nor biological).

9. B100, as a biolubricant, applies to all types of 4T (4-Stroke) internal combustion engines, whether they are vehicular, commercial, or industrial; including all diesel engines, but also all spark-ignition engines. Video 2; test vehicle spark-ignition engines; https://bit.ly/3HgVXJi

10. The Bio-Circular Engine described is a model that is inserted in the concept of the Circular Economy, in that it uses in one (1) single substance or product, for two (2) totally different functions (Bio-Lubricant + Bio-Fuel) successively and simultaneously.

## Acknowledgements

To God, source of love, peace, and inspiration; to my parents Luz Alba and Pedro Pablo; to my beloved wife Blanca Aurora; to my dear daughter María Piedad; to my brothers Pedro, Ferney, Carlos and Linda Piedad. Bio D S. A, Colombia, who generously donated the PREMIUM GOLD BIODIESEL to carry out all the laboratory tests and start up the functional model, and the test vehicle. Eng. Erika Díaz; Eng. Danisa Leguizamón. http://www.biodsa.com.co/?lang=es Technological School Central Technical Institute, Bogotá D.C. Colombia. Eng. Jim Landinez Cañón; M.Sc. Alejandro Martínez. https://etitc.edu.co/en/; Colombian Federation of Biofuels. Eng. Alfonso Santos; Eng. Carlos Graterón; Eng. Jorge Bendeck. https://www.fedebiocombustibles.com/ Special thanks to: Professor PhD Luis Eduardo Benítez Hernández; National University of Colombia. Professor PhD Carlos Alberto Guerrero Fajardo; National University of Colombia. Professor PhD John Ramiro Agudelo Santamaría; University of Antioquia, Colombia. M. Sc Yordanka Reyes Cruz; Petroleum Research Center, City of Havana, Cuba; Federal University of Rio de Janeiro, Brazil.

## Additional information

There is an authoritative summary, published in the proceedings of the EUBCE 2021 event (ETA Florence Renewable Energies. www.etaflorence.it); From that summary I have taken some elements of my authorship (the same author of this chapter) that are part of the preliminary investigation of this chapter, in which some additional findings are presented. From what was published in the EUBCE minutes, the use of elements was authorized. The cited abstract was not peer reviewed.

## Author details

Cesar Bautista Sterling
Engineer in Machine Design, Independent Researcher, Technological School Central Technical Institute, Bogotá, DC, Colombia

*Address all correspondence to: cesarbautistasterling@gmail.com

IntechOpen

© 2022 The Author(s). Licensee IntechOpen. This chapter is distributed under the terms of the Creative Commons Attribution License (http://creativecommons.org/licenses/by/3.0), which permits unrestricted use, distribution, and reproduction in any medium, provided the original work is properly cited.

## References

[1] Placeholder Text Inter Governmental Panel on Climate Change. Available from: https://www.ipcc.ch/

[2] World Health Organization (WHO). Available from: https://www.who.int/es

[3] Colombian Petroleum Association "Manual of Oils Used". Available from: https://acp.com.co/web2017/images/pdf/combustiblesylubricantes/FAU/Manual-AU-final-14.pdf

[4] JE Murillo. Production of Biodiesel from Palm Oil. Chemical Engineering Department, University of Colombia, Manizales. 2003

[5] Vargas FAA. Production and Characterization of Palm Biodiesel [Doctoral Thesis]. Amsterdam, Netherlands: Elsevier; ISBN: 978-84-693-4594-8/DL: T.1007-2010

[6] Available from: http://noria.mx/lublearn/volatilidad-vs-punto-de-inflamacion-lo-que-debe-saber/?utm_source=LubeTips+en+espa%C3%B1ol&utm_campaign=011b2a5657-Lube_Tips_N_23_2019&utm_medium=email&utm_term=0_b3e8ef5031011b2a5657-211997503&ct=t(Lube_Tips_N_23_2019)&mc_cid=011b2a5657&mc_eid=c366601c8f

[7] Molina BA. Prediction of the effect of temperature on the viscosity of palm oil and its blends with conventional diesel. Energy Magazine. 2006;**35**:33-38 National University of Colombia, Medellin Campus

[8] Campi AR, et al. Depósito de Patente Brasileira-PI9903413-1. Aditivo para melhorar as propriedades de combustível automotivo. 4 de agosto de. 1999

[9] Campi AR., et al. Depósito de Patente Brasileira-PI9903414-0. Mistura alcoólica solúvel em óleo diesel. 4 de agosto de

[10] Glasstone S. Tratado de Química-Física. La Habana: Universidad de Oklahoma, Instituto Cubano del Libro, Ed. Científico-Técnica; 1972

[11] Guibet J, Faure-Birchem E. Fuel and Engines. Francia: Editions Technip; 1999. p. 1999

[12] SÁ SR 2002. Apostila de Físico-Química, Termodinâmica das Superfícies. pp. 1-4, 19, 23-24

[13] Available from: www.bretis.com module 6 synthetic lubricants

[14] Available from: https://noria.mx/lublearn/la-importancia-de-la-resistencia-de-la-pelicula-lubricante/

[15] Design, Selection and Production of New Biolubricants. Available from: https://www.tdx.cat/bitstream/handle/10803/48759/Tesi%20Doctoral%20-%20Albert%20Garcia%20Colomer%20%28FINAL%29.pdf?sequence=1&isAllowed=y

Chapter 12

# Pressure Fluctuation Characteristics of High-Pressure Common Rail Fuel Injection System

*Yun Bai, Zhaoyang Chen, Wei Dou, Xiangdong Kong, Jing Yao, Chao Ai, Fugang Zhai, Jin Zhang and Liu Yang*

## Abstract

In high-pressure common rail fuel injection system, fuel pressure wave propagates back and forth in the system during fuel injection, and the cycle fuel injection volume is affected by the fluctuation of fuel injection pressure. Therefore, to reduce the influence of pressure fluctuation on the cycle fuel injection volume fluctuation, it is of great theoretical significance to analyze the mechanism of pressure fluctuation and its influence law. In this chapter, the dynamic pressure fluctuation characteristics of the high-pressure common rail fuel injection system are analyzed based on the injector inlet pressure, and experimental research and theoretical analysis are carried out for the time domain and frequency domain characteristics of injector inlet pressure fluctuation, aiming at revealing the pressure fluctuation mechanism and its influence law, and providing theoretical support for improving the control accuracy of multiple injection cycle fuel injection volume.

**Keywords:** diesel engine, high pressure common rail, fuel injection, pressure wave, time domain characteristics, frequency domain characteristics

## 1. Introduction

As the most advanced fuel system, the high-pressure common rail fuel injection system can realize the flexible, accurate and stable control of fuel injection pressure, fuel injection timing and cycle fuel injection volume, which can not only make the diesel engine power performance and economy best but also meet the increasingly strict requirements of emission regulations [1–5]. The existence of common rail separates the fuel supply process and fuel injection process of high-pressure common rail fuel injection system. The high pressure fuel pump only provides high pressure fuel to the common rail according to the working condition of the system. The electrical control unit (ECU) drives the high speed solenoid valve to control the injector to inject high pressure fuel into the cylinder. The two parts work independently. This is the main characteristic of high-pressure common rail fuel injection system different from traditional fuel injection systems [6–11]. The flow characteristics of high pressure fuel in high-pressure common rail fuel injection systems have an important effect on cycle fuel injection volume. The fuel injector is the main executing part of the fuel injection system. Due to the fuel inertia, fuel in the control chamber and nozzle volume does not immediately stop flowing when the control valve and needle of the injector are suddenly closed. The fuel kinetic energy near the control valve and needle is converted into local pressure gain, then this conversion propagates at the speed of sound to the control chamber and nozzle volume. Finally, the fuel compression wave or expansion wave is reflected back. Because of the energy imbalance, the dynamic pressure wave propagates and oscillates repeatedly in the system until the system reaches a stable state again due to the dissipative effect [12–18].

The pressure fluctuation has a significant effect on fuel injection rate, which affects the cycle fuel injection volume of the high-pressure common rail fuel injection system. Reference [19] proposed a simplified physical model to predict the fluctuation of fuel injection pressure. Reference [20] designed a fuel acceleration pipeline at the nozzle and two sets of control systems were added to control the fuel flow state in the pipeline. A numerical model was established to predict the fluctuation of fuel injection pressure in theory. To explore the relationship between pressure fluctuation frequency and system structure during fuel injection, Ref. [21] established an LC zero-dimensional equivalent model of common rail, high pressure fuel pipeline and injector. Aiming at the influence of pressure fluctuation of high-pressure common rail fuel injection system on cycle fuel injection volume characteristics, Ref. [22] studied the interrelationship between geometrical dimensions of high pressure fuel pipeline between common rail and injector and pressure fluctuation characteristics and cycle fuel injection volume. The research results show that the size change of high pressure pipeline has a significant impact on the characteristics of single injection cycle fuel injection volume. Reference [23] analyzed the characteristics of cycle fuel injection volume of high-pressure common rail fuel injection system under two working conditions. The research results show that the fuel pressure fluctuation in the injector internal pipeline has a more significant effect on the change of cycle fuel injection volume compared with the pressure fluctuation in the common rail. Reference [24] established a simulation model of high-pressure common rail fuel injection system. The simulation analysis shows that the pressure fluctuation in the fuel chamber during fuel injection is the main reason for the high frequency characteristics of fuel injection rate variation. However, the low frequency characteristics of fuel injection rate variation are determined by the fluctuation of fuel injection pressure.

Multiple injections is one of the main technical means for diesel engines to meet increasingly strict emission regulations. Many scholars have studied the influence of pressure fluctuation on multiple injection cycle fuel injection volume characteristics of high-pressure common rail fuel injection systems. Reference [25] studied the characteristics of cycle fuel injection volume under different injection modes of high-pressure common rail fuel injection systems. The research results show that the fuel pressure fluctuation has an important influence on cycle fuel injection volume because it affects the injection timing of pilot injection, main injection and post-injection. Pressure fluctuation generated after the main injection will cause the needle to be difficult to open during post-injection, which results in post-injection volume fluctuations. Reference [26] changed the injection interval between pilot injection and main injection of high-pressure common rail fuel injection system. It is found that the pulse width of the main injection fluctuates periodically with the increase of the injection interval between pilot injection and main injection when the actual pilot injection cycle fuel injection volume and the actual main injection cycle fuel injection volume are fixed. The pulse width fluctuation frequency only depends on the structural parameters and is independent of diesel engine speed, cycle fuel injection volume, fuel injection pulse width and fuel injection pressure. Reference [27] simulated and analyzed the influence of different fuel properties on the pressure wave and cycle fuel injection volume in high pressure fuel pipeline during three injection processes of high-pressure common rail fuel injection system. The results show that the post-injection fuel volume is affected by the pressure fluctuation caused by the main injection. The change of fuel properties leads to the different phases of pressure fluctuation, which affects the opening of the needle. Thus, the cycle fuel injection volume decreases with the increase of the bulk modulus of elasticity of the fuel. Reference [28] studied the influence of pilot injection timing and pilot injection fuel volume on soot, NOx, combustion noise and fuel consumption rate of diesel engines. The results show that the pressure wave caused by specific pilot injection timing in common rail and high pressure fuel pipeline leads to the dramatic change of main injection cycle fuel injection volume, especially when the injection interval between pilot injection and main injection changes. It has important influence on soot and NOx. In order to reduce the repeated reflection and propagation of fuel pressure fluctuation in high pressure fuel circuits, Ref. [29–31] designed a pressure storage chamber at the outlet end of high pressure fuel pump and developed a new type of high-pressure common rail fuel injection system. By studying the cycle fuel injection volume of the system and the conventional high-pressure common rail fuel injection system under different common rail pressures, it was found that the different arrangement of the two fuel injection systems leads to the difference in fuel pressure wave propagation and reflection, which leads to the variation of fuel pressure fluctuation characteristics and causes the cycle fuel injection volume to be different.

In this chapter, the pressure fluctuation of high-pressure common rail fuel injection system will be investigated theoretically. The dynamic pressure wave fluctuation mechanism of the system will be analyzed through the experiments. On this basis, the influence of different parameters on dynamic pressure wave during fuel injection will be studied. The results will provide the support for revealing the cycle fuel injection volume fluctuation and its generation mechanism of high-pressure common rail fuel injection system.

## 2. Composition and working principle of high-pressure common rail fuel injection system

High-pressure common rail fuel injection system is mainly composed of the low pressure fuel supply part, including tank, low pressure pump, the high pressure fuel injection part, including high pressure pump, common rail, electrical control injector, the fuel return circuit, which transfers excess fuel from each part back to the tank, the electrical control part, including ECU and various sensors, as shown in **Figure 1**. During the work process of high-pressure common rail fuel injection system, the plunger of high pressure pump moves downward with the rotation of the camshaft under the action of spring. Fuel is sucked into the high pressure pump plunger chamber from the tank by the low pressure pump through the fuel filter to complete the fuel absorption process. The plunger of the high pressure pump moves upward with the rotation of the camshaft driven by the cam. Fuel in the plunger chamber is compressed and the pressure increases. The pressurized fuel is pumped to the common rail through the high pressure pipeline to complete the process of fuel pressurization and fuel supply. The high pressure fuel in the common rail is distributed to the

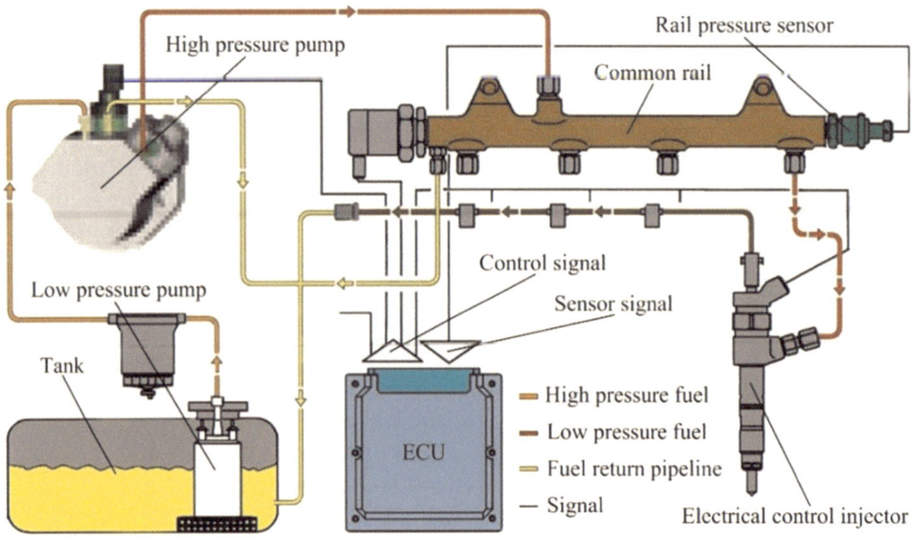


**Figure 1.**
*Schematic diagram of the high-pressure common rail fuel injection system.*

injector of each cylinder by the high pressure pipeline. Fuel is injected into the cylinder through the nozzle when the solenoid valve coil of the injector is energized. The needle closes the nozzle hole to finish fuel injection when the coil is de-energized, which completes a fuel injection process. The fuel metering valve on high pressure pump connects the plunger chamber of the low pressure pump and high pressure pump, which is opened a triangle fuel metering hole. The ECU controls fuel supply volume by adjusting the opening of the fuel metering hole through outputs pulse width modulation signal, thus realizing the adjustment of common rail pressure. Through the feedback signals of various sensors, ECU outputs corresponding control signals according to the working state of the diesel engine and drives the high speed solenoid valve on the injector to realize the control of injection timing, injection duration and injection times, to complete the real-time control of the fuel injection system.

## 3. Theoretical study on pressure fluctuation of high-pressure common rail fuel injection system

The fluctuation characteristics of fuel pressure in the pipeline of high-pressure common rail fuel injection system can be represented as one-dimensional partial differential equations of unstable compressible flow, as shown in Eq. (1).

$$\begin{cases} \dfrac{\partial p}{\partial t} + u\dfrac{\partial p}{\partial x} + a^2\rho\dfrac{\partial u}{\partial x} = 0 \\ \dfrac{\partial p}{\partial x} + \rho\dfrac{\partial u}{\partial t} + \rho u\dfrac{\partial u}{\partial x} + 2\kappa\rho u = 0 \end{cases} \tag{1}$$

where $u$ is the fuel flow velocity, $a$ is the fuel pressure wave propagation velocity and $\kappa$ is the fuel flow resistance coefficient.

The above-mentioned partial differential equations can be converted into ordinary differential equations as follows.

$$\begin{cases} \left(\frac{dx}{dt}\right)_R = u + a, & \frac{dp}{dt} + a\rho\frac{du}{dt} + 2a\kappa\rho u = 0 \\ \left(\frac{dx}{dt}\right)_L = u - a, & \frac{dp}{dt} - a\rho\frac{du}{dt} - 2a\kappa\rho u = 0 \end{cases} \tag{2}$$

where $(dx/dt)_R$ and $(dx/dt)_L$ are the trace lines of pressure waves propagating to the right and left inside the pipeline. The above equations directly reflect the propagation relationship of the pressure wave in the pipeline of high-pressure common rail fuel injection system. When the fuel pressure wave propagates to a specific position in the pipeline, the pressure at this position rises, that is, $dp$ is positive, and the pressure wave is a compression wave. Conversely, the pressure drops. $dp$ is negative, and it is an expansion wave. Assuming that the right direction of $x$ axis is the positive direction, the fuel pressure wave propagating along the forward direction is called the right–traveling wave, while the fuel pressure wave propagating along the negative direction is called the left–traveling wave. The right–traveling wave is represented by $dp_R$ and $du_R$, and it has the following relationship.

$$dp_R = a\rho du_R \tag{3}$$

While the left–traveling wave is represented by $dp_L$ and $du_L$, which has the following relationship.

$$dp_L = -a\rho du_L \tag{4}$$

The pressure wave in the pipeline of high-pressure common rail fuel injection system is divided into left-traveling wave and right-traveling wave according to the direction of propagation and divided into compression wave and expansion wave according to the change of fuel pressure caused by propagation. Therefore, pressure waves in the pipeline can be divided into the following four types.

a. Right-traveling compression wave

   This kind of fuel pressure wave propagates along the positive direction of $x$-axis, where $dp_R$ is positive and $du_R$ is also positive, that is, the fuel pressure and velocity along the x-forward propagating direction increase.

b. Right-traveling expansion wave

   This kind of fuel pressure wave propagates along the positive direction of the $x$ axis, but $dp_R$ is negative and $du_R$ is also negative, that is, the fuel pressure and velocity along the x-forward propagating direction decrease.

c. Left-traveling compression wave

   This kind of fuel pressure wave propagates backward along the $x$-axis, where $dp_L$ is positive and $du_L$ is negative, that is, the fuel velocity decreases along the x-forward propagating direction. However, the fuel pressure and velocity of left-traveling compression wave along with the propagating direction increase.

d. Left-traveling expansion wave

This kind of fuel pressure wave propagates backward along the $x$-axis, where $dp_L$ is negative and $du_L$ is positive, that is, the fuel velocity along the x-forward propagating direction increases.

Supposing the right-traveling wave arriving at $x$ position of the fuel pipeline at moment $t$ is $dp_R$ and $du_R$, and the left-traveling wave is $dp_L$ and $du_L$. According to the pressure wave synthesis theory, the total pressure variation $dp$ and velocity variation $du$ are

$$\begin{cases} dp = dp_R + dp_L \\ du = du_R + du_L \end{cases} \tag{5}$$

where $dp$ and $du$ are called synthetic pressure waves at $(x, t)$. In fact, after the fuel pressure wave is synthesized, each single-traveling wave continues to propagate along its determined direction. Therefore, the above equation can also be understood as the pressure fluctuation in a certain position in the pipeline can be decomposed into left-traveling waves and right-traveling waves.

When the pressure wave propagates to the boundary surface, another returned pressure wave can be obtained based on the pressure wave and the boundary condition at the moment, which is the reflected wave. Assuming that the right end of the high-pressure common rail fuel injection system is closed, the boundary condition is $u = 0$ and $du = 0$. At this time, if there is a right-traveling wave $dp_R$ arriving on the left, there must be a left-traveling reflected pressure wave $dp_L$, so that $du = du_R + du_L = 0$. It can be seen that the fuel reflected pressure wave from the pipeline end boundary surface is the result of coupling the boundary condition with the propagating fuel pressure wave.

There are three boundary conditions for pressure wave propagation and reflection in the pipeline of high-pressure common rail fuel injection system. The boundary type at common rail and high pressure fuel pipeline is outlet isobaric end (outlet opening end). The boundary type is the closed-end when the nozzle needle valve is closed. The boundary type is the orifice flow ends when the needle opening nozzle and injector injection.

a. Outlet isobaric end

There are the following equations when the right-traveling wave $dp_R$ reaches the open end of the right end.

$$\begin{cases} dp_R = a\rho du_R \\ dp_L = -a\rho du_L \\ dp = dp_R + dp_L = 0 \\ du = du_R + du_L \end{cases} \tag{6}$$

Thus, $dp_L = -dp_R$, $du_L = du_R$, $du = 2du_R$, $dp = 0$.

It can be seen that the signs of the incident pressure wave and reflected pressure wave are opposite when the pressure wave propagates to the common rail. However, the absolute value of the amplitude of the pressure wave is the same and the wave velocity is the same. The changing of the pressure at the pipeline end is zero and the changing of the velocity is twice that of the incident pressure wave. The reflection with the property of incident wave and reflected wave is opposite (an expansion wave, the other is compression wave) and the absolute value of the amplitude of the pressure wave is the same is called complete negative reflection. Therefore, the pressure wave reflection at common rail and high pressure fuel pipelines is a complete negative reflection.

b. Closed-end

There are the following equations when the right-traveling wave $\mathrm{d}p_{\mathrm{R}}$ reaches the closed end of the right end.

$$\begin{cases} \mathrm{d}p_{\mathrm{R}} = a\rho \mathrm{d}u_{\mathrm{R}} \\ \mathrm{d}p_{\mathrm{L}} = -a\rho \mathrm{d}u_{\mathrm{L}} \\ \mathrm{d}p = \mathrm{d}p_{\mathrm{R}} + \mathrm{d}p_{\mathrm{L}} \\ \mathrm{d}u = \mathrm{d}u_{\mathrm{R}} + \mathrm{d}u_{\mathrm{L}} = 0 \end{cases} \tag{7}$$

Thus, $\mathrm{d}p_{\mathrm{L}} = \mathrm{d}p_{\mathrm{R}}$, $\mathrm{d}u_{\mathrm{L}} = -\mathrm{d}u_{\mathrm{R}}$, $\mathrm{d}p = 2\mathrm{d}p_{\mathrm{R}}$, $\mathrm{d}v = 0$.

It can be seen that the amplitude of the incident pressure wave and reflected pressure wave is the same after the pressure wave propagates to the nozzle when the needle is closed. However, the velocity disturbance value is the opposite. The velocity change at the nozzle is zero and the pressure variation is twice the amplitude of the incident pressure wave. The reflection with the property of incident wave and reflected wave is the same, that is, an expansion wave or a compression wave at the same time, and the amplitude of pressure disturbance is the same is called complete positive reflection. Therefore, the pressure wave reflection when the needle valve is closed is a complete positive reflection.

c. Orifice flow end

As shown in **Figure 2**, the fuel flow from the C-C boundary surface of the pipeline with section $F$ through the orifice of section $F_t$ to the space where the backpressure is $p_c$. Assuming the initial state of fuel is $p_0 = p_c$ and $u_0 = 0$, the Bernoulli equation and continuity equation from C-C boundary surface to the minimum throat section t-t can be established as follows.

$$\begin{cases} p + \frac{1}{2}\rho u^2 = p_c + \frac{1}{2}\rho u_t^2 \\ Fu = \alpha_F F_t u_t \end{cases} \tag{8}$$

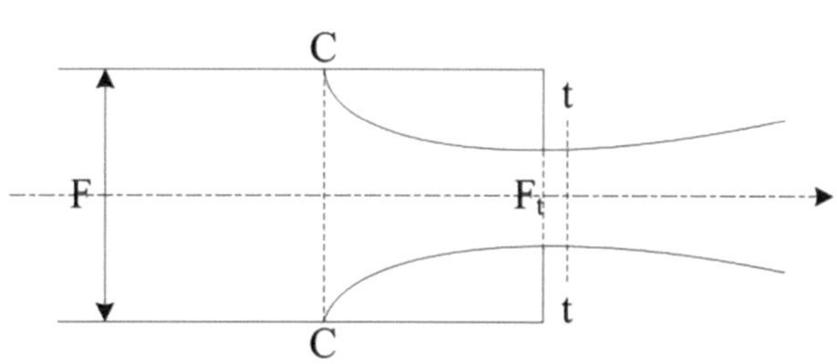


**Figure 2.**
*Schematic of the orifice flow end.*

where $p$ and $u$ are the fuel pressure and flow rate at C-C boundary surface, $\alpha_F$ is the orifice flow coefficient, and $u_t$ is the fuel flow velocity at t-t of minimum throat section.

Thus,

$$p = \frac{1}{2}\rho u^2 \frac{1-\phi_F^2}{\phi_F^2} + p_c \tag{9}$$

where, $\varphi_F = \alpha_F F_t / F$, which is the effective flow section ratio at the orifice end.

The above equation is the boundary condition equation of the orifice outflow of the high-pressure common rail fuel injection system. It can be seen from the above equation that when $\varphi_F = 0$, $F_t = 0$, is a closed-end, and when $\varphi_F = 1$, $F_t = F$, is a outlet flow opening end. However, the reflection at the boundary surface transitions from complete positive reflection to complete negative reflection when $\varphi_F$ changes from 0 to 1.

## 4. Test bench of dynamic pressure fluctuation for high-pressure common rail fuel injection system

The test bench mainly includes the high pressure common rail injection system test stand, the operation stand and the water-cooling unit. It can measure the dynamic injection characteristics of a six-cylinder engine, such as the fuel injection rate, the fuel injection volume and the fuel injection duration for each cycle at most. The main functions of the high-pressure common rail fuel injection system test bench are to drive high pressure fuel pump, supply fuel to high pressure fuel pump at specified temperature and pressure, drive fuel injection system and real-time measure the dynamic injection characteristics of the system. The operation stand includes monitoring system, electronic control device for single injection instrument, driving equipment of fuel injector, programmable synchronous timing pulse generator, common rail pressure regulator, torque analyzer and DC power supply, etc. It can realize the real-time control and monitor of the experimental process. All of the experimental data results and the environmental parameters of each injector can be recorded after the experiment.

The high-pressure common rail fuel injection system test bench is shown in **Figure 3**. It consists of a driving motor, high pressure fuel pump, common rail, common rail pressure sensor, oscilloscope, single injection instrument, mechanical part of injection flow and rate (IFR), injector, pressure sensor, high pressure pipeline, electronic part of IFR, ECU, computer terminal and tank. The high pressure fuel pump is driven by the driving motor, which can provide a stable speed input for the system. The common rail pressure sensor installed on the common rail measures the common rail pressure in real-time and feeds the common rail pressure signal back to the ECU. The ECU adjusts the fuel flow into the common rail by adjusting the flow control valve on the high pressure fuel pump to stabilize the common rail pressure. In order to measure the dynamic pressure fluctuation at the injector inlet, a piezoresistive high pressure sensor is installed near the injector end on the high pressure pipeline between the common rail and the injector. The oscilloscope receives the pressure signal at the injector inlet measured by the pressure sensor and the common rail pressure signal measured by the common rail pressure sensor and stores

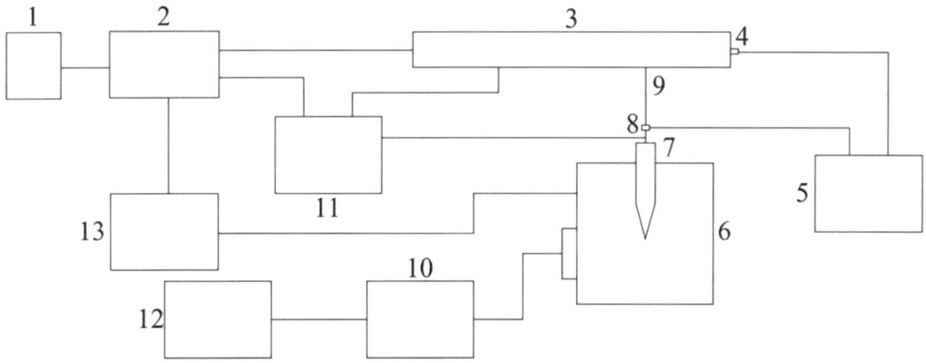


**Figure 3.**
*Schematic diagram of the high-pressure common rail fuel injection system test bench. 1. Drive motor 2. High pressure fuel pump 3. Common rail 4. Common rail pressure sensor 5. Oscilloscope 6. Mechanical part of IFR 7. Injector 8. Pressure sensor 9. High pressure pipeline 10. Electronic part of IFR 11. ECU 12. Computer terminal 13. Tank.*

the signal data. The ECU provides control current to the injector according to experiment conditions to complete fuel injection under different conditions.

## 5. Fluctuation mechanism of dynamic pressure wave for high-pressure common rail fuel injection system

The fuel pressure wave reciprocating propagates within high-pressure common rail fuel injection system during fuel injection. The cycle fuel injection volume is affected by the fuel injection pressure. Therefore, it has important theoretical and practical significance on system optimization design and taking effective method to reduce the adverse impact of pressure fluctuations on cycle fuel injection volume characteristics by thorough analysis on the fluctuation mechanism and influence rule of dynamic pressure wave for high-pressure common rail fuel injection system. Theoretically, the fuel injection pressure refers to the fuel pressure near the nozzle hole. However, due to the fuel pressure in the nozzle is high and the size of the fuel cavity in the nozzle is small, it is difficult to install the pressure sensor near the nozzle hole. More importantly, the installation of a pressure sensor close to the nozzle hole will cause the change of flow field distribution in the nozzle, thus affecting the dynamic pressure fluctuation characteristics of the system. Since the fuel pressure wave propagation in the system with a limited speed, the fuel pressure of the injector inlet only slightly lags behind the nozzle volume pressure at time sequence (fuel injection pressure). It is easily measured and can actually represent the dynamic pressure fluctuation characteristics of the system. In this chapter, the fuel pressure of the injector inlet is used instead of injection pressure to analyze the dynamic pressure wave of the system [32].

**Figure 4** shows the characteristics of solenoid valve drive current, fuel injection rate and injector inlet pressure before and after fuel injection of high-pressure common rail fuel injection system when the cycle fuel injection volume is 30 $mm^3$. As shown in the figure, there is a delay characteristic between solenoid valve energized and fuel injection due to the hydraulic delay of the system. In addition, the fuel injection duration is longer than the solenoid valve energized time. The opening of the

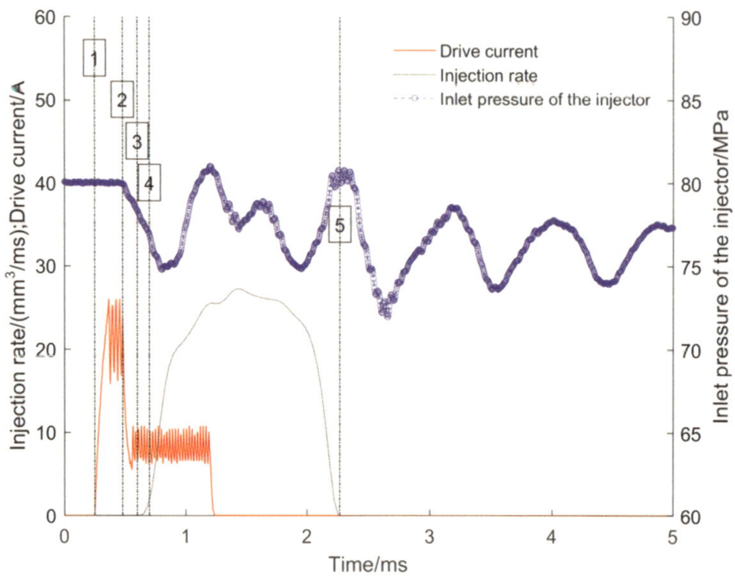


**Figure 4.**
*Drive current of the solenoid valve, injection rate and inlet pressure of the injector at a fuel injection volume of 30 mm³.*

solenoid valve and needle causes the pressure drop at the injector inlet. The fuel injection duration is consistent with the opening time of the needle due to the dynamic response characteristics of the needle.

According to the various characteristics of driving current of the solenoid valve, fuel injection rate and injector inlet pressure at different moments, as shown in **Figure 4**, the curves can be divided into five different stages as follows.

1–2 stages. The solenoid valve coil of the injector is energized, the current gradually increases and the electromagnetic force increases. But the electromagnetic force of the solenoid valve is less than the pretightening force of the control valve reset spring. The control valve is pressed against the control seat and the outlet orifice is not open. The control chamber is filled with high pressure fuel. The resultant of fuel hydraulic pressure on the upper end surface of the needle and pretightening force of reset spring of the needle is larger than the fuel hydraulic pressure on the lower end surface of the needle. The needle is pressed against the needle seat and the injector does not injection fuel. Thus, the injector inlet pressure remains unchanged at 80 MPa.

2–3 stages. The injector inlet pressure drops. The reason is analyzed as follows. The control valve overcomes the pretightening force of the control valve reset spring and moves upward under the action of electromagnetic force of the solenoid valve, and the outlet orifice is opened. The high pressure fuel in the control chamber is discharged to the tank through the low pressure return fuel circuit. The fuel pressure in the control chamber drops rapidly. However, the resultant of fuel hydraulic pressure on the upper end surface of the needle and pretightening force of reset spring of the needle is still larger than the fuel hydraulic pressure on the lower end surface of the needle. The needle is still pressed against the needle seat. The nozzle hole is closed and the injector does not inject fuel. The control valve opens the outlet orifice suddenly arousing an instantaneous expansion wave, which starts between the control valve and the control valve seat.

3–4 stages. The injector inlet pressure continues to drop, but the pressure drop gradient increases. The reason is that the resultant of fuel hydraulic pressure on the upper end surface of the needle and pretightening force of reset spring of the needle is less than the fuel hydraulic pressure on the lower end surface of the needle as the decreasing of the control chamber fuel pressure. The needle moves upward and opens the nozzle hole. The injector starts fuel injection and the fuel injection rate appears. The sudden opening of the needle also arouses an instantaneous expansion wave, which starts between the needle and the needle seat and propagates upward. Due to the existence of the needle channel orifice on the injector body, the pressure drop at the injector inlet is not significant, however, the pressure drop gradient is larger than that when the control valve is opened alone.

4–5 stages. Complete fuel injection process. The expansion wave aroused by the moving parts working processes of the system propagates upward along the fuel circuit in the injector. When it propagates to the common rail, reflecting back a compression wave. This compression wave attempts to recover the fuel pressure in the fuel circuit to the initial value. When it propagates to the injector inlet causes the inlet pressure increasing, as shown in **Figure 4**. In fact, there is no expansion wave generated between the needle and the needle seat when the needle reaches its maximum lift, and the size of the nozzle hole becomes the main factor limiting fuel injection.

Stage after 5. The nozzle hole is closed by the needle and fuel injection is stopped. The closing of the needle will cause a water hammer effect in the system, a compression wave in the nozzle aroused and propagates upward along the fuel circuit in the injector. The inlet pressure increases when it propagates to the injector inlet, as shown in **Figure 4**. Since then, the needle and control valve shut down completely. The pressure wave propagates repeatedly in the system. Because the hydraulic shear resistance restrains the pressure wave oscillation, the amplitude of the fuel pressure wave decreases gradually, and the pressure at the injector inlet shows an attenuation oscillation characteristic.

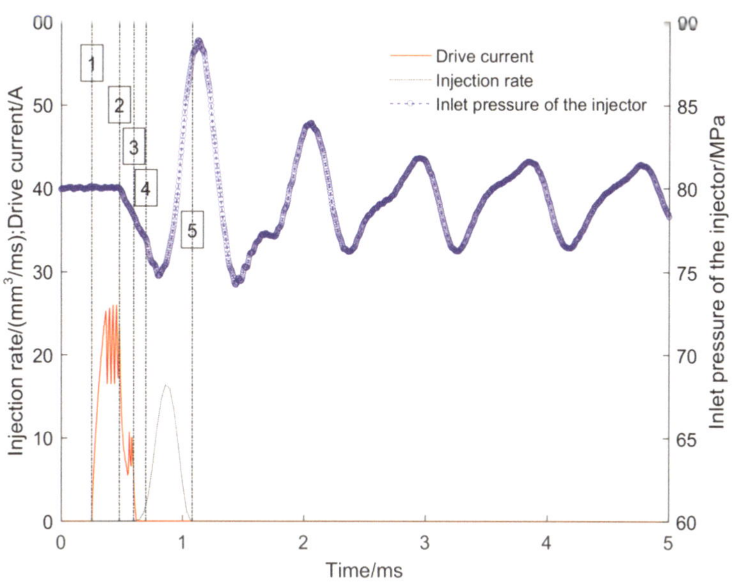


**Figure 5.**
*Drive current of the solenoid valve, injection rate and inlet pressure of the injector at a fuel injection volume of 3 mm$^3$.*

It can be seen from the above analysis that the pressure fluctuation characteristic of the injector inlet before the opening of the needle is independent of the energized time of the solenoid valve coil or the fuel injection duration since this phenomenon will be caused whenever the control valve or the needle starts to move. The pressure fluctuation characteristic of the injector inlet caused by the water hammer effect is obviously dependent on the energized time of the solenoid valve coil since it is generated after the needle valve is closed. Therefore, when the energized time of the solenoid valve coil is shorter, the time interval between the two pressure peaks in **Figure 4** is small, the third pressure peak and the subsequent pressure oscillation peak depend on the energized time of the solenoid valve coil due to the pressure wave interaction. As shown in **Figure 5**, the pressure fluctuation amplitude of the injector inlet is significant when the cycle fuel injection volume of the system is 3 $mm^3$. The fusion of the two pressure peaks is called hydraulic resonance as shown in **Figure 4**.

## 6. Study on the influence factors of dynamic pressure wave in high-pressure common rail fuel injection system

According to the wave mechanism of dynamic pressure wave for high-pressure common rail fuel injection system, the fuel injection rate and fuel injection duration are different with different fuel injection pulse widths, which results in different cycle fuel injection volumes. The change of injection pulse width has a different influence on pressure fluctuation characteristics in the system when the fuel is injected. In addition, the cycle fuel injection volume is different even if the injection pulse width is the same when the high-pressure common rail fuel injection system is under different common rail pressures. Therefore, this section mainly analyzes the influence rule of two key control parameters of the system, namely injection pulse width, and common rail pressure, on the dynamic pressure wave in the system, which provides support for the study of the fluctuation characteristics of cycle fuel injection volume of the system.

**Figure 6** shows the pressure fluctuation characteristics of injector inlet during fuel injection of high-pressure common rail fuel injection system with injection pulse width of 400, 600, 800 and 1000 μs, respectively. It can be seen from the figure that under the same high pressure pipeline size and common rail pressure, the injector inlet pressure with different injection pulse widths shows attenuation fluctuation characteristics. The smaller the injection pulse width, the larger the pressure fluctuation amplitude during the injection duration. With the increase of injection pulse width from 400 μs to 1000 μs, the change rate of pressure fluctuation amplitude at injector inlet decreases, and the average injector inlet pressure increases after injection. This is because the system circulates less fuel injection with small pulse width under the same common rail pressure. After the needle is seated and the nozzle is closed, the high pressure fuel in common rail immediately flows through the high pressure pipeline to replenish that injected in the injector. The larger the injection pulse width, the more fuel needed to replenish and the longer the time required. In addition, with the increase of injection pulse width, the needle gradually reaches its maximum lift. At this time, the injection pulse width only affects the moment when the needle closes the nozzle but has no influence on the needle from opening to reaching its maximum lift. Therefore, as shown in the figure, there is no significant difference between the injector inlet pressure from the first trough to the first crest when the injection pulse width is 800 μs and 1000 μs.

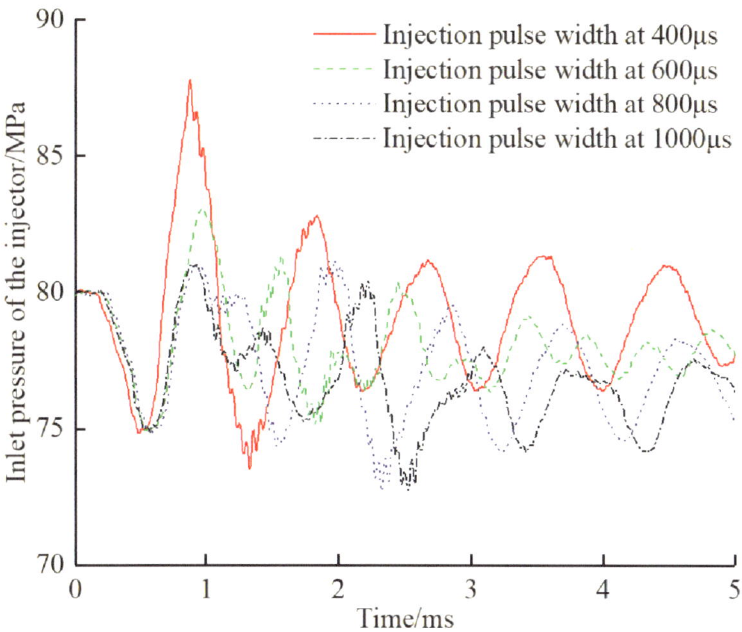


**Figure 6.**
*Inlet pressure of the injector at different injection pulse widths.*

**Figure 7** shows the pressure fluctuation characteristics of injector inlet during fuel injection of high-pressure common rail fuel injection system with rail pressures of 40, 80, 120 and 160 MPa, respectively. Each point in the figure is the difference between the injector inlet pressure at specific common rail pressure and the set common rail pressure when the injection pulse width is 800 μs, which reflects the pressure fluctuation characteristics of the system under different common rail pressures more intuitively. As shown in the figure, the inlet pressure of the injector decreases to a certain extent under different common rail pressures when the size of the high pressure pipeline and injection pulse width is constant. The inlet pressure fluctuation of the injector under the common rail pressure of 40 MPa is obviously different from that under the other three common rail pressures. The average pressure fluctuation of the injector inlet under this common rail pressure is higher than that under the other three common rail pressures. The inlet pressure fluctuation rules are consistent when the common rail pressure increases from 80 MPa to 160 MPa. The higher the common rail pressure, the larger the inlet pressure drop amplitude and the lower the average value of pressure fluctuation. When the size of the high pressure pipeline and injection pulse width is the same, the injection pressure increases with the increase of common rail pressure, and the injection pulse width required by the needle to reach the maximum lift decreases. When the common rail pressure is 40 MPa, due to the low common rail pressure, the moment when the control valve fully opens outlet orifice lags behind, and the moment when the control valve closes outlet orifice is advanced. The fuel pressure relief time in the control chamber is shortened and the needle does not reach its maximum lift. The nozzle is closed again before it is fully opened. At this time, the maximum fuel injection rate and the fuel injection duration of the system are small. Therefore, the average value of pressure fluctuation at the injector inlet is high. With the increase of common rail pressure, the difference of control valve opening outlet orifice decreases. But the pressure difference between the control chamber and low

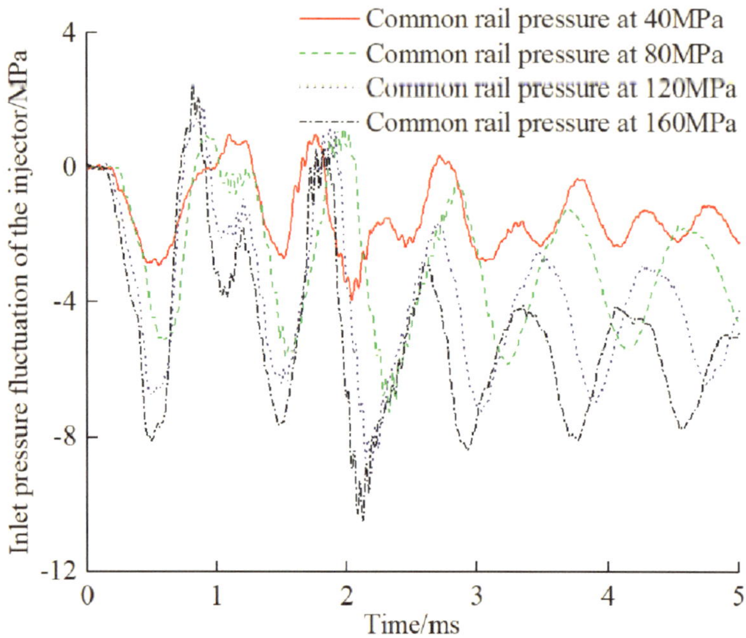


**Figure 7.**
*Inlet pressure fluctuation of the injector at different common rail pressures. (a) Common rail pressure at 40 MPa. (b) Common rail pressure at 80 MPa. (c) Common rail pressure at 120 MPa. (d) Common rail pressure at 160 MPa.*

pressure fuel circuit and nozzle volume and cylinder is larger when the common rail pressure is high. The fuel discharge rate of the control chamber and the fuel injection rate of the nozzle hole are accelerated, which results in the system pressure drop gradient increases. In addition, the higher the common rail pressure, the longer the needle is maintained at the maximum lift position. This is the main reason why the higher the common rail pressure, the larger the injector inlet pressure drop amplitude, the lower the average pressure fluctuation.

It can be seen from the above analysis that the dynamic pressure wave of the system shows different fluctuation characteristics under different injection pulse widths and common rail pressures when the size of the high pressure pipeline is constant. Therefore, the dynamic pressure fluctuation frequency and amplitude of the system are further analyzed with the injection pulse width of 400, 600, 800 and 1000 μs and common rail pressure of 40, 80, 120 and 160 MPa, respectively, to reveal the dynamic pressure wave variation rule of the system under different injection pulse width and common rail pressure.

The area enclosed below the power spectrum density curve represents the amount of energy generated by the fluctuation in the frequency range [33]. **Figure 8** shows the power spectrum density obtained by the fast Fourier transform of injector inlet pressure fluctuation under different injector pulse widths and common rail pressures. As shown in **Figure 8(a)**, when the common rail pressure is 40 MPa and the injection pulse width is 400, 600 and 1000 μs, the dynamic pressure wave energy of the system is mainly in the frequency band of 799 Hz–1199 Hz, and the crest characteristics of power spectrum density are significant, and all reach the main crest at the frequency of 999 Hz. At this time, the dynamic pressure wave in the system shows obvious periodic fluctuation characteristics, which mainly fluctuates in the frequency of the

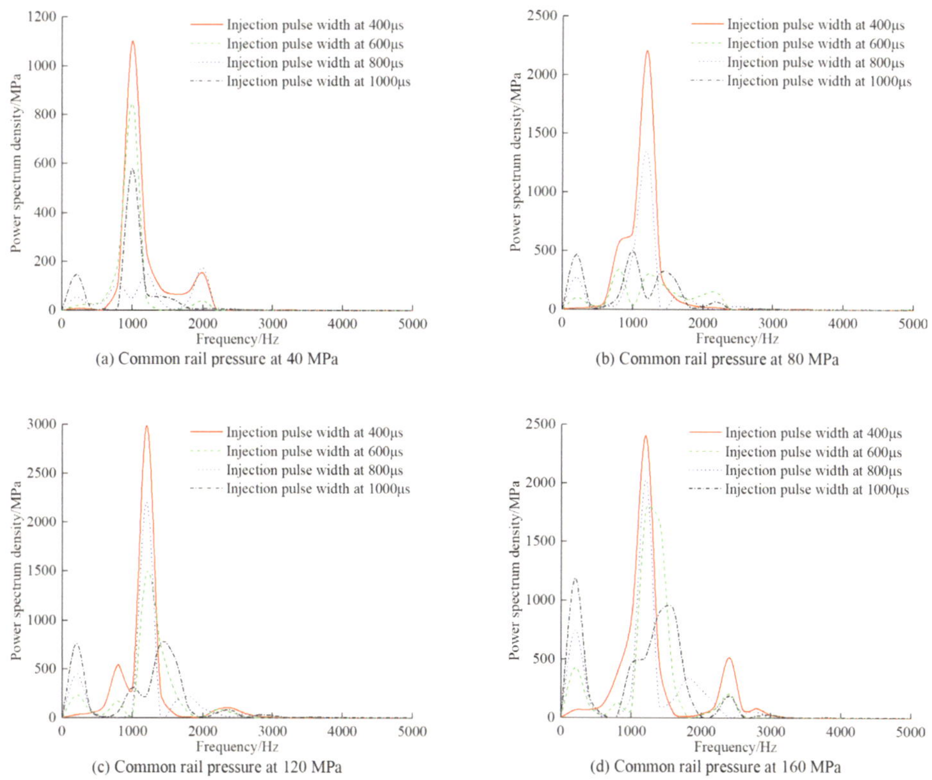


**Figure 8.**
*Power spectrum density of the injector inlet pressure at different common rail pressures and injection pulse widths.*

main crest. When the fuel injection pulse width is 800 μs, the crest characteristics of the dynamic pressure wave power spectrum density are not obvious, and the fuel pressure fluctuation does not show significant periodic fluctuation characteristics. The pressure wave mainly fluctuates at the frequency of 799, 1199 and 1998 Hz. This may be because the pressure wave frequency aroused by the control valve and needle movement in the system reaches the resonance frequency under this injection pulse width, and all kinds of pressure waves propagate repeatedly and superimpose in the system, which changes the pressure wave frequency characteristics. At this time, the fluctuation characteristic of the system is the most complex, and the influence on the dynamic injection characteristic of the system is the most serious.

The injector inlet pressure power spectrum density differs greatly under the four injector pulse widths when the common rail pressure is 80 MPa. The dynamic pressure wave energy of the system is between the frequency band of 599 Hz to 1398 Hz and 799 Hz to 1398 Hz, respectively when the injection pulse width is 400 μs and 800 μs. The power spectrum densities of pressure waves at the injector inlet under the two injection pulse widths have significant crest characteristics, both of which show obvious periodic fluctuation characteristics at the main crest frequency of 1199 Hz. The dynamic pressure wave of the system does not show periodic fluctuation when the injection pulse width is 600 μs and 1000 μs, which shows multi-frequency characteristics. As shown in **Figure 8(b)**, the pressure wave at the injector inlet mainly fluctuates at the frequency of 799 Hz and 1199 Hz when the injection pulse width is

600 μs. The dynamic pressure wave mainly fluctuates at the frequency of 199 Hz and 999 Hz when the injection pulse width is 1000 μs.

As shown in **Figure 8(c)**, the power spectrum density of pressure wave at the injector inlet shows obvious crest characteristics when the common rail pressure is 120 MPa and the injection pulse width is 400, 600 and 800 μs, respectively. The main crest frequency of the pressure wave power spectrum density is 1199 Hz under three injection pulse widths. However, the energy frequency bands are different. The dynamic pressure wave energy is mainly between the frequency band of 799 Hz–1398 Hz when the injection pulse width is 400 μs and 800 μs. The main frequency band of the dynamic pressure wave becomes wider when the injection pulse width is 600 μs, ranges from 799 Hz to 1798 Hz. At the same time, the dynamic pressure wave of the system shows the characteristics of multi-frequency fluctuation when the injection pulse width is 1000 μs, which mainly fluctuates at the frequency of 199 Hz and 1398 Hz.

As shown in **Figure 8(d)**, the variation rule of pressure wave power spectrum density at the injector inlet is similar to that of common rail pressure is 120 MPa when the common rail pressure is 160 MPa and the injection pulse width is 400, 600 and 800 μs, respectively. The dynamic pressure wave mainly fluctuates periodically at the main crest frequency of 1199 Hz. But the energy bands of dynamic pressure wave are different under three injection pulse widths, and the energy band of dynamic pressure wave becomes smaller with the increase of injection pulse width. The main energy bands of the dynamic pressure wave are 599 Hz–1598 Hz, 999 Hz–1798 Hz and 999 Hz–1398 Hz, respectively when the injection pulse width increases from 400 μs to 800 μs. The dynamic pressure wave of the system also shows the multi-frequency fluctuation characteristics when the injection pulse width is 1000 μs, which mainly fluctuates at the frequency of 199 Hz and 1598 Hz.

The fuel density increases with the increase of pressure, which results in the acceleration of pressure wave propagation in the system. Comparing the pressure wave power spectrum density of injector inlet at different common rail pressures under the same injection pulse width in **Figure 8**, it can be seen that the crest characteristics of pressure wave power spectrum density at injector inlet under different common rail pressures are significant, except for the pressure wave multi-frequency fluctuation operating points. The main crest frequency of the injector inlet pressure wave power spectrum density is the lowest when the common rail pressure is 40 MPa, that is, the pressure wave frequency in the system is low when the common rail pressure is low.

To sum up, the dynamic pressure wave of the system has different frequency characteristics under different injection pulse widths and common rail pressure when the size of the high pressure pipeline is constant. It either fluctuates at the main crest frequency or shows the characteristics of multi-frequency fluctuation. While the dynamic pressure wave of the system shows low frequency fluctuation under a low common rail pressure at the same injection pulse width.

As shown in **Figures 6** and 7, the average injector inlet pressure varies with different common rail pressure and injection pulse width. The difference between average injector inlet pressure and setted common rail pressure not only reflects the decreased amplitude of injection pressure in the fuel injection process but also reflects the average amplitude of pressure fluctuation in the system after fuel injection. Therefore, the average pressure drop is defined in this chapter as the difference between the setted common rail pressure and the average injector inlet pressure which locating the moment of injector solenoid valve energized time coordinates from 0 to

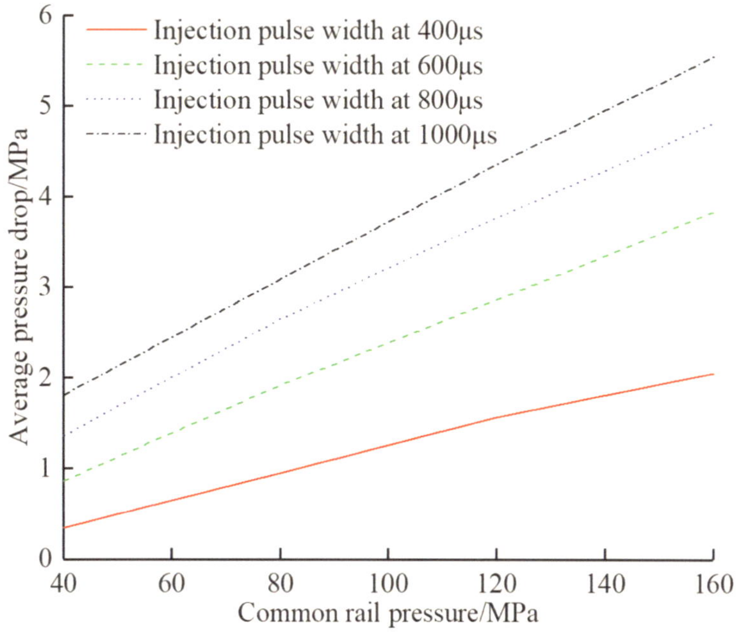


**Figure 9.**
*Variation characteristics of the average pressure drop caused by common rail pressure at different injection pulse widths.*

10 ms. **Figure 9** shows the average pressure drop with common rail pressure under different injection pulse widths. As shown in the figure, the average pressure drop increases linearly with the increase of common rail pressure from 40 MPa to 160 MPa when the size of the high pressure fuel pipeline and injection pulse width is constant. The larger the injection pulse width, the faster the average pressure drop increase rate. In addition, the average pressure drop increases with the increase of injection pulse width from 400 μs to 1000 μs when the size of high pressure fuel pipeline and common rail pressure are constant, and the average pressure drop amplitude increases with the increase of common rail pressure at the two same injection pulse widths. It can be seen that the higher the common rail pressure and the injection pulse width, the larger the average pressure drop.

The first trough of the pressure wave at the injector inlet is the minimum pressure of the system after the fuel injection when the nozzle hole is opened, which reflects the maximum pressure drop after stable fuel injection. The first crest of injector inlet pressure is the maximum compression wave returned in the system when the injection pulse width is large. It reflects the pressure fluctuation amplitude of hydraulic resonance when the injection pulse width is small. Therefore, the first trough and the first crest of the injector inlet pressure are the characteristic parameters reflecting the dynamic pressure wave characteristics of the system. **Tables 1** and **2** show the trough and crest values of injector inlet pressure fluctuation when the common rail pressure is 40, 80, 120 and 160 MPa and the injection pulse width is 400, 600, 800 and 1000 μs, respectively. For comparative analysis, all values in **Table 1** are the difference between the setted common rail pressure and the injector inlet pressure wave trough, and all values in **Table 2** are the difference between injector inlet pressure wave crest and setted common rail pressure.

| Common rail pressure/MPa | Injection pulse width/μs | | | |
|---|---|---|---|---|
| | 400 | 600 | 800 | 1000 |
| 40 | 2.91 | 2.89 | 2.91 | 2.84 |
| 80 | 5.17 | 5.15 | 5.14 | 5.15 |
| 120 | 6.77 | 6.74 | 6.75 | 6.79 |
| 160 | 7.98 | 8.14 | 8.13 | 8.05 |

**Table 1.**
*Trough of the injector inlet pressure fluctuation at different common rail pressures and injection pulse widths.*

| Common rail pressure/MPa | Injection pulse width/μs | | | |
|---|---|---|---|---|
| | 400 | 600 | 800 | 1000 |
| 40 | 5.34 | 2.32 | 0.98 | 0.10 |
| 80 | 7.82 | 3.06 | 0.96 | 1.04 |
| 120 | 6.12 | 3.77 | 1.83 | 1.87 |
| 160 | 6.75 | 3.84 | 2.46 | 2.50 |

**Table 2.**
*Crest of the injector inlet pressure fluctuation at different common rail pressures and injection pulse widths.*

As shown in **Table 1**, the troughs of injector inlet pressure fluctuation increase approximately linearly with the increase of common rail pressure under different injection pulse widths when the high pressure fuel pipeline size is constant. The change of injector pulse width has little effect on the trough of injector inlet pressure fluctuation under the same common rail pressure. The trough of injector inlet pressure fluctuation between four injector pulse widths has a maximum difference of 0.16 MPa under the same common rail pressure. It can be seen that the trough of dynamic pressure fluctuation is independent of injection pulse width, but increases with the increase of common rail pressure. The reasons are as follows. The higher the common rail pressure, the faster the opening response of the needle. The fuel injection rate increases after the needle is opened during the same time, the effective flow area at the nozzle hole increases and the fuel in nozzle volume injects through the nozzle hole more quickly. The pressure drop of the system increases and the trough of the pressure wave increases. The increase of injection pulse width under the same common rail pressure does not affect the early opening of the needle. That is, the needle motion state is the same before the injection pulse width is 400 μs. Therefore, the change of injection pulse width does not affect the trough of system pressure fluctuation.

As shown in **Table 2**, the crests of injector inlet pressure fluctuation increase with the increase of common rail pressure under the same injection pulse width when the size of high pressure fuel pipeline is constant. The smaller the injection pulse width, the higher the crest of injector inlet pressure fluctuation under the same common rail pressure. This is because the higher the common rail pressure with the same injection pulse width, the faster the fuel pressure wave propagates in the system. The superposition time of the large amplitude compression wave reflected from the common rail and the expansion wave aroused by the opening of the needle is advanced, which leads to the increase of the crest of pressure fluctuation at the injector inlet. The smaller the injection pulse width, the shorter the injection duration when the common rail pressure is the same. The short of opening and closing time of the needle will lead to a decrease in the encounter time of the first pressure crest and the third pressure crest as shown in **Figure 4**. The hydraulic resonance effect of the fuel pressure wave is more significant. Therefore, the crest of injector inlet pressure fluctuation decreases with the increase of injection pulse width at the same common rail pressure.

## 7. Conclusions

In this chapter, the pressure fluctuation of high-pressure common rail fuel injection system is studied theoretically. On the basis of revealing the wave mechanism of dynamic pressure wave, the influence of different parameters on the dynamic pressure wave of the fuel injection is investigated. The conclusions are as follows.

1. The theoretical study of pressure fluctuation for high-pressure common rail fuel injection systems shows that the reflected fuel pressure wave returned from the boundary surface of the pipeline end is the result of the coupling of boundary conditions and propagated pressure wave. The reflection of the pressure wave at common rail and high pressure fuel pipeline is a complete negative reflection. The reflection of the pressure wave when the needle closing is a complete positive reflection. The boundary condition type of needle opening nozzle hole is an orifice flow outlet end.

2. The dynamic pressure wave mechanism in high-pressure common rail fuel injection system is revealed. The results show that there is a delay characteristic from the solenoid valve energizing to fuel injection, and the fuel injection duration is longer than the solenoid valve energized time. The opening of the solenoid valve and needle causes the drop of injector inlet pressure. The fuel injection duration is consistent with the opening time of the needle. The fluctuation characteristics of injector inlet pressure before the opening of the needle are independent of the solenoid valve energized time or fuel injection duration. However, the fluctuation characteristics of injector inlet pressure caused by the water hammer effect when the needle closing obviously dependent on the solenoid valve energized time.

3. The influence rules of injection pulse width and common rail pressure on dynamic pressure wave of high-pressure common rail fuel injection system are analyzed. The results show that the inlet pressure of the injector fluctuates in attenuation mode when the injection pulse width is different. The smaller the injection pulse width, the larger the amplitude of pressure fluctuation during fuel injection duration. The change rate of inlet pressure fluctuation amplitude decreases with the increase of injection pulse width. The average injector inlet pressure increases after fuel injection. The inlet pressure of the injector decreases to some extent under different common rail pressure. The average inlet pressure fluctuation of the injector is higher than that of the other three common rail pressures when the common rail pressure is 40 MPa. The inlet pressure fluctuation rules are consistent when the common rail pressure increases from 80 MPa to 160 MPa. The higher the common rail pressure, the larger the injector inlet pressure drop amplitude and the lower the average pressure fluctuation. The dynamic pressure wave of the system has different frequency characteristics

under different injection pulse widths and common rail pressure. It either fluctuates at the main crest frequency or shows the characteristics of multi-frequency fluctuation. However, the dynamic pressure wave of the system shows low frequency fluctuation characteristics under a low common rail pressure at the same injection pulse width. The average pressure drop increases linearly with the increase of common rail pressure, and the increase rate of average pressure drop is faster with the increase of injection pulse width. In addition, it increases with the increase of fuel injection pulse width, and the higher the common rail pressure between the two same fuel injection pulse widths, the larger the increased amplitude of the average pressure drop. Both the trough and crest of dynamic pressure waves increase with the increase of common rail pressure. The smaller the fuel injection pulse width, the higher the crest of the pressure wave at the injector inlet.

## Acknowledgements

The authors gratefully acknowledge the financial support from Hebei Provincial Key Laboratory of Heavy Machinery Fluid Power Transmission and Control.

## Conflict of interest

The authors declare no conflict of interest.

## Author details

Yun Bai*, Zhaoyang Chen, Wei Dou, Xiangdong Kong, Jing Yao, Chao Ai,
Fugang Zhai, Jin Zhang and Liu Yang
School of Mechanical Engineering, Yanshan University, Qinhuangdao, China

*Address all correspondence to: baiyun@ysu.edu.cn

IntechOpen

© 2022 The Author(s). Licensee IntechOpen. This chapter is distributed under the terms of the Creative Commons Attribution License (http://creativecommons.org/licenses/by/3.0), which permits unrestricted use, distribution, and reproduction in any medium, provided the original work is properly cited.

## References

[1] Kim K, Si W, Jin D, Kim J-H, Cho J, Baek S, et al. Characterization of engine oil additive packages on diesel particulate emissions. Journal of Mechanical Science and Technology. 2020;**34**(2):931-939

[2] Wang TJ. Effects of insulation on exhaust temperature and subsequent SCR efficiency of a heavy-duty diesel engine. Journal of Mechanical Science and Technology. 2019;**33**(2):923-929

[3] Ismael MA, Heikal MR, Aziz ARA, Syah F, Zainal EZA, Crua C. The effect of fuel injection equipment on the dispersed phase of water-in-diesel emulsions. Applied Energy. 2018; **15**(222):762-771

[4] Balz R, von Rotz B, Sedarsky D. In-nozzle flow and spray characteristics of large two-stroke marine diesel fuel injectors. Applied Thermal Engineering. 2020;**180**:115809

[5] Yu H, Goldsworthy L, Brandner PA, Li J, Garaniya V. Modelling thermal effects in cavitating high-pressure diesel sprays using an improved compressible multiphase approach. Fuel. 2018; **15**(222):125-145

[6] Ghiji M, Goldsworthy L, Brandner PA, Garaniya V, Hield P. End of injection process in a single-hole diesel injector. At Sprays. 2018;**28**(1): 23-45

[7] Wu X, Deng J, Cui H, Xue F, Zhou L, Luo F. Numerical simulation of injection rate of each nozzle hole of multi-hole diesel injector. Applied Thermal Engineering. 2016;**108**:793-797

[8] Qiu T, Song X, Lei Y, Liu X, An X, Lai M. Influence of inlet pressure on cavitation flow in diesel nozzle. Applied Thermal Engineering. 2016;**25**(109): 364-372

[9] Soriano JA, Mata C, Armas O, Ávila C. A zero-dimensional model to simulate injection rate from first generation common rail diesel injectors under thermodynamic diagnosis. Energy. 2018;**1**(158):845-858

[10] Oerley F, Hickel S, Schmidt SJ, Adams NA. Large-Eddy simulation of turbulent, cavitating fuel flow inside a 9-hole diesel injector including needle movement. International Journal of Engine Research. 2017;**18**(3):195-211

[11] Ghiji M, Goldsworthy L, Garaniya V, Brandner PA, Hield P, Novozhilov V, et al. Effect of residual air bubbles on diesel spray structure at the start of injection. Fuel. 2019;**1**(241):25-32

[12] Ferrari A, Paolicelli F. Modal analysis of fuel injection systems and the determination of a transfer function between rail pressure and injection rate. Journal of Engineering for Gas Turbines and Power. 2018;**140**(11): 112808-112808-112808-112811

[13] Luo T, Jiang S, Moro A, Wang C, Zhou L, Luo F. Measurement and validation of hole-to-hole fuel injection rate from a diesel injector. Flow Measurement and Instrumentation. 2018;**61**:66-78

[14] Rehman KU, Liu X, Wang H, Zheng L, Rehman RU, Cheng X, et al. Effects of black soldier fly biodiesel blended with diesel fuel on combustion, performance and emission characteristics of diesel engine. Energy Conversion and Management. 2018;**173**:489-498

[15] Lee Y, Lee CH. An uncertainty analysis of the time-resolved fuel

injection pressure wave based on BOSCH method for a common rail diesel injector with a varying current wave pattern. Journal of Mechanical Science and Technology. 2018;**32**(12):5937-5945

[16] Piano A, Boccardo G, Millo F, Cavicchi A, Postrioti L, Pesce FC. Experimental and numerical assessment of multi-event injection strategies in a solenoid common-rail injector. SAE International Journal of Engines. 2017 [cited 2017 Sep 18];**10**(4):2129-2140. Available from: http://papers.sae.org/2017-24-0012/

[17] Yu H, Goldsworthy L, Brandner PA, Garaniya V. Development of a compressible multiphase cavitation approach for diesel spray modelling. Applied Mathematical Modelling. 2017; **45**:705-727

[18] Rounthwaite NJ, Williams R, McGivery C, Jiang J, Giulliani F, Britton B. A chemical and morphological study of diesel injector nozzle deposits - insights into their formation and growth mechanisms. SAE International Journal of Fuels and Lubricants. 2017;**10**(1): 106-114

[19] Ubertini S. Injection pressure fluctuations model applied to a multidimensional code for diesel engines simulation. Journal of Engineering for Gas Turbines and Power. 2006;**128**(3):694

[20] Catalano LA, Tondolo VA, Dadone A. Dynamic Rise of Pressure in the Common-Rail Fuel Injection System. In 2002 [cited 2017 Aug 10]. Available from: http://papers.sae.org/2002-01-0210/

[21] Catania AE, Ferrari A, Manno M, Spessa E. Experimental investigation of dynamics effects on multiple-injection common rail system performance. Journal of Engineering for Gas Turbines and Power. 2008;**130**(3):032806

[22] Beierer P, Huhtala K, Vilenius M. Experimental Study of the Hydraulic Circuit of a Commercial Common Rail Diesel Fuel Injection System. In 2007 [cited 2017 Aug 12]. Available from: http://papers.sae.org/2007-01-0487/

[23] Bianchi GM, Falfari S, Brusiani F, Pelloni P, Osbat G, Parotto M. Numerical Investigation of Critical Issues in Multiple-Injection Strategy Operated by a New C.R. Fast-Actuation Solenoid Injector. In 2005 [cited 2019 Apr 20]. p. 2005-01–1236. Available from: http://papers.sae.org/2005-01-1236/

[24] Seykens XLJ, Somers LMT, Baert RSG. Modelling of common rail fuel injection system and influence of fluid properties on injection process. [cited 2017 Aug 9]; Available from: http://citeseerx.ist.psu.edu/viewdoc/summary?doi=10.1.1.475.2416

[25] Henein NA, Lai M-C, Singh IP, Zhong L, Han J. Characteristics of a Common Rail Diesel Injection System under Pilot and Post Injection Modes. In 2002 [cited 2019 Apr 19]. p. 2002-01–0218. Available from: http://papers.sae.org/2002-01-0218/

[26] Herfatmanesh MR, Peng Z, Ihracska A, Lin Y, Lu L, Zhang C. Characteristics of pressure wave in common rail fuel injection system of high-speed direct injection diesel engines. Advances in Mechanical Engineering. 2016;**8**(5):168781401664824

[27] Boudy F, Seers P. Impact of physical properties of biodiesel on the injection process in a common-rail direct injection system. Energy Conversion and Management. 2009;**50**(12):2905-2912

[28] Badami M, Millo F, D'Amato DD. Experimental Investigation on Soot and NOx Formation in a DI Common Rail Diesel Engine with Pilot Injection. In

2001 [cited 2019 Apr 20]. p. 2001-01–0657. Available from: http://papers.sae.org/2001-01-0657/

[29] Ferrari A, Paolicelli F, Pizzo P. Hydraulic performance comparison between the newly designed common feeding and standard common rail injection Systems for Diesel Engines. J Eng Gas Turbines Power-Trans Asme. 2016;**138**(9):092801

[30] Ferrari A, Mittica A. Response of different injector typologies to dwell time variations and a hydraulic analysis of closely-coupled and continuous rate shaping injection schedules. Applied Energy. 2016;**169**:899-911

[31] Catania AE, Ferrari A. Development and performance assessment of the new-generation CF fuel injection system for diesel passenger cars. Applied Energy. 2012;**91**(1):483-495

[32] Li P, Zhang Y, Li T, Xie L. Elimination of fuel pressure fluctuation and multi-injection fuel mass deviation of high pressure common-rail fuel injection system. Chin J Mech Eng. 2015;**28**(2):294-306

[33] Ismail MY, Mamat R, Ali O, Aziz A, Mohd A, Kamarulzaman M, et al. The combustion of N-butanol-diesel fuel blends and its cycle to cycle variability in a modern common-rail diesel engine. Journal of Engineering and Applied Science. 2016;**1**(11):2297-2301